Advances in Intelligent and Soft Computing 89

Editor-in-Chief: J. Kacprzyk

Advances in Intelligent and Soft Computing

Editor-in-Chief

Prof. Janusz Kacprzyk
Systems Research Institute
Polish Academy of Sciences
ul. Newelska 6
01-447 Warsaw
Poland
E-mail: kacprzyk@ibspan.waw.pl

Further volumes of this series can be found on our homepage: springer.com

Vol. 76. T. Bastiaens, U. Baumöl, and B.J. Krämer (Eds.)
On Collective Intelligence, 2010
ISBN 978-3-642-14480-6

Vol. 77. C. Borgelt, G. González-Rodríguez, W. Trutschnig, M.A. Lubiano, M.Á. Gil, P. Grzegorzewski, and O. Hryniewicz (Eds.)
Combining Soft Computing and Statistical Methods in Data Analysis, 2010
ISBN 978-3-642-14745-6

Vol. 78. B.-Y. Cao, G.-J. Wang, S.-Z. Guo, and S.-L. Chen (Eds.)
Fuzzy Information and Engineering 2010
ISBN 978-3-642-14879-8

Vol. 79. A.P. de Leon F. de Carvalho, S. Rodríguez-González, J.F. De Paz Santana, and J.M. Corchado Rodríguez (Eds.)
Distributed Computing and Artificial Intelligence, 2010
ISBN 978-3-642-14882-8

Vol. 80. N.T. Nguyen, A. Zgrzywa, and A. Czyzewski (Eds.)
Advances in Multimedia and Network Information System Technologies, 2010
ISBN 978-3-642-14988-7

Vol. 81. J. Düh, H. Hufnagl, E. Juritsch, R. Pfliegl, H.-K. Schimany, and Hans Schönegger (Eds.)
Data and Mobility, 2010
ISBN 978-3-642-15502-4

Vol. 82. B.-Y. Cao, G.-J. Wang, S.-L. Chen, and S.-Z. Guo (Eds.)
Quantitative Logic and Soft Computing 2010
ISBN 978-3-642-15659-5

Vol. 83. J. Angeles, B. Boulet, J.J. Clark, J. Kovecses, and K. Siddiqi (Eds.)
Brain, Body and Machine, 2010
ISBN 978-3-642-16258-9

Vol. 84. Ryszard S. Choraś (Ed.)
Image Processing and Communications Challenges 2, 2010
ISBN 978-3-642-16294-7

Vol. 85. Á. Herrero, E. Corchado, C. Redondo, and Á. Alonso (Eds.)
Computational Intelligence in Security for Information Systems 2010
ISBN 978-3-642-16625-9

Vol. 86. E. Mugellini, P.S. Szczepaniak, M.C. Pettenati, and M. Sokhn (Eds.)
Advances in Intelligent Web Mastering – 3, 2011
ISBN 978-3-642-18028-6

Vol. 87. E. Corchado, V. Snášel, J. Sedano, A.E. Hassanien, J.L. Calvo, and D. Ślęzak (Eds.)
Soft Computing Models in Industrial and Environmental Applications, 6th International Workshop SOCO 2011
ISBN 978-3-642-19643-0

Vol. 88. Y. Demazeau, M. Pěchouček, J.M. Corchado, and J.B. Pérez (Eds.)
Advances on Practical Applications of Agents and Multiagent Systems, 2011
ISBN 978-3-642-19874-8

Vol. 89. J.B. Pérez, J.M. Corchado, M.N. Moreno, V. Julián, P. Mathieu, J. Canada-Bago, A. Ortega, and A.F. Caballero (Eds.)
Highlights in Practical Applications of Agents and Multiagent Systems, 2011
ISBN 978-3-642-19916-5

Javier Bajo Pérez, Juan M. Corchado,
María N. Moreno, Vicente Julián,
Philippe Mathieu, Joaquin Canada-Bago,
Alfonso Ortega, and
Antonio Fernández Caballero (Eds.)

Highlights in Practical Applications of Agents and Multiagent Systems

9th International Conference on Practical Applications of Agents and Multiagent Systems

Springer

Editors

Prof. Javier Bajo Pérez
Universidad Pontificia de Salamanca
Escuela Universitaria de Informática
Compañía 5, 37002 Salamanca, Spain

Prof. Juan M. Corchado
Universidad de Salamanca
Departamento de Informática y Automática
Facultad de Ciencias, Plaza de la Merced
S/N, 37008 Salamanca, Spain
E-mail: corchado@usal.es

Prof. María N. Moreno
Universidad de Salamanca
Departamento de Informática y Automática
Facultad de Ciencias, Plaza de la Merced
S/N, 37008 Salamanca, Spain

Prof. Vicente Julián
Universidad Politécnica de Valencia
Departamento de Sistemas Informáticos
y Computación, Camino de Vera s/n,
46022 Valencia, Spain

Prof. Philippe Mathieu
Université des Sciences et Technologies
de Lille, LIFL (Laboratoire d'Informatique
Fondamentale de Lille) UMR CNRS 8022,
Bâtiment M3 (extension), 59655
Villeneuve d'Ascq Cedex, France

Prof. Joaquin Canada-Bago
Universidad de Jaén, Área de
Ingeniería Telemática, Departamento
de Ingeniería de Telecomunicación
E.P.S. de Linares, C/ Alfonso X el
Sabio, 28, 23700 Linares (Jaén), Spain

Prof. Alfonso Ortega
Ciudad Universitaria de Cantoblanco
Departamento de Ingeniería
Informática, Calle Francisco Tomás
y Valiente, 11, 28049 Madrid, Spain

Prof. Antonio Fernández Caballero
Universidad de Castilla-La Mancha
Escuela de Ingenieros Industriales
de Albacete, Departamento de
Sistemas Informáticos
Campus Universitario s/n, 02071, Spain

ISBN 978-3-642-19916-5 e-ISBN 978-3-642-19917-2

DOI 10.1007/978-3-642-19917-2

Advances in Intelligent and Soft Computing ISSN 1867-5662

Library of Congress Control Number: 2011923213

Typeset & Cover Design: Scientific Publishing Services Pvt. Ltd., Chennai, India.

Printed on acid-free paper
5 4 3 2 1 0
springer.com

Preface

PAAMS'11 Special Sessions are a very useful tool in order to complement the regular program with new or emerging topics of particular interest to the participating community. Special Sessions that emphasized on multi-disciplinary and transversal aspects, as well as cutting-edge topics were especially encouraged and welcome.

Research on Agents and Multi-Agent Systems has matured during the last decade and many effective applications of this technology are now deployed. An international forum to present and discuss the latest scientific developments and their effective applications, to assess the impact of the approach, and to facilitate technology transfer, has become a necessity.

PAAMS, the International Conference on Practical Applications of Agents and Multi-Agent Systems is an evolution of the International Workshop on Practical Applications of Agents and Multi-Agent Systems. PAAMS is an international yearly tribune to present, to discuss, and to disseminate the latest developments and the most important outcomes related to real-world applications. It provides a unique opportunity to bring multi-disciplinary experts, academics and practitioners together to exchange their experience in the development of Agents and Multi-Agent Systems.

This volume presents the papers that have been accepted for the 2011 edition in the special sessions: Special Session on Agents Behaviours for Artificial Markets, Special Session on Multi-Agent Systems for safety and securit, Special Session on Web Mining and Recommender Systems, Special Session on Adaptative Multi-Agent System, Special Session on Integration of Artificial Intelligence Technologies in Resource-Constrained Devices, Special Session on Bio-Inspired and Multi-Agents Systems: Applications to Languages and Special Session on Agents for smart mobility.

We would like to thank all the contributing authors, as well as the members of the Program Committees of the Special Sessions and the Organizing Committee for their hard and highly valuable work. Their work has helped to contribute to the success of the PAAMS'11 event. Thanks for your help, PAAMS'11 wouldn't exist without your contribution.

Juan Manuel Corchado
Javier Bajo
PAAMS'11 Organizing Co-chairs

Acknowledgements: This conference has been supported by Junta de Castilla y León (Spain).

Organization

Special Sessions

SS1 – Special Session on Agents Behaviours for Artificial Markets
SS2 – Special Session on Multi-Agent Systems for safety and security
SS3 – Special Session on Web Mining and Recommender Systems
SS4 – Special Session on Adaptative Multi-Agent System
SS5 – Special Session on Integration of Artificial Intelligence Technologies in Resource-Constrained Devices
SS6 – Special Session on Bio-Inspired and Multi-Agents Systems: Applications to Languages
SS7 – Special Session on Agents for smart mobility

Special Session on Agents Behaviours for Artificial Markets Committee

Mathieu Philippe (Co-chairman)	Lille1 University (France)
Florian Hauser	University of Innsbruck (Austria)
Olivier Brandouy	University Paris 1 Panthéon Sorbonne (France)
Silvano Cincotti	University of Genoa (Italy)

Special Session on Multi-Agent Systems for Safety and Security Committee

Antonio Fernández Caballero	University of Castilla-La Mancha (Spain)
Elena María Navarro Martínez	University of Castilla-La Mancha (Spain)
Francisco J. Garijo	Telefónica (Spain)
Javier Jaen	Universidad Politécnica de Valencia (Spain)
José Manuel Gascueña	Universidad de Castilla La Mancha (Spain)
Maria Teresa López	Universidad de Castilla La Mancha (Spain)

Special Session on Web Mining and Recommender Systems Committee

María N. Moreno García (Chairman)	University of Salamanca (Spain)
Luis Alonso Romero	University of Salamanca (Spain)
Angel Luis Sánchez Lázaro	University of Salamanca (Spain)
Vivian López Batista	University of Salamanca (Spain)
María José Polo Martín	University of Salamanca (Spain)
María Dolores Muñoz Vicente	University of Salamanca (Spain)
Ana Belén Gil González	University of Salamanca (Spain)
Antonio Garrote Hernández	University of Salamanca (Spain)
Joel P. Lucas	University of Salamanca (Spain)
Maguelonne Teisseire	University of Montpellier 2 (France)
Ana María Almeida	Institute of Engineering of Porto (Portugal)
César Hervás	University of Córdoba (Spain)
Chris Cornelis	Ghent University (Belgium)
Carlos Alonso	University of Valladolid (Spain)
Anne Laurent	University of Montpellier 2 (France)
María José del Jesús	University of Jaen (Spain)
Rafael Corchuelo	University of Sevilla (Spain)
Constantino Martins	Institute of Engineering of Porto (Portugal)
Yolanda Blanco	University of Vigo (Spain)

Special Session on Adaptative Multi-Agent System Committee

Vicente Julian	Polytechnic University of Valencia (Spain)
Alberto Fernández	University of Rey Juan Carlos (Spain)
Juan M. Corchado	University of Salamanca (Spain)
Vicente Botti	Polytechnic University of Valencia (Spain)
Sascha Ossowski	University of Rey Juan Carlos (Spain)
Sara Rodríguez González	University of Salamanca (Spain)
Carlos Carrascosa	Polytechnic University of Valencia (Spain)
Javier Bajo	Pontifical University of Salamanca (Spain)
Martí Navarro	Polytechnic University of Valencia (Spain)
Stella Heras	Polytechnic University of Valencia (Spain)
Carlos A. Iglesias	Technical University of Madrid (Spain)
Holger Billhardt	University King Juan Carlos I

Javier Palanca	Polytechnic University of Valencia (Spain)
Juan Francisco de Paz Santana	University of Salamanca (Spain)
Álvaro Herrero	University of Burgos
Victor Sanchez	Polytechnic University of Valencia (Spain)

Special Session on Integration of Artificial Intelligence Technologies in Resource-Constrained Devices Committee

Juan Antonio Botía Blaya	University of Murcia (Spain)
Joaquin Canada-Bago	University of Jaen (Spain)
Jose Angel Fernandez-Prieto	University of Jaen (Spain)
Manuel Angel Gadeo-Martos	University of Jaen (Spain)
Alicia Triviño-Cabrera	University of Malaga (Spain)
Juan Ramon Velasco	University of Alcala (Spain)

Special Session on Bio-Inspired and Multi-Agents Systems: Applications to Languages Committee

Gemma Bel Enguix (Co-chairman)	GRLMC - Rovira i Virgili University (Spain)
Alfonso Ortega de la Puente (Co-chairman)	Universidad Autónoma de Madrid (Spain)
María de la Cruz Echeandía (Co-chairman)	Universidad Autónoma de Madrid (Spain)
Alejandro Echeverría Rey	Universidad Autónoma de Madrid (Spain)
Peter Leupold	Rovira i Virgili University (Spain)
Robert Mercas	Rovira i Virgili University (Spain)
Diana Pérez Marín	Universidad Rey Juan Carlos (Spain)
Leonor Becerra-Bonache	Rovira i Virgili University (Spain)
Luis Fernando de Mingo López	Universidad Politécnica de Madrid (Spain)

Special Session on Agents for Smart Mobility Committee

Miguel Ángel Sánchez (chairman)	Pontifical University of Salamanca (Spain)
Montserrat Mateos	Pontifical University of Salamanca (Spain)
Encarnación Beato	Pontifical University of Salamanca (Spain)

Ana M. Fermoso	Pontifical University of Salamanca (Spain)
Alberto Pedrero	Pontifical University of Salamanca (Spain)
Vidal Alonso	Pontifical University of Salamanca (Spain)
Roberto Berjón	Pontifical University of Salamanca (Spain)

Organizing Committee

Juan M. Corchado (Chairman)	University of Salamanca (Spain)
Javier Bajo (Co-Chairman)	Pontifical University of Salamanca (Spain)
Juan F. De Paz	University of Salamanca(Spain)
Sara Rodríguez	University of Salamanca (Spain)
Dante I. Tapia	University of Salamanca (Spain)
Emilio Corchado	University of Salamanca (Spain)
Fernando de la Prieta Pintado	University of Salamanca (Spain)
Davinia Carolina Zato Domínguez	University of Salamanca (Spain)

Contents

Special Session on Web Mining and Recommender Systems

Special Session on Adaptative Multi-Agent System

Special Session on Bio-Inspired and Multi-Agents Systems: Applications to Languages

Special Session on Agents for Smart Mobility

Calibration of an Agent Based Model for Financial Markets

Annalisa Fabretti

Abstract. Agent based model are very widely used in different discipline. In financial markets they can explain very well known features called stylized facts and fit statistical properties of data. For such a reason in predicting future price movements they could perform better than standard models using gaussianity. At this scope calibration and validation in order to choose the model and the model parameters are very essential issues. However calibrating such models is a hard issue to tackle and not yet very well considered in literature. The present paper presents the attempt to calibrate the Farmer Joshi model by a Nelder Mead algorithm with threshold. Different objective function are considered in order to identify the best choice.

1 Introduction

Agent based models (ABM) are able to reproduce some of the statistical properties of returns seen in real stock markets [2]. These features include a distribution of returns that is more peaked than the Gaussian distribution, periods of persistent high volatility, period of persistent high trading volume and correlation between volatility and trading volume. Traditional economic models adopt gaussian distribution considering extreme events as outliers or build statistical process able to reproduce these features without providing any understanding of how these features are observed over all financial markets. The agent-based approach considers a population of intelligent adaptive agents and let them acting in order to maximize their financial performance. These kind of model are able to put attention on interactions, learning dynamics or herding behaviour very commons in markets. Heterogeneity is one of the key issue in modelling financial markets. ABM focus on describing different approaches to trade [9] [3], sometimes using oversimplified technical trading rules [4]. Some models put attention on the price formation, i.e. the way of matching

Annalisa Fabretti
Dept. SEFEMeQ University of Rome Tor Vergata, Via Columbia 2 Rome Italy
e-mail: annalisa.fabretti@uniroma2.it

J.B. Pérez et al. (Eds.): Highlights in PAAMS, AISC 89, pp. 1–8.
springerlink.com

orders, which matters in reproducing stylized facts [11]. A comprehensive view of the literature can be found in [7] and [8].

Since ABM are useful in reproducing market statistics, validation and calibration cover an important role. According to Tesfatsion [13] there are three alternative ways of validating computational models: a descriptive output validation; a predictive output validation; and an input validation. But how can these be done in a scientifically meaningful manner? At the moment there are only a limited number of contributions dealing with it. An interesting and wide discussion of the literature can be found in [15]. A model can be validated if the statistical properties of simulated data match those of real data and the two series can be said to belong to the same distribution, but how can the model parameters be chosen? It is difficult to justify why to choose one specification of parameters and not another; this recalls the problem of calibration. Calibrating is also useful in selecting the best model in reproducing the real data. ABMs can have different ingredients and not any of them can fit well the qualitative and quantitative stylized fact [14]. Main methods of calibration presented so far in the literature are descriptive or heuristic based on similarity of generated sample [10]. Unfortunately classical calibrating model are subject to fail due to the irregularity of the simulated data, however Gilli and Winker in [6] propose a Nelder-Mead simplex algorithm with a local search heuristic called threshold accepting. Others methods, as genetic algorithms used in social science, are also proposed to calibrate agent based model [12] and could give useful improvement to the issue.

The aim of the present work is to study the issue of calibrating an ABM. The standard idea for calibrating this kind of model is to fit statistical properties as moments. For this reason another question addressed in this work is: which moments are to be preferred in the construction of the objective function? In [6] authors combined kurtosis and Arch(1)-effect in the objective function, in [12] mean and standard deviation, while in [16] a more wide discussion on the choice of the objective function is provided. In the present work using the Nelder Mead algorithm with threshold proposed in [6] different moments in the objective function are compared in calibrating the model in [4]. The Farmer Joshi model is choosn for its simplicity but at the same time for its capacity to reproduce market dynamics. The mechanism of price formation, if compared with more complex and realistic mechanism can be considered trivial, however it is out of the scope of the present work to discuss the goodness of Farmer Joshi model, instead it is an aim of the present work to verify the utility of the method proposed by Gilli and Winker in [6].

The paper is organized as follows: Sect. 2 presents briefly the Farmer Joshi Model; Sect. 3 describes data and explains the methodology; Sect. 4 explains results and finally concludes focusing on future researches.

2 The Farmer Joshi Model

The model [4] considers two type of agents characterized by common strategies in the market, fundamentalist and chartist; to aggregate demands the model also includes a risk neutral market maker. At the single time step t the traders observe the most recent prices $P_t, P_{t-1}, ..P_{t-d}$ and the information I_t and submit an order ω_t^i; then the market maker fills all the orders at the new price P_{t+1}. The market maker bases price formation only on the net order $\omega_t = \sum_{i=1}^{N} \omega_t^i$, where N is the number of agents in the market. The market maker fills price for the net order using the so called market impact function $P_{t+1} = P_t \cdot e^{\frac{\omega_t}{\lambda}}$, where λ is called the liquidity parameters. Letting $p_t = \log P_t$, and adding a noise term ξ_{t+1} the logarithm of the price is

$$p_{t+1} = p_t + \frac{1}{\lambda} \sum_{i=1}^{N} \omega_t^i + \xi_{t+1}. \quad (1)$$

The quantity ω_t^i consists in the difference $x_t^i - x_{t-1}^i$, where $x_t^i = x_t^i(p_t, p_{t-1}.., I_t)$ is the position at time t of the i-th trader.

Trend followers invest based on the belief that price changes have inertia. A trend strategy takes a positive (long) position if prices have recently been going up, and a negative (short) position if they have recently been going down. Their position is $x_{t+1}^i = c(p_t - p_{t-d})$, where c is a positive constant and d is the time lag.

Value investors make a subjective assessment of the value in relation to price. They believe that their perceived value may not be fully reflected in the current price, and that future prices will move toward their perceived value. They attempt to make profits by taking positive (long) positions when they think the market is undervalued and negative (short) positions when they think the market is overvalued.
Let the logarithm of the value v_t be a random walk: $v_{t+1} = v_t + \eta_{t+1}$, where η_t is a noise process IID with mean μ_η and standard deviation σ_η. The value investor takes the position $x_{t+1}^i = c(v_t - p_t)$, where $c > 0$ is a constant proportional to the trading capital.

From the point of view of a practitioner, a concern with the simple position-based value strategies and the simple trend follower strategies is excessive transaction costs. Trades are made whenever the mispricing changes. To reduce trading frequency strategies a threshold for entering a position and another threshold for exiting it can be used. The goal is to trade only when the expected price movement is large enough to beat transaction costs. Assume that a short position $-c$ is entered when the mispricing exceeds a threshold T and exited when it goes below a threshold τ . Similarly, a long position c is entered when the mispricing drops below a threshold $-T$ and exited when it exceeds $-\tau$. In general, different traders will choose different entry and exit thresholds, the entry threshold T^i and exit threshold τ^i are extracted from uniform distributions of entry thresholds ranging from T_{min} to T_{max} and exit thresholds ranging from τ_{min} to τ_{max}, respectively. Finally, parameter c is chosen so that $c = a(T - \tau)$, where a is a positive constant called scale parameter for capital assignment.

3 Data and Methodology

An objective function considering moments not always well behaves to guarantee a global solution, hence standard methods are subject to fail. In such a situation in different fields of science and engineering, heuristics optimization have been applied successfully. Such heuristics includes simulated anealing, threshold accepting heuristic, neural networks and genetic algorithms [5]. These kind of methods require a strong computational power and for this are really time consuming. Due to the complexity of the problems and the stochastic elements of the algorithm and the model, they cannot pretend to produce exact solution in every case with certainty. However, a stochastic high-quality approximation of a global optimum is probably more useful than a deterministic poor-quality local minimum provided by a classical method or no solution at all. Here the methodology proposed by [6] is implemented to solve the problem

$$\min_{\theta \in \Theta} f(\theta), \tag{2}$$

where θ is the vector of parameters, Θ is the space of feasible parameters, and $f(\theta)$ is the objective function. The objective function is a weighted combination of estimating errors on some chosen moments. Let be $\mathbf{w}$ a vector of weights, the function takes the form

$$f(\theta) = \mathbf{w}'|\mathbf{m}^e - \mathbf{m}^m|, \tag{3}$$

where $\mathbf{m}^j$, with $j = e, m$, is the moments vector $\mathbf{m} = (k, m, \sigma, \alpha)'$, where k is the kurtosis, m the mean, σ the standard deviation and α is the Arch(1) effect. The index e identify the empirical value of data and m the value of the simulated time series, finally $'$ indicates transposition.

The vector of parameters to calibrate is $\theta = [\lambda, \mu_\eta, \sigma_\eta]$, the liquidity parameter, the drift and the volatility of the random walk driving the fundamental value v_t, respectively. Actually, the model have many parameters to be set, but a too many parameters estimation would easily fail, for this reason this first attempt of calibration is focused only on 3 parameters and keeps constant the others, see table 1. Note that the parameter σ_ξ is not mentioned, because the calibration is done on the "pseudo deterministic "part of the model, that is the log-price in (1) without the noise ξ. Hence the σ_ξ can be found fitting the residuals.

The calibration method consists of two phases. First a Nelder-Mead algorithm is run, later the neighbourhood of the solution found at the first step is explored by a threshold algorithm. For a full description of the algorithm see [6]. The stopping criterium for the Nelder-Mead phase imposes a stop if the number of maximum iteration N_{iter}^{max} or the maximum admissible error err are reached. The first case is considered a failure because often occurs when the algorithm is not able to find the direction in which the objective function decreases. The second case instead is a success because the algorithm is able to find a θ^* minimizing the objective function; when this situation occurs the calibration procedure passes to the second phase, the threshold algorithm, exploring n_R rounds (with n_S steps for each round) of the neighbourhood of θ^*. In the trials presented N_{iter}^{max} has been set to 100, err to $1 \cdot 10^{-1}$, R to 20 and both n_R and n_S to 20.

Data considered are closing prices of S&P500 Composite index from 14/10/2005 to 28/09/2006. In Figure 1(a) logarithmic returns and (b) their histogram are shown. In table 2 a summary of the statistics of interest are reported.

Table 1 Parameters of the model

Name of Parameter	Symbol	Value
Number of agents	N^{trend} N^{fund}	100
Maximum threshold for entering positions	T^{trend}_{max} T^{fund}_{max}	4
Minimum threshold for entering positions	T^{trend}_{min} T^{fund}_{min}	0.2
Maximum threshold for exiting positions	τ^{trend}_{max} τ^{fund}_{max}	0
Minimum threshold for exiting positions	τ^{trend}_{min} τ^{fund}_{min}	-0.2
Maximum offset for log of perceived value	v_{max}	0.5
Minimum offset for log of perceived value	v_{min}	-0.5
Scale parameter for capital assignment	a^{trend} a^{fund}	0.001
Time delay for trend followers	d	5

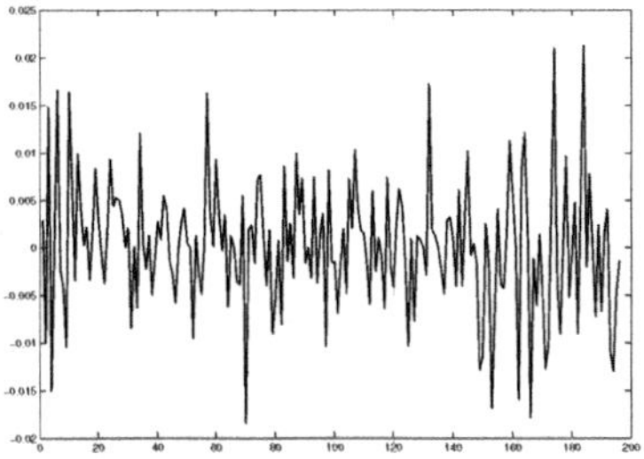

(a) real log returns

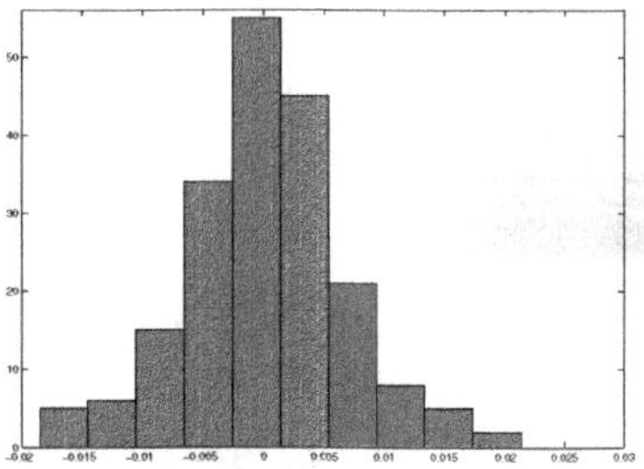

(b) histogram of real log returns

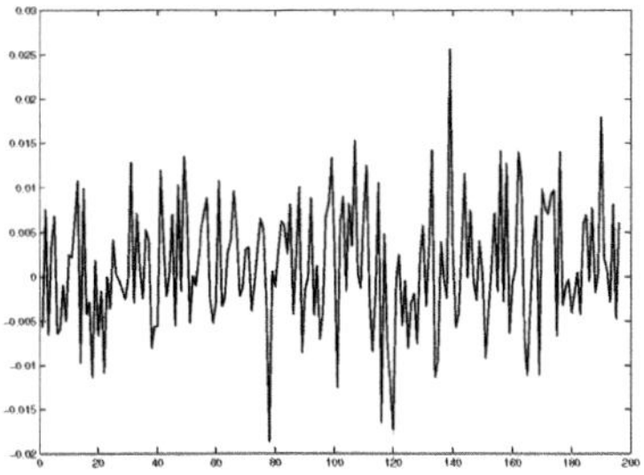

(c) simulated log returns

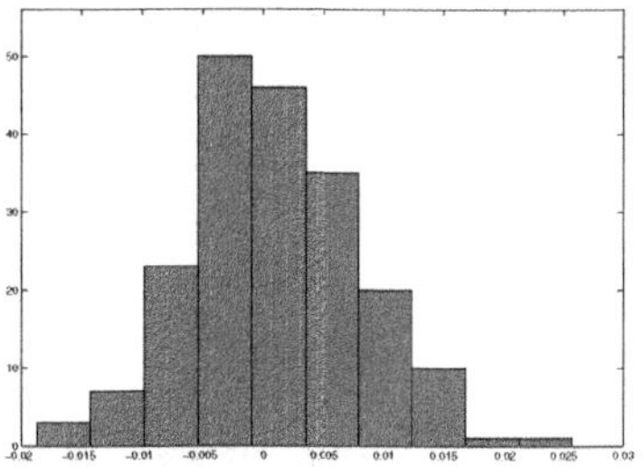

(d) histogram of simulated log returns

Fig. 1 Log returns of real data (a) and relative histogram (b) from 14/10/2005 to 28/09/2006 and simulated log returns (c) and relative histogram obtained with parameters $[1.0003, 0.0103, 0.0203]$.

Table 2 Descriptive statistics of Data. S&P500 Composite Index from 14/10/2005 to 28/09/2006

Variable	Price	Log Return
Maximum Value	1339.2	0.0213
Minimum Value	1177.8	-0.0185
Kurtosis	3.7553	3.948
Mean	1273.6	4.8582e-004
Standard Deviation	31.6027	0.0066
α arch(1) effect	1.0005	0.99972

4 Results, Comments and Further Research

In the following the results are reported and commented. In Figure 2 objective functions considering different moments are shown plotted for μ_η in $[0.02, 0.6]$ and σ_η in $[0.02, 0.5]$ and λ fixed to 1. More precisely the objective functions considered are a linear combination of the kurtosis, the mean and the standard deviation (a),

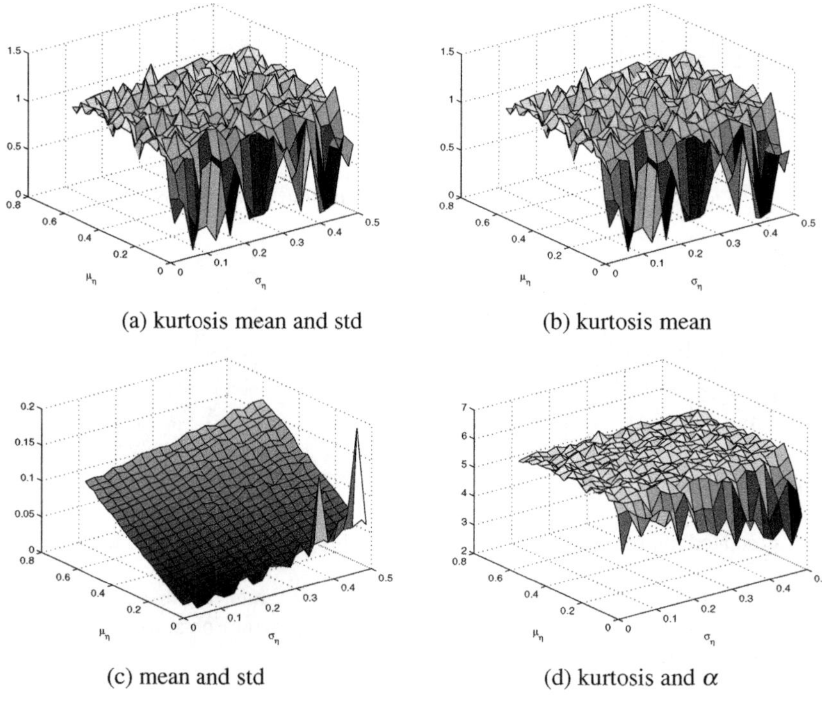

(a) kurtosis mean and std

(b) kurtosis mean

(c) mean and std

(d) kurtosis and α

Fig. 2 Different objective functions are plotted for μ_η in $[0.02, 0.6]$ and σ_η in $[0.02, 0.5]$ and $\lambda = 1$. The objective functions considered are a combination of the kurtosis, the mean and the standard deviation (a), of the kurtosis and the mean (b), of the mean and the standard deviation (c) and of the kurtosis and α(d).

of the kurtosis and the mean (b), of the mean and the standard deviation (c) and of the kurtosis and α effect (d). It is notably to observe that where the kurtosis is considered the surface is very rough with higher values respect with the smoothest objective function obtained by the couple m and σ, that unfortunately is not able to catch the well known stylized facts of financial data.

Many trials have been done for the objective function with weights vector $\mathbf{w} = [1,2,1,5]$ and all of them have failed. Indeed the algorithm is unable to find the direction of descending of the objective function. This failure can be attributed either to the algorithm or to the model. In fact some ABMs are able to reproduce qualitative feature of the market but could be not adapt to reproduce the quantitative feature of the market as observed in [14] and here it could be the case. The Farmer Joshi model considers, for example, a very simple price formation mechanisms where more complex market microstructure better can reproduce stylized facts. More investigations are needed to understand how to improve the calibration or why it fails. In table 3 two successful calibration trials are presented, for each successful trial the moments of simulated data are provided. A successful calibration is obtained for the objective function with $\mathbf{w} = [1,0,0,5]$, while the threshold algorithm alone has been used in minimizing the objective function with $\mathbf{w} = [0,1,1,0]$. Data analysis is necessary to test the goodness of the resulting calibrated parameters; it can be seen comparing moments in table 3 with those in table 2 that only $\theta^* = [1.0003, 0.0103, 0.0203]$ fits real data, while $\theta^* = [0.7921, 0.9752, 0.6337]$ gives unsatisfactory results, suggesting that the algorithm stopped in a local minimum. In Figure 1 simulated data (c) for $\theta^* = [1.0003, 0.0103, 0.0203]$ are plotted in comparison with real data (a), see also their histograms (b) (d).

Some observations are due: first, it is worth to note that the choice of weights of the objective function would require a more rigorous criterium; second, attempts to calibrate data on a longer time window (14/10/2005 - 14/10/2010) has been done without success. A so long data interval in fact includes events like the recent financial crisis, the total kurtosis and the standard deviation of these data are so high that even a long time interval simulation is not able to catch completely the oscillations. This drawback may be bypassed evaluating moments using bootstrap.

In conclusion, even if the results presented are still preliminary they can offer a starting point for future investigations. The procedure implemented is still very slow to converge and often fails. The minimization of the objective function must be improved with a reasonable choice of the weights and a robust estimation, like the bootstrap, of the moments. By this first study it can be concluded that, when used for rough objective function, the Nelder Mead algorithm, even if coupled with an heuristic, can easily fail reaching a local minimum instead than a global one. A further test can be done by using this method for a model which it is possible to derive a closed form solution for the distribution of returns as in [1]. Other heuristic calibration methods can also successfully replace the one studied here. In particular interesting results are given by the genetic algorithms, which also allow the simultaneous calibration of many parameters. Future research are intended to be in such directions. The hope is to present in future more exhaustive studies on methods and their comparison.

Table 3 Calibration results of successful trials. The time is indicated in minutes.

Obj Fun	$\theta^* = [\lambda^*, \mu_\eta^*, \sigma_\eta^*]$	min Obj Fun	time	mean	std	kurtosis
k, α	$[0.7921, 0.9752, 0.6337]$	0.9235	167.12	0.1103	0.0968	2.9650
m, σ	$[1.0003, 0.0103, 0.0203]$ [1]	0.0012	44.43	0.0010	0.0070	3.2990

[1]Obtained with the threshold algorithm alone.

References

1. Alfarano, S., Wagner, F., Lux, T.: Estimation of Agent-Based Models: The Case of an Asymmetric Herding Model. Computational Economics 26, 19–49 (2005)
2. Cont, R.: Empirical properties of asset returns:stylized facts and statistical issues. Quant. Fin. 1, 223–236 (2001)
3. Day, R.H., Huang, W.: Bulls, Bears and Market Sheep. J. of Econ. Behav. and Org. 14, 299–329 (1990)
4. Farmer, J.D., Joshi, S.: The price dynamics of common trading strategies. J. of Econ. Behav. and Org. 49, 149–171 (2002)
5. Fonseca, C.M., Fleming, P.J.: Genetic Algorithms for multi-objective optimization: Formulation, Discussion, Generalization. In: Proceedings Fifth International Conference on Genetic Algorithm, pp. 416–423 (1993)
6. Gilli, M., Winker, P.: A global optimization heuristic for estimating agent based models. Comp. Stat. & Data Anal. 42, 299–312 (2003)
7. Hommes, C.H.: Heterogeneous Agent Models in Economics and Finance. In: Tesfatsion, L., Judd, K.L. (eds.) Handbook of Computational Economics, vol. 2, pp. 10–13. Elsevier Science B.V., Amsterdam (2006)
8. LeBaron, B.: Agent based computational finance: Suggested readings and early research. J. Econ. Dyn. Contr. 24, 679–702 (2000)
9. Lux, T., Marchesi, M.: Scaling and criticality in a stochastic multi-agent model of a financial market. Nature 397, 498–500 (1999)
10. Palin, J.: Agent-Based Stockmarket Models: Calibration Issues and Application. PhD thesis, University of Sussex, UK (2002)
11. Pellizzari, P., Dal Forno, A.: A comparison of different trading protocols in an agent based market. J. of Econ. Inter. and Coord. 2, 27–43
12. Rogers, A., Von Tessin, P.: Multi-Objective calibration for and agent-based model. In: Agent-Based Simulation, Lisbon, Portugal (May 2004)
13. Tesfatsion, L.: Website on ACE: Verification and Empirical Validation of Agent-Based Computational Models, `http://www2.econ.iastate.edu/tesfatsi/empvalid.htm`
14. Veryzhenko, I., Brandouy, O., Mathieu, P.: Agent's Minimal Intelligence Calibration for Realistic Market Dynamics. In: Li Calzi, M., Milone, L., Pellizzari, P. (eds.) Progress in Artificial Economics, pp. 3–14. Springer, Heidelberg (2002)
15. Windrum, P., Fagiolo, G., Moneta, A.: Empirical Validation of Agent-Based Models: Alternatives and Prospects. J. of Art. Soc. and Social Sim. 10(2) (2007)
16. Winker, P., Gilli, M., Jeleskovic: An objective function for simulation based inference on exachange rate data. J. Econ. Interac. Coord. 2, 125–145 (2007)

A Multiagent System for Web-Based Risk Management in Small and Medium Business

Juan F. De Paz, Javier Bajo, M. Lourdes Borrajo, and Juan M. Corchado

Abstract. Business Intelligence has gained relevance during the last years to improve business decision making. However, there is still a growing need of developing innovative tools that can help small to medium sized enterprises to predict risky situations and manage inefficient activities. This article present a multiagent system especially conceived to detect risky situations and provide recommendations to the internal auditors of SMEs. The core of the multiagent system is a type of agent with advanced capacities for reasoning to make predictions based on previous experiences. This agent type is used to implement an evaluator agent specialized in detect risky situations and an advisor agent aimed at providing decision support facilities. Both agents incorporate innovative techniques in the stages of the CBR system. An initial prototype was developed and the results obtained related to small and medium enterprises in a real scenario are presented.

Keywords: Hybrid neural intelligent system, CBR, MAS, Business Intelligence, business risk prediction.

1 Introduction

Nowadays, the organization systems employed in enterprises are increasing in complexity. In the present financial context, it is increasilly relevant to provide innovative tools and decision support systems that can help the small-medium

Javier Bajo
Facultad de Informática. Universidad Pontificia de Salamanca Compañía 5, 37002, Salamanca, Spain
e-mail: jbajope@upsa.es

M. Lourdes Borrajo
Dept. Informática, University of Vigo, Edificio Politécnico, Campus As Lagoas s/n, Ourense, 32004, Spain
e-mail: lborrajo@uvigo.es

Juan F. De Paz · Juan M. Corchado
Departamento Informática y Automática. University of Salamanca, Plaza de la Merced s/n, 37008, Salamanca, Spain
e-mail: {fcofds,corchado}@usal.es

J.B. Pérez et al. (Eds.): Highlights in PAAMS, AISC 89, pp. 9–17.
springerlink.com

enterprises (SMEs) to improve their functioning (Khashman 2009, Li and Sun in press, Li et al. in press, Sun and Li 2009a, Sun and Li 2009b). These tools and methods can contribute to improve the existing business control mechanisms, reducing the risk by predicting undesiderable situations and providing recommendations based on previous experiences (Chi-Jie 2009, Li and Su 2008, Sun and Li 2008a, Sun and Li 2008b).

The processes carried out inside a company are grouped into functional areas (Corchado et al. 2000) denominated "Functions". A Function is a group of coordinated and related activities that are systematically and iteratively carried out during the process of reaching the company's objectives (Li and Sun 2009). The functions that are usually carried out within a company, as studied within the framework of this research, are: Purchases, Cash Management, Sales, Information Technology, Fixed Assets Management, Compliance to Legal Norms and Human Resources. Each one of these functions is broken down into a series of activities. For example, the Information Technology function is divided into the following activities: Computer Plan Development, Study of Systems, Installation of Systems, Treatment of Information Flows, and Security Management.

This article propose an innovative approach, based on multiagent systems (Bajo et al. 2009), to propose a model for risk management and prediction in SMEs. Multiagent systems are the most prevalent solution to construct Artificial Intelligence distributed systems. Agents are computational entities that can be characterized through their capacities in areas such as autonomy, reactivity, pro-activity, social abilities, reasoning, learning and mobility (Bajo et al. 2009). These capacities make the multi-agent systems very appropriate for constructing intelligent environments. An agent can act as an interface between the user and the rest of the elements of the intelligent environment. Moreover, intelligent agents can incorporate advanced artificial intelligence models to predict risky situations.

The article is structured as follows: the next section briefly introduces the problem that motivates this research. Section 3 presents the multi-agent system for managing small and medium enterprises. Section 4 presents the results obtained after testing the system and the conclusions of this study.

2 Application of Business Web Intelligence to Enterprise Risk Assessment and Management

"Risk Management" is a broad term for the business discipline that protects the assets and profits of an organization by reducing the potential for loss before it occurs, mitigating the impact of a loss if it occurs, and executing a swift recovery after a loss occurs. It involves a series of steps that include risk identification, the measurement and evaluation of exposures, exposure reduction or elimination, risk reporting, and risk transfer and/or financing for losses that may occur. All organizations practice risk management in multiple forms, depending on the exposure being addressed (Calderon and Cheh 2002, Risk and Insurance Management Society 2008).

The economic environment has increased the pressure on all companies to address risk at the highest levels of the organization. Companies with a strategic approach to risk management use more tools and have more structured and frequent reporting on risk management than do firms with other approaches. As such, they are in a better position to ensure that risk management provides relevant and applicable information that meets the needs of the organization and executive team. But no matter what an organization's approach is, the tools used must be backed up by solid, actionable reporting addressed (Calderon and Cheh 2002, Risk and Insurance Management Society 2008). It's not always necessary for the risk managers to be conducting their own studies for their voices to be heard. Forging a strong relationship with internal auditors and other departments can allow risk practitioners to supplement their reports with the risk manager's own analysis (Colbert 1995).

Enterprise Risk Management (ERM) is defined as "a process, effected by an entity's board of directors, management and other personnel, applied in strategy-setting and across the enterprise, designed to identify potential events that may affect the entity, and manage risk to be within its risk appetite, to provide reasonable assurance regarding the achievement of entity objectives." (Committee of Sponsoring Organizations of the Treadway Commission -COSO 2009).

The managing of risks and uncertainties is central to the survival and performance of organizations. Enterprise risk management (ERM) is an emerging approach to managing risks across different business functions in an organization that represents a paradigm shift from specialized, silo-ed approaches in managing specific risks (Ding et al. 2008, Huang et al. 2008, Li and Sun 2009, Lin et al. 2009, Ramamoorti et al. 1999). This paper provides a web intelligent model to Enterprise risk assessment, which will subsequently lead to better organizational performance.

3 Multiagent Web System for Risk Management

Agent and multi-agent systems (SMEs) have become increasily relevant during the last decades and have gained relevance in different areas (Bajo et al. 2009, Sun and Li 2008).

In this article we propose a distributed approach where the components of a SME are modeled as intelligent agents that collaborate to create models that can evolve over the time and adapt to the changing conditions of the environment. The multiagent system provides a web system interface to facilitate the remote interaction with the human users involved in the risk management process.

Thus, making possible to detect risky situations for the SMEs and providing suggestions and recommendations that can help to avoid possible undesiderable situations. The agents in the system allow the users to access the system through distributed applications, which run on different types of devices and interfaces (e.g. computers, cell phones, PDA). Figure 1 shows the basic schema of the proposed architecture, where all requests and responses are handled by the agents in the platform.

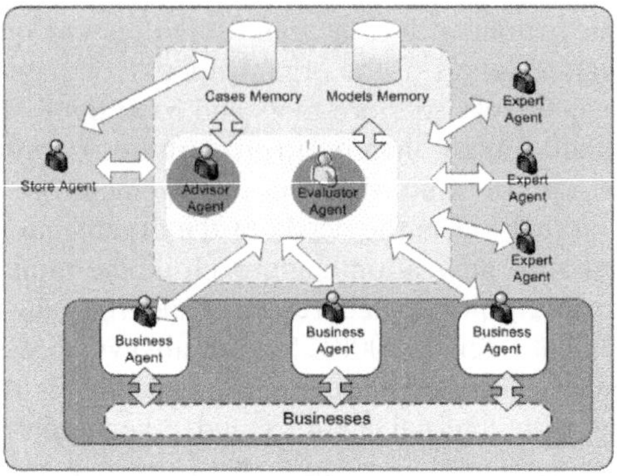

Fig. 1 Crisis multiagent architecture basic schema.

There are different kinds of agents in the architecture, each one with specific roles, capabilities and characteristics:

- Business Agent. This agent was assigned for each enterprise in order to collect new data and allow consultations. The enterprise can interact with the system by means of this agent, introducing information and receiving predictions.
- Evaluator Agent. It is responsible for the evaluation and predictions of potential risky situations.
- Advisor agent. The objective of this agent is to carry out recommendations to help the internal auditor decide which actions to take in order to improve the company's internal and external processes.
- Expert Agent. This agent helps the auditors and enterprise control experts that collaborate in the project to provide information and feedback to the multiagent system. These experts generate prototypical cases from their experience and they receive assistance in developing the Store agent case-base.
- Store Agent. This agent has a memory that has been fed with cases constructed with information provided by the enterprise (through its agent) and with prototypical cases identified by 34 enterprises control experts, using personal agents who have collaborated and supervised the developed model.

The core of the multiagent system are the evaluator and advisor agents, that incorporate new techniques to analyze the data from enterprises, extract the relevant information, and detect possible failures or inefficiencies in the operation processes.

The evaluator and advisor agents are CBR-BDI agents (Corchado and Laza 2003) that make use to past experiences to resolve new problems. As such, it is perfectly suited for solving the problem at hand. In addition, CBR (Case-Based Reasoning) makes it possible to incorporate the various stages of expression analysis into the reasoning cycle of the CBR (Kolodner 1993), thus facilitating the creation of strategies similar to the processes followed in small and medium enterprises.

On one hand the evaluator agent is specialized in detecting risky situations. The recovery of information from previous experiences simplifies the prediction process by detecting and eliminating relevant and irrelevant patterns detected in previous analyses. The retrieve phase of the hybrid neural intelligent system incorporates a novel Expectation Maximization clustering technique (Dellaert 2002). The reuse stage incorporates an innovative mixture of experts that makes use of multilayer perceptron, support vector regression and radial basis function neural network. The revise and retain stages implement a decision support system for experts. Moreover, the knowledge obtained during the prediction process is of great importance for subsequent predictions. On the other hand, the advisor agent is specialized in providing recommendations to avoid risky situations and improve the overall functioning of the SME. The retrieve phase recovers similar cases and their corresponding solutions. The reuse phase incorporates a novel approach based on decision trees and probabilistic gain functions to ponder efficient and inefficient tasks. The revise and retain stages also implement a decision support system for experts. There are various artificial intelligence techniques such as artificial neural networks (Bianchia et al. 2007, Sawa and Ohno-Machado 2003), Bayesian networks (Baladandayuthapani et al. 2005), and fuzzy logic (Avogadri and Valentini 2009) which have been applied to business failure prediction. While these techniques can be applied to failure detection and prediction, the knowledge obtained cannot be incorporated into successive tests and included in subsequent analyses. The approach presented in this article is an evolution of our previous works and proposes an innovative perspective (Borrajo et al. 2005, Corchado et al. 2005). The new approach proposes a multiagent system to model the organizational structure of a SME. Moreover, the core of the system is a CBR-BDI agent type with the ability to adapt to the changes in the environment. In (Borrajo et al. 2005) we presented a system composed of two case-based reasoning systems to detect the associate risk in the activities of SMEs in the textile sector and generate recommendations to improve the erroneous processes. In (Corchado et al. 2005) we presented a decision support tool based on a case-based reasoning system that automates the internal control processes of a SME. The new approach proposes new methods for the retrieval stage of the CBR systems, as the Expectation Maximization clustering, that notably improves the case's recovery reducing the final quantity of cases stored and making it easier to recover the most similar cases to the problem introduced. Moreover, the approach proposes very innovative reuse mechanisms, based on mixture of experts and probabilistic decision trees.

4 Results and Conclusions

A case study aimed at providing innovative web business intelligence tools for the management of SMEs was carried out in the Castilla y León region, in Spain. The experiment consisted on the construction of the initial prototype of memory of cases and then in predicting potential risky situations for the enterprises taken into considerations and providing recommendations. The case study presented in this work was oriented to detect possible risky situations in SMEs, taken into account the crisis that affects the market. A multiagent system was implemented and 22 SMEs participated in the experiment and were assigned a personal business agent. The enterprises were situated in different sectors and located in the Spanish region of Castilla y León. The economic context is the same for all the SMEs. The system was tested during 24 months, from January 2008 to January 2010, tuned and improved taking into account the experience acquired using a total of 238 cases. The evolution of the enterprise is monitored through its internal activities and the predictions are made based on the previous experiences and on the status of the market (the possible crisis that affect the market). To provide information about the Enterprise, the experts have to complete a survey.

To validate the overall functioning of the system it was necessary to individually evaluate the Evaluator and Advisor agents. These agents provide predictions on the performance of the activities and detect those tasks that can be improved for each activity in order to get an overall improvement of the activity. To validate the performance of the Evaluator agent, an estimation of the efficiency of the predictions provided by the Evaluator agent was carried out. To evaluate the significance of the different techniques integrated within the Evaluator agent, a cross validation was established, following the Dietterich's 5x2- Cross-Validation Paired t-Test algorithm (Dietterich 1998). The value 5 in the algorithm represents the number of replications of the training process and value 2 is the number of sets in which the global set is divided. Thus, for each of the techniques, the global dataset S was divided into two groups S_1 and S_2 as follows: $S = S_1 \cup S_2$ y $S_1 \cap S_2 = \phi$. Then, the learning and estimation processes were carried out. This process was repeated 5 times and had the following steps: the system was trained using S_1 and then it was used to classify S_1 y S_2. In a second step, the system was trained using S_2 and then it was used to classify S_1 y S_2. The results obtained by the evaluator agent using the mixture of experts were compared to the results obtained using an individual RBF and an individual MLP to the same dataset and the same 5x2- Cross-Validation process. Table 1 shows the error rate obtained for each of the techniques, using the test in each of the 5 repetitions. As can be seen in Table 1, the estimated error was lower for the Evaluator agent than for the rest of the evaluated techniques.

Table 1 Absolute error for the estimation of the status of the activities.

Method	S_2	S_1	S_2	S_1	S_2	S_1	S_2	S_1	S_2	S_1
Evaluator agent	0,297	0,309	0,210	0,281	0,207	0,355	0,226	0,343	0,239	0,302
MLP	0,677	0,669	0,489	0,507	0,513	0,806	0,530	0,696	0,506	0,485
RBF	1,009	0,833	0,656	0,985	0,878	0,959	0,620	0,831	0,643	0,783

A Paired t-Test was applied to check that the difference between the methods can be considered as significative if a value $\alpha = 0.05$ is established.

To evaluate the Advisor agent it is necessary to take into account that the aim of this agent is to detect inefficient tasks by means of gain functions, as explained in Section 3. The evaluation of the functioning of the Advisor agent was carried out by selecting those tasks with higher values for the gain function. The selected tasks were used to estimate the different scenarios for different execution values for the task. The estimation was performed using the values provided by the Evaluator agent, obtaining a concrete value for the task. In this way, Figure 2a presents the evolution of the system for the average status of 5 activities along 12 months. As shown, the evolution for the 5 activities can be considered as positive.

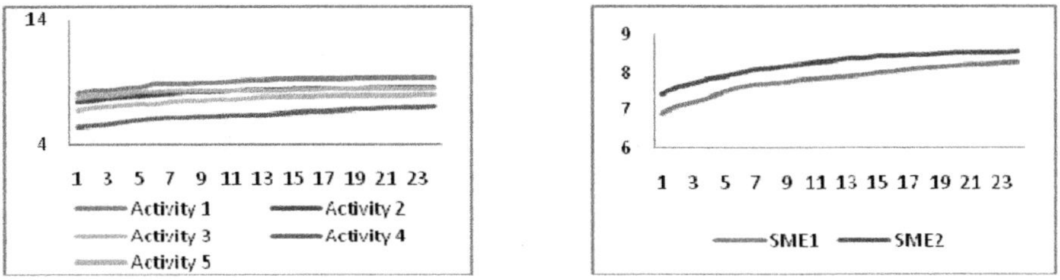

Fig. 2 a) Evolution of the average status of 5 activities during 12 months. b) Evolution of the average status of 2 SMEs during 12 months.

Looking at the evolution of the global efficiency for the activities analyses for two SMEs, shown in Figure 2b, it is possible to observe a growing tendency in the average status of the business along the time, which indicates a reduction of inefficient tasks in each of the activities. The results obtained demonstrate that the multiagent system caused a positive evolution in all enterprises. This evolution was reflected in the reduction of inefficient processes. The indicator used to determine the positive evolution of the companies was the state of each of the activities analyzed. After analyzing one of the company's activities, it was necessary to prove that the state of the activity (valued between 1 and 100) had increased beyond the state obtained in the previous three month period. The system considers small changes in the tasks performed in the SMEs, and all the experts that participated in the experiments considered 3 months as a significant time to evaluate the evolution of a SME related to these changes.

We had certain problems implementing the system, partly because the management and experts were not familiar with the use of computational devices and multiagent systems, so some courses were given to introduce them to these technologies and teach them how to use the system interface. The proposed novel multiagent system is a unique system useful for dynamic environments and open enough to be used in other enterprise environments. The experts noted that the behavior of the system improves as the number of cases in the memory of cases grows. This is a typical behavior in CBR-based systems. We found some difficulties with the surveys, because some of the SMEs were reticent to complete surveys.

The users indicated that the use of a web-based interface facilitates the use of the new system. However, some of them were reticent about trusting the system because they were reluctant to facilitate their internal data and because updating the information about the enterprise requires specialized human resources and time. However, the auditors and experts believe that the CBR-BDI agents may support their work and provide a highly appreciated decision support tool. They believe that this hybrid neural intelligent architecture has more advantages than disadvantages and that the system helped them to detect inconsistent processes in the enterprises. They tend to argue that the hybrid neural intelligent architecture should incorporate a shared memory of cases to compare data from different firms, but with the guarantee of data privacy.

As a conclusion, we can say that the use of innovative tools in business intelligence can notably help to detect potential risky situations and have a better understanding of their internal functioning. This article presented a multiagent-based tool aimed at modeling the behavior of SMEs and providing a decision support tool that can contribute to detect potential risky situations and to avoid them by acting on the tasks that compose each of the activities of the business.

Acknowledgements

This work has been supported by the Spanish Ministry of Science and Innovation project TIN2009-13839-C03.

References

Avogadri, R., Valentini, G.: Fuzzy ensemble clustering based on random projections for DNA microarray data analysis. Artificial Intelligence in Medicine 45(2-3), 173–183 (2009)

Bajo, J., De Paz, Y., De Paz, J.F., Corchado, J.M.: Integrating Case Planning and RPTW Neuronal Networks to Construct an Intelligent Environment for Health Care. Expert Systems with Applications 36(3), 5844–5858 (2009)

Baladandayuthapani, V., Ray, S., Mallick, B.K.: Bayesian Methods for DNA Microarray Data Analysis. Handbook of Statistics 25(1), 713–742 (2005)

Bianchia, D., Calogero, R., Tirozzi, B.: Kohonen neural networks and genetic classification. Mathematical and Computer Modelling 45(1-2), 34–60 (2007)

Borrajo, M.L., Corchado, J.M., Yáñez, J.C., Fdez-Riverola, F., Díaz, F.: Autonomous internal control system for small to medium firms. In: Muñoz-Ávila, H., Ricci, F. (eds.) ICCBR 2005. LNCS (LNAI), vol. 3620, pp. 106–121. Springer, Heidelberg (2005)

Calderon, T.G., Cheh, J.J.: A roadmap for future neural networks research in auditing and risk assessment. International Journal of Accounting Information Systems 3(4), 203–236 (2002)

Chi-Jie, L., Tian-Shyug, L., Chih-Chou, C.: Financial time series forecasting using independent component analysis and support vector regression. Decision Support Systems 47(2), 115–125 (2009)

Colbert, J.L.: Risk. Internal Auditor, 36–40 (1995)

Committee of Sponsoring Organizations of the Treadway Commission (COSO), Guidance on Monitoring Internal Control Systems (COSO's Monitoring Guidance) (2009)

Corchado, J.M., Laza, R.: Constructing Deliberative Agents with Case-based Reasoning Technology. International Journal of Intelligent Systems 18(12), 1227–1241 (2003)

Corchado, J.M., Borrajo, M.L., Pellicer, M.A., Yanez, C.Y.: Neuro-symbolic system for business internal control. In: Perner, P. (ed.) ICDM 2004. LNCS (LNAI), vol. 3275, pp. 1–10. Springer, Heidelberg (2004)

Dellaert, F.: The Expectation Maximization Algorithm, Technical Report. Georgia Institute of Technology (2002)

Ding, Y., Song, X., Zen, Y.: Forecasting financial condition of Chinese listed companies based on support vector machine. Expert Systems with Applications: An International Journal 34(4), 3081–3089 (2008)

Huang, S., Tsai, C., Yen, D.C., Cheng, Y.: A hybrid financial analysis model for business failure prediction. Expert Systems with Applications: An International Journal 35(3), 1034–1040 (2008)

Khashman, A.: A neural network model for credit risk evaluation. International Journal of Neural Systems 19(4), 285–294 (2009)

Kolodner, J.: Case-Based Reasoning. Morgan Kaufmann, San Francisco (1993)

Li, H., Sun, J.: Ranking-order case-based reasoning for financial distress prediction. Knowledge-Based Systems 21(8), 868–878 (2008)

Li, H., Sun, J.: Gaussian case-based reasoning for business failure prediction with empirical data in China. Information Sciences 179(1-2), 89–108 (2009)

Li, H., Sun, J.: On performance of case-based reasoning in Chinese business failure prediction from sensitivity, specificity, positive and negative values. Applied Soft Computing (in Press)

Li, H., Huang, H.B., Sun, J.: On sensitivity of case-based reasoning to optimal feature subsets in business failure prediction. Expert Systems with Applications (in Press)

Li, H., Liang, Y., Xu, Q.: Support vector machines and its applications in chemistry. Chemometrics and Intelligent Laboratory Systems 95(2), 188–198 (2009)

Lin, R., Wang, Y., Wu, C., Chuang, C.: Developing a business failure prediction model via RST, GRA and CBR. Expert Systems with Applications 36(2), 1593–1600 (2009)

Ramamoorti, S., Bailey, A.D., Traver, R.O.: Risk assessment in internal auditing: a neural network approach. Int. J. IntellSyst. Account Finance Manage (8), 159–180 (1999)

Risk and Insurance Management Society, Inc., Financial Crisis- A Wake up call for Enterprise Risk Management (2008)

Sawa, T., Ohno-Machado, L.: A neural network based similarity index for clustering DNA microarray data. Computers in Biology and Medicine 33(1), 1–15 (2003)

Sun, J., Li, H.: Data mining method for listed companies' financial distress prediction. Knowledge-Based Systems 21(1), 1–5 (2008a)

Sun, J., Li, H.: Listed companies' financial distress prediction based on weighted majority voting combination of multiple classifiers. Expert Systems with Applications 35(3), 818–827 (2008b)

Sun, J., Li, H.: Financial distress early warning based on group decision making. Computers & Operations Research 36(3), 885–906 (2009a)

Sun, J., Li, H.: Financial distress prediction based on serial combination of multiple classifiers. Expert Systems with Applications 36(4), 8659–8666 (2009b)

Drawing Technical Figures in Double Auction Artificial Stock Markets

Albert Meco, Javier Arroyo, and Juan Pavón

Abstract. This paper presents an artificial stock market that is able to reproduce both technical figures, such as the price bouncing between support and resistance lines, and some typical stylized facts of financial time series. The proposed market implements a continuous double auction system where agents can either follow a fundamental or a chartist investment approach. Some novelties are introduced to this setting. Firstly, the fundamental strategy is based on a time-varying fundamental value, which does not exhibit the stylized facts of financial time series. The fundamental strategy tries to make the price follow this fundamental value. Secondly, chartist agents can adopt either a trend-following or a bouncing strategy. Interestingly, chartist strategies are implemented taking into account the support and resistance lines that each agent draws. As a result of both investment approaches, some typical technical figures can be spotted on the time series of the prices.

1 Introduction

Agent-based simulation can be used to reproduce and analyze phenomena that arise in financial markets [8]. Different simulation models have been proposed to explain phenomena such as volatility, market efficiency and the behavior of the investors [9, 5, 7, 2]. Regarding investors' behavior, it is widely acknowledged that two investment approaches coexist in the markets: fundamentalism and chartism. Fundamentalism is usually associated with long-term expectations that try to stabilize the price; while chartism is associated with shorter-term expectations that act as a destabilizing force that tries to push the price away from the fundamental value [3].

Many artificial stock markets (ASMs) include both chartist and fundamentalist investment approaches. Both approaches usually are combined by means of weights

A. Meco · J. Arroyo · J. Pavón
Grupo de Investigación en Agentes Software, Universidad Complutense Madrid, 28040 Madrid, Spain
e-mail: `albert.meco@totemcat.com`, `{javier.arroyo,jpavon}@fdi.ucm.es`

J.B. Pérez et al. (Eds.): Highlights in PAAMS, AISC 89, pp. 19–26.
springerlink.com

[2] or switches that allow agents to change from one approach to the other [7]. Behavior heterogeneity is a key factor to reproduce financial stylized facts such as returns unpredictability and volatility clustering among others. In ASMs, chartist behavior is usually reduced to a trend following strategy that makes possible to reproduce departures from the fundamental value in the price time series. However, the chartist toolbox includes more than just that strategy.

This paper presents an ASM that is inspired by the continuous double auction system in [1] and by the agents' switching behavior in [6]. The proposed ASM aims to reproduce chartist patterns in the prices time series by means of a sophisticated chartist behavior. This behavior includes both trend following and bouncing strategies. The bouncing strategy consists of drawing support and resistance lines that contain the recent past of the prices time series and considering that the future price will not cross them. As a result, prices bounce from one line to the other. In addition, the fundamental strategy tries to follow a time-varying fundamental value, which is slightly different for each agent, and that is represented by a time series whose returns are unpredictable and do not exhibit volatility clustering.

2 Description of the Artificial Stock Market

The ASM is a double auction market as many real-life stock markets. This kind of markets has two queues one for buy orders and the other for sell orders. Agents can place limit and market price orders. Limit orders are sorted by their respective price in its corresponding queue. When a market price (buy or sell) order arrives, the most competitive order in the opposite queue is matched. The price is given by the price of the last orders matched. If no transaction occurs, the price is given the average of the lowest ask price and the highest bid price stored in the public order book. If no fixed price orders are stored in the bid or ask lists of the order book, then the resulting price would be the last observed price.

In our market, each trading session t' is divided into 25 cycles or trading periods and at each cycle only one market order is executed. Thus, each session consists of at most 25 transactions. Market-price orders are executed in the sequence they arrive to the market. If several market-price orders have the same priority, the double auction mechanism executes the order sent by the agent with higher priority ρ_t^i, which is a random value renewed in each cycle. It is also important to remark that orders are stored in the public order book for 400 cycles and are discarded after this period. Order size is one share. Agents with a pending order in the market will not place a new order until its current order is either matched or discarded.

Agents will decide whether to place a buy or a sell order taking into account the information they have about the state of the market. As they are rationally bounded, their orders are based on their perceptions and beliefs about the market behavior. These perceptions and beliefs may be different between agents.

At each trading period t, each agent i makes an expectation about the return of the next trading period, $\hat{r}_{t+1}^i$. This expectation can be based on either fundamental or chartists beliefs. Fundamental beliefs essentially follow the premise that the asset price should be close to its fundamental value. On the other hand, chartist beliefs are rooted in the idea that markets price movements tend to repeat themselves because investors collectively tend toward patterned behavior. In the proposed ASM, agents switch their beliefs between fundamental and chartist ideas, similarly to what is done in [6].

The switch is driven by the difference between the fundamental value and the asset price in each instant of time, f_t and p_t respectively. If an agent considers that the difference between them is too large, then it trades the asset assuming that the market will eventually recognize and correct the "misprice" (fundamental behavior). However, if an agent considers that the price suitably reflects the fundamental value, then it tries to earn some money by analyzing the price patterns ignoring the fundamental value (chartist behavior). In our market, f_t remains constant within each trading session t. The deviation perceived for agent i at time t is represented by

$$u_t^i = \ln(g_1^i f_t) - \ln p_t, \tag{1}$$

where g_1^i is a distortion factor that represents the different perception of each agent i of the fundamental value f_t. This factor follows a random variable $N(1, \sigma_1^2)$ with $g_1^i >= 0$. The value u_t^i the difference perceived by agent i at time t. If $u_t^i > 0$ the agent believes that the actual price is under the fundamental value, while if $u_t^i < 0$ the asset is overpriced.

The rule that triggers the switch of behavior between chartist and fundamentalist uses the observed deviation s_t^i and a threshold v^i as follows:

- if $s_t^i > v^i$, then the agent will estimate its expectation of the return for the next trading period $t+1$ using fundamental information;
- otherwise, its expectation will be based on chartist information;

where the deviation is $s_t^i = e^{|u_t^i|}$ with u_t^i shown in (1), and the threshold is $v_i = 1 + |N(0, \sigma_2^2)|$. The greater the threshold value is the more reluctant is the agent to adopt a fundamental behavior. As the deviation from the fundamental value increases more agents estimate their expectations using a fundamentalist approach and vice versa.

2.1 The Fundamental and Chartist Investment Approaches

The future return expected in t by an agent i with fundamental behavior is

$$\hat{r}_{t+1}^i = u_t^i + \varepsilon_t^i, \tag{2}$$

where u_t^i is the deviation between the fundamental value perceived by agent i and the actual price in t, see Eq. (1), and ε_t^i is a noise component that follows $N(0, \sigma_3^2)$ and changes every period for each agent. Fundamental investors consider this deviation as a source of arbitrage opportunity. Consequently, they expect the market to reverse to the fundamental value. In our ASM, the fundamental value f_t is represented by a time series similar to a random walk, whose returns are uncorrelated and do not exhibit clusters of volatility. It is given by

$$f_{t+1} = f_t e^{\gamma_t}, \tag{3}$$

where γ factor follows a random variable $N(0, \sigma_7^2)$. The time-varying fundamental value is inspired by the model proposed in [4], where a real-life time series is used to represent an international factor that affects the market price along with the fundamental and the chartist components.

On the other hand, agents with chartist behavior will use two well-known technical strategies: trend following and support and resistance bouncing. Under normal circumstances, a chartist agent will behave as a trend follower. However, if its expected price estimated with the trend following strategy falls out of the channel that it perceives, then the agent will ignore this price and will follow a support and resistance bouncing strategy.

The **trend following** behavior is implemented in a similar way to that in [1]. Each agent estimates the future return using an average of the last returns

$$\hat{r}^i_{t+1} = g^i_2 \bar{r}_{L_i} + \varepsilon^i_t, \tag{4}$$

where ε^i_t is a noise component that follows $N(0, \sigma_3^2)$, g^i_2 represents the bias that the agent introduces in the average trend estimation that follows a random variable such as $N(1, \sigma_4^2)$ with $g^i_2 >= 0$, and $\bar{r}_{L_i}$ is the mean of the last L_i returns,

$$\bar{r}_{L_i} = \frac{1}{L_i} \sum_{j=0}^{L_i} \ln p_{t-j} - \ln p_{t-j-1}, \tag{5}$$

where L_i is different for each agent i and follows a discrete uniform distribution $U(1, L_{max})$.

The **support and resistance bouncing strategy** requires that each agent draws its own support and resistance line. This is done as follows

$$S^i_t = S_t(1 - g^i_3), \text{ and } R^i_t = R_t(1 + g^i_3), \tag{6}$$

where S_t and R_t are the support and resistance levels at t and g^i_3 follows a random variable $N(1, \sigma_5^2)$ with $0 <= g^i_3 <= 0.5$. In our market, S_t and R_t remains constant for each trading session. For the sake of simplicity it is assumed that all the agents estimate the levels S_t and R_t using the same procedure and for the same time horizon. These levels are estimated using an algorithm that uses exponential weighted moving averages and looks for local maxima and minima, which are used to determine the support and resistance levels, respectively. However, some changes are introduced to update the local maxima and minima lists in a faster way. As g^i_3 is different for each agent, it makes possible for the agents to use S_t and R_t to implement the bouncing strategy in its own way.

The bouncing strategy is implemented as follows. If the trend-following expected price crosses the support or the resistance level, chartist agents believe that the price will bounce and will stay within the channel defined by the support and the resistance. A chartist agent i adopts this strategy if these three conditions are satisfied

$$p_{t'-1} > S_t^i,\ p_{t'-1} < R_t^i, \text{ and } \frac{R_t}{S_t} >= \mu, \tag{7}$$

where μ is a constant ensuring a minimum distance between support and resistance, and $p_{t'-1}$ represents the closing price of the past trading session $t' - 1$.

The future price estimated by chartist agents with bouncing strategy is given by

$$\begin{aligned} &\text{if } p_t e^{\hat{r}_{t+1}^i} < S_t \text{ then } \hat{p}_{t+1}^i = R_t^i, \\ &\text{if } p_t e^{\hat{r}_{t+1}^i} > R_t \text{ then } \hat{p}_{t+1}^i = S_t^i, \end{aligned} \tag{8}$$

where $\hat{r}_{t+1}^i$ is the trend-following expected return for $t+1$ given in eq.(4). The strategy expects the price to bounce, so the suitable idea is to place the order in close to the opposite boundary of the channel.

2.2 *The Bid and Ask Orders*

The future price expected by agent i follows the method used in [1], given by

$$\hat{p}_{t+1}^i = p_t e^{\hat{r}_{t+1}^i}, \tag{9}$$

where p_t is the current price and $\hat{r}_{t+1}^i$ is the return expected according to its strategy, which can be fundamentalist (2) or chartist with trend-following behavior (4). For the case of agents with a chartist bouncing-based strategy, they estimate the future price using the rules in (8), as has been explained.

Given the future price expected $\hat{p}_{t+1}^i$, each agent i decides whether bidding or asking and whether placing a market-price or a limit order. If the agent expects a price increase, it decides to bid. Otherwise, it decides to ask. The agent bids (resp. asks) at a price b_t^i (resp. a_t^i) lower (resp. higher) than his expected future price $\widehat{p_t^i + 1}$. These prices are determined by

$$b_t^i = \hat{p}_{t+1}^i (1 - k^i) \tag{10}$$

$$a_t^i = \hat{p}_{t+1}^i (1 + k^i), \tag{11}$$

where $k^i \sim N(0, \sigma_6^2), 0 <= k^i <= 0.5$.

If b_t^i (resp. a_t^i) is smaller (resp. greater) than the current quoted ask (resp. bid) the agent submits a limit order at b_t^i (resp. a_t^i). While if b_t^i (resp. a_t^i) is larger (resp. smaller) than or equal to the current quoted ask (resp. bid) the agent submits a market order and trade at a_t^q (resp. b_t^q).

3 Simulation and Results

The simulation was carried out with 5000 traders during 1000 sessions. The number of traders was adjusted after several simulations in order to ensure liquidity in the market and heterogeneity of orders. A lower number of traders makes the bid-ask spread to increase dramatically. The agent parameters were $\sigma_1^2 = 0.01$, $\sigma_2^2 = 0.04$, $\sigma_3^2 = 1E^{-6}$, $\sigma_4^2 = 1$, $\sigma_5^2 = 1E^{-4}$ and $\sigma_6^2 = 1$. The parameter $\sigma_7^2 = 2.89E^{-4}$ represents the variance of the SP500 daily returns from March 2008 to October 2009. Other simulation parameters were $Lmax = 200$, $\mu = 1.001$, $m = 5$ and $n = 3$. The warm-up period was 1000 trading sessions. This period is required to initialize the algorithm that draws the support and resistance lines.

3.1 Price Evolution and Technical Patterns

Figure 1 shows the simulation from $t' = 550$ to $t' = 1000$. First of all, it can be seen that the support and resistance lines drawn by the agents actually enclose the simulated price. During long periods, the price bounces between these lines, e.g. $t' = [640, 770]$ and $t' = [800, 960]$. This is the self-fulfilling prophecy effect. However, when the pressure made by fundamental agents is big enough, i.e. when the distance between the fundamental value and the price is deemed excessive for most agents, these lines are broken and the price converges towards the fundamental value; e.g., around $t' = 775$ and $t' = 965$. It is interesting to see how the simulated price reflects the trend of the fundamental value, even if it does it from a certain distance and while bouncing within the channel drawn by the support and resistance lines; e.g., see the upwards and downwards trends in $t' = [800, 950]$.

It is interesting to remark that resistance and support lines are estimated in real time. At the beginning of each session t', the support and resistance levels, $S_{t'}$ and $R_{t'}$, are updated taking into account only previous information. As a result, these levels are step-wise. However, in many cases, these lines are quite similar to those that can be drawn taking into account the whole time series, e.g., in $t' = [800, 900]$.

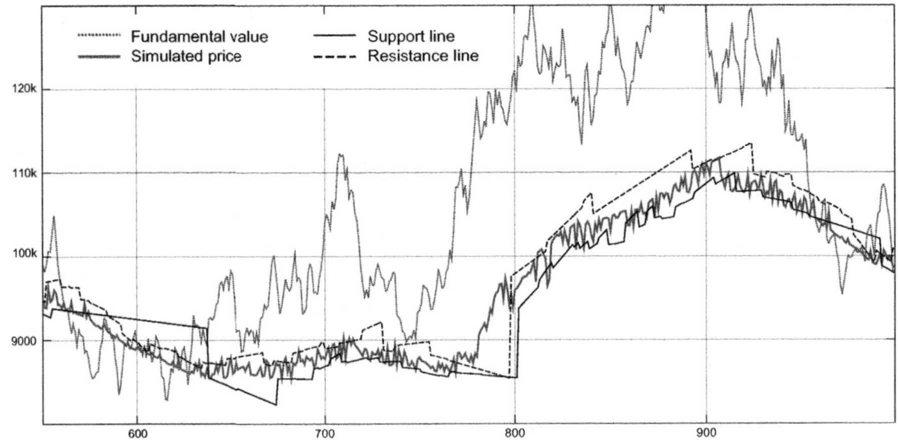

Fig. 1 Close price series of the simulation and the fundamentalist reference.

However, when a change in the trend occurs, the algorithm takes some time to detect that the change is significant enough to correct the levels.

3.2 *Stylized Facts in the Simulated Prices*

Simulated stock market prices should reproduce some statistical facts that are present in real-life prices. Some of those stylized facts are: unpredictable asset returns with almost no autocorrelation and slow decay of autocorrelations of square returns and absolute returns [6, 7]. Figure 2 shows that our simulated prices reproduce these stylized facts, while the fundamental value used shows no autocorrelation in the returns, absolute returns and squared returns time series.

In our ASM, the autocorrelation function (ACF) of the returns reveals no linear structure. Moreover, the autocorrelation of the first lag is negative and greater than for the rest of the lags, as in real-life financial market returns. The ACFs of the

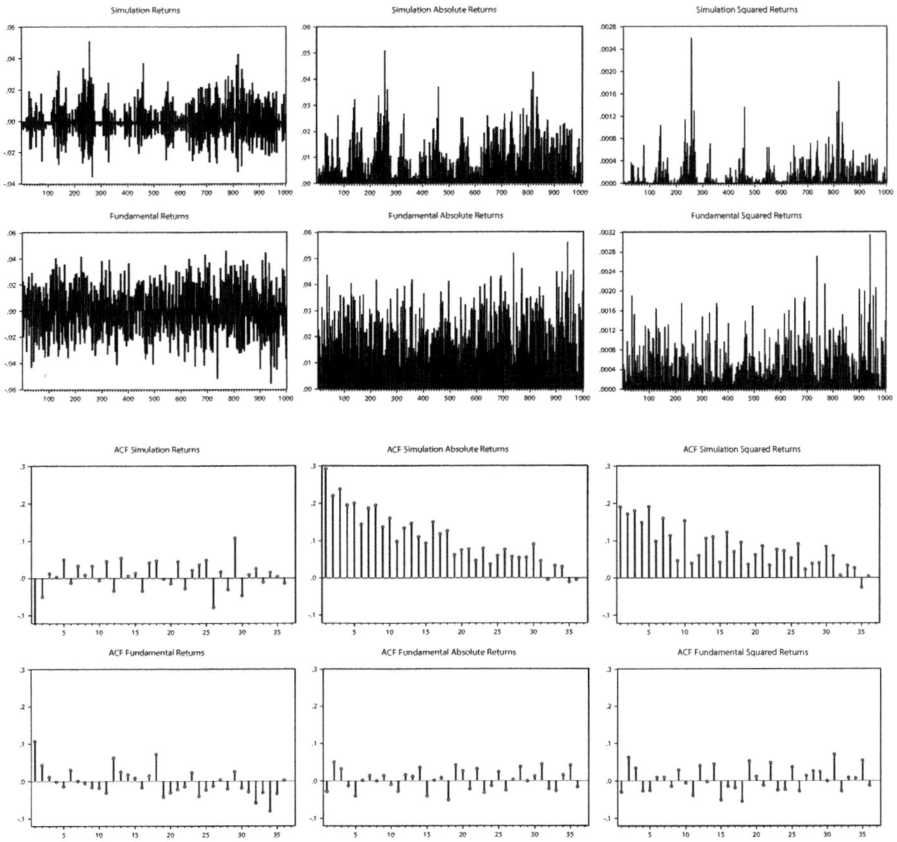

Fig. 2 Returns and autocorrelation functions of simulated market and fundamental data

absolute and the squared returns show positive and slow decaying autocorrelation values that characterize long volatility clustering, as in real-life financial markets.

4 Conclusions

This ASM can be considered as a starting point that can be used to reproduce more sophisticated behavior. For example, agents could be guided by expected utility maximization and risk averse behavior. Agents could also adopt strategies with different time horizons. Regarding the auction system, the market could be extended to allow agents to trade more than one stock at a time. These new features should help to reproduce different stylized facts and technical patterns.

Acknowledgments. Work supported by grants TIN2008-06464-C03-01 (Spanish Council for Science and Innovation) and Banco Santander - UCM GR58/08.

References

1. Chiarella, C., Iori, G.: A simulation analysis of the microstructure of double auction markets. Quantitative Finance 2(5), 346–353 (2002)
2. Chiarella, C., Iori, G., Perelló, J.: The impact of heterogeneous trading rules on the limit order book and order flows. Journal of Economic Dynamics and Control 33(3), 525–537 (2009)
3. Frankel, J., Froot, K.: Chartists, fundamentalists and the demand for dollars. In: Private Behaviour and Government Policy in Interdependent Economies, pp. 73–126. Oxford University Press, New York (1990)
4. de Jong, E., Verschoor, W.F., Zwinkels, R.C.: Behavioural heterogeneity and shift-contagion: Evidence from the asian crisis. Journal of Economic Dynamics and Control 33(11), 1929–1944 (2009)
5. LeBaron, B., Arthur, W.B., Palmer, R.: Time series properties of an artificial stock market. Journal of Economic Dynamics and Control 23(9-10), 1487–1516 (1999)
6. Lux, T., Marchesi, M.: Scaling and criticality in a stochastic multi-agent model of a financial market. Nature 397(6719), 498–500 (1999)
7. Lux, T., Marchesi, M.: Volatiliy clustering in financial markets: A micro-simulation of interacting agents. International Journal of Theoretical and Applied Finance 3(4), 675–702 (2000)
8. Mathieu, P., Brandouy, O.: A generic architecture for realistic simulations of complex financial dynamics. In: 8th Int. Conference on Practical Applications of Agents and Multiagent Systems, PAAMS 2010. AISC, vol. 70, pp. 185–197. Springer, Heidelberg (2010)
9. Palmer, R.G., Arthur, W.B., Holland, J.H., LeBaron, B., Tayler, P.: Artificial economic life: a simple model of a stockmarket. Physica D 75, 264–274 (1994)

VigilAgent Methodology: Modeling Normal and Anomalous Situations

José Manuel Gascueña, Elena Navarro, and Antonio Fernández-Caballero

Abstract. This paper describes the experience and the results of using the *VigilAgent* agent methodology to develop surveillance systems. This work is based on the suitability of agent characteristics for developing surveillance systems. It is usual to develop them following an ad-hoc approach instead of using a methodology to guide stakeholders in achieving quality standards expected from commercial software. The experimental setting is based on the development of an application for monitoring and controlling the normal and anomalous situations that happen when humans access to a specific area.

Keywords: Surveillance systems, Agent methodology.

1 Introduction

In the last few decades, it is usual to install security systems [9] in environments such as bank, parking, motorway and underground to protect humans from attacks or burglaries. The development of surveillance systems is very complex as they must work in highly dynamic and heterogeneous environments. They deploy, in the observed scenarios, several kinds of sensors that perform actions with certain degree of autonomy to collect information about their surrounding area, and to cooperate in the recognition of special situations in a semi-automatic way. The characteristics of autonomy and cooperation are often cited as the rationale of why multi-agent systems (MAS) are especially appropriate for surveillance tasks [8], [6]. In fact, agent technology has already been used in several surveillance systems [3]. However, to the best of our knowledge, they are usually developed following an ad-hoc approach without using a methodology that guides stakeholders in achieving quality standards

José Manuel Gascueña · Elena Navarro · Antonio Fernández-Caballero
Universidad de Castilla-La Mancha, Departamento de Sistemas Informáticos & Instituto de Investigación en Informática de Albacete, 02071-Albacete, Spain
e-mail: jmanuel@dsi.uclm.es

J.B. Pérez et al. (Eds.): Highlights in PAAMS, AISC 89, pp. 27–35.
springerlink.com

expected from commercial software. So, this paper proposes the introduction of an Agent-Oriented Software Engineering (AOSE) methodology, named *VigilAgent*, to carry out well documented surveillance applications throughout the different phases that form the development process.

Specifically, it is described how *VigilAgent* is applied to face a particular surveillance task supported in these systems such as monitoring and controlling the access to a specific area. Assigning human personal to carry out this task is a common solution. However, this is an expensive solution where human's attention decreases after some elapsed time working or watching video. Thus, it sounds interesting to perform this task automatically via software which will be responsible for collecting information to inform when some anomalous situation occurs. In addition, it is relevant to record normal situations to guarantee some usual activities; for instance, that the current occupation capacity does not exceed above a threshold. So, a company will take knowledge about economical losses arising for cunning users, where it is necessary to have more personal to prevent risky situations. The cost of the personnel is decreased because it is not necessary to have a guard in every access control point.

2 VigilAgent Methodology

The five phases of *VigilAgent* are briefly described next. (1) *System specification* - the analyst identifies the system requirements and the environment of the problem, which are obtained after several meetings arranged with the client; (2) *Architectural design* - the system architect determines what kind of agents the system will have and how the interaction between them will be; (3) *Detailed design* - the agent designer and application designer collaborate to specify the internal structure of each entity that will make up the system overall architecture produced in the previous phase; (4) *Implementation* - the software developer generates and completes the application code; and (5) *Deployment* - the deployment manager deploys the application according to a specified deployment model.

At this point, it is worthy of notice several issues about this development process. The first one is that phases named system specification and architectural design in *VigilAgent* are the two first phases of Prometheus methodology [5]. Another detail is that the third phase of *VigilAgent* (detailed design) uses models of INGENIAS [7]. Finally, notice that code is generated and deployed for ICARO-T framework [2]. Several reasons that are introduced in the following have conducted to this integration.

Prometheus is significant because of the guidelines it offers to identify which the agents and their interactions are. Another advantage of Prometheus is the explicit use of the concept *scenario* which is closely related to the specific language used in the surveillance domain. Indeed, a surveillance application is developed to deal with a collection of scenarios. Nevertheless, notice that Prometheus last phase has not been integrated in *VigilAgent* because it focuses on BDI agents and how the entities obtained during the design are transformed into concepts used in a specific agent-oriented programming language named JACK [1]; this supposes, in principle, a loss of generality. On the contrary, INGENIAS does facilitate a general process to transform models specified during the design phase into executable code. However, INGENIAS does not offer guidelines to identify the entities of the model; the developer's experience is necessary for their identification. Therefore, *VigilAgent* methodology is not developed from scratch but integrates facilities of both Prometheus and INGENIAS to take advantage of both of them.

Regarding implementation, the ICARO-T framework has been selected because it provides high level software components that facilitate the development of agent applications. Moreover, it is independent of the agent architecture; that is, the developer can develop new architectures and incorporate them in the framework. This is a clear difference regarding other agent frameworks such as JACK or JADE [1], which provide a middleware, instead of an extensible architecture, to establish the communications among agents. An additional advantage are the functionalities already implemented in the framework to automatically carry out component management, application initialization and shutdown, reducing the developers' amount of work and guarantying that the components are under control. These last functionalities are usually not provided by other frameworks.

3 Case Study: Access Control

Access control is the usual and basic term used for monitoring and controlling entrances to and exits from a specific area. This section illustrates how to use *VigilAgent* in order to develop an intelligent system that automatically controls entrances/exits of humans to/from an enclosure throughout the installed modules. Specifically, each module facilitates the entrances and exits according to its configuration and is composed of the following components: a reader device, an automatic door, a contact sensor and an infrared sensor.

In order to go in/out of the enclosure throughout a module, first, the user inserts a ticket into the reader device that the system verifies against the users' database. Then, a LED illuminates in green if the user is authorized, otherwise it illuminates in red. Moreover, if the user is authorized then the door is opened, and closed once the user has crossed or some time has elapsed.

In addition, the system collects and shows the guard statistics about the number of humans crossing each door and the number of humans located inside the enclosure by using the infrared sensor located in each module. It should also control if any anomalous situation happens, such as tailgating or if a door is blocked by a human when the system opens it. A *tailgating* situation is detected when some cunning human crosses a door that has been opened by a user correctly authenticated. The system also shows the state of the devices and offers the guard the possibility of disabling a module if its door remains closed despite having correctly authenticated a user.

3.1 System Specification

Usually, the system specification phase begins with the *analysis overview diagram*, which shows the interactions between the system and the environment (see Fig. 1a). At this level, firstly, an *actor* for each device of a module (reader, door, infrared and contact sensors) has been identified; there is also a *Central_A* actor representing the user interface that supports the human interaction with the system, that is, it shows the monitored activities to the security guard, and the commands he/she can send to the system to disable modules. Moreover, on the one hand, the information that comes from the environment is identified as *percepts*. For example, the code associated to the card introduced by the user into the reader (*Ticket_p*) and the signal captured by the infrared device when its beam is broken (*BrokenBeam_p*). On the other hand, every operation performed by the system on the actors is identified as an *action*. For example, the commands issued to open and close a door (*OpenDoor_a* and *CloseDoor_a*). Finally, relations with the scenarios identified to control the entrance, the exit and anomalous situations are established (see *Tailgating* and *Sabotage Door* scenarios).

A scenario is a sequence of structured steps - labeled as action (A), percept (P), goal (G), or other scenario (S) - that represents a possible execution way of the system. As an example, Fig. 1b illustrates the process performed by the system to control the crossing of authorized users through a module. For every scenario a goal is identified that represents the goal to be achieved by the scenario. In our multi-agent system approach, several agents communicate and coordinate to pursue the common general goal *Control entrance and exit*. In the *goal overview diagram*, this general goal has been refined into five goals (*Control Entrance*, *Control Exit*, *Tailgating*, *Sabotage Door*, and *Disable Module*) related to the scenarios identified. Similarly, some of these goals have also been decomposed into several sub-goals to denote how to achieve each parent goal. Finally, roles are identified by clustering goals, and linking perceptions and actions. Table 1 summarizes the roles identified.

Table 1 Roles description

Role	Description
User identification_R	It manages the user identification process to gain access to or come out of enclosure.
Facilitate access_R	It manages the door actuator to provide the users authenticated correctly with access.
Control infrared_R	It aims to update statistics when the infrared sensor detects a person located in front of it.
Control contact_R	Its objective is to monitor the door state thanks to the information perceived by the contact sensor installed in the door.
Management tailgating_R	It is responsible for detecting and notifying when the anomalous situation named tailgating happens.
Management sabotage door_R	It is responsible for detecting and notifying when someone, different from the security guard, blocks the door.
Intercommunication_R	It is responsible for managing the communication between software entities for person access control.

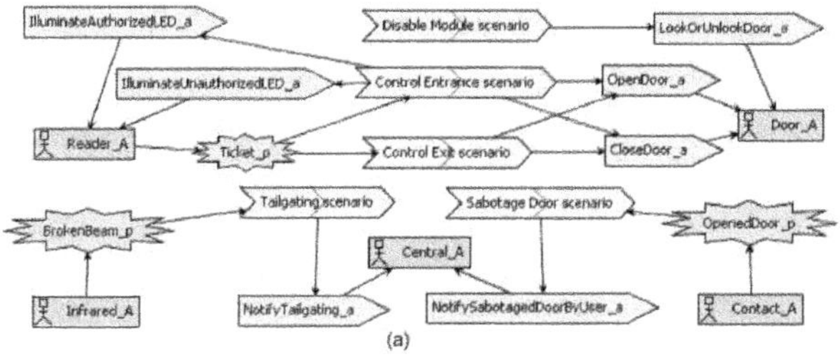

(a)

	Type	Name	Role
1	P	ClosedDoor_p	Control contact_R
2	P	Ticket_p	User identification_R
3	G	Check identification	User identification_R
4	A	IlluminateAuthorizedLED_a	User identification_R
5	A	IlluminateAuthorizedLEDCentral_a	User identification_R
6	G	Manage opening process	Facilitate acces_R
7	G	Control door state	Control contact_R
8	A	DelayClosingDoor_a	Facilitate acces_R
9	P	BrokenBeam_p	Control infrared_R
10	G	Update entrance traffic	Control infrared_R
11	G	Manage closing process	Facilitate acces_R
12	G	Control door state	Control contact_R

(b)

Fig. 1 (a) Analysis overview diagram; (b) Steps of Control Entrance scenario.

3.2 Architectural Design

Usually, applications need also reading and/or writing data to achieve the functionality related with some roles. This information is specified during the architectural design phase in the *data coupling diagram*. A data is information that will be managed by agents or beliefs representing agent knowledge about the environment or itself. For example, a data (*BDD*) should be available to validate every ticket introduced. Furthermore, there are data to count how many people cross each module (*Traffic*), to have control on the enclosure capacity level (*OccupationLevel*) and determine if a user goes in or goes out the enclosure (*ModuleConfiguration*).

Once the functionality of the roles has been described through their relations with other entities (percepts, actions, goals and data), the guidelines offered by Prometheus to decide the types of agents that will be included in the system are applied during the second phase namely architectural design. In general terms, these guidelines consist of grouping roles to obtain the system. The cohesion and coupling criteria are used to decide which the best groupings are to ease their maintenance. These two concepts are essential to obtain a suitable software development that will be distinguished by its maximum cohesion and minimum coupling. In our case study, to achieve the system requirements, an agent is introduced for each role identified.

Another relevant task performed in architectural design phase is to define the agent conversations (Interaction Protocols, IP) in order to describe what should happen to realize the specified goals and scenarios. Interaction Protocols are a graphical representation that shows (i) interactions among agents, and (ii) interactions among agents and the environment. For example, Fig. 2a details the *Tailgating_IP* interaction protocol internal structure, a sub-protocol of *Access_IP*. As can be noticed, it involves two agents and two actors (identified by the dotted squares in the diagram) to detect tailgating situations. This situation happens when the door remains open after an authorized person has passed through it and the related infrared sensor has been activated once. From this moment, whenever the *InfraredSensorAg* agent perceives that the infrared sensor beam has been broken again, a new crossing through the module is counted (*IncreaseTrafficCounter_a*), the number of people inside the enclosure is updated (*IncreaseOccupationCounter_a*), and the detection is

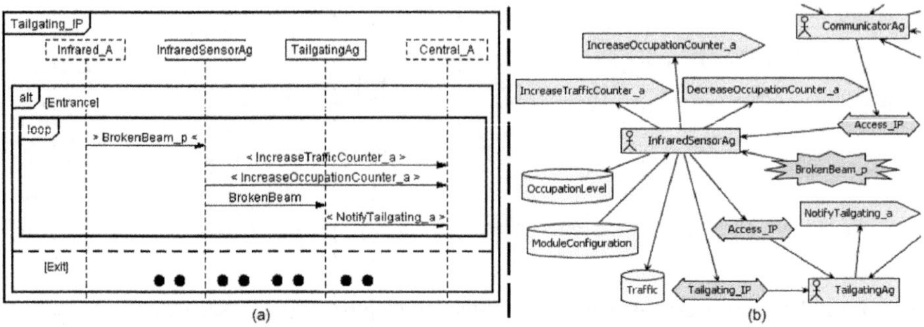

Fig. 2 (a) Tailgating interaction protocol; (b) Fragment of system overview diagram.

communicated to the *TailgatingAg* agent sending a *BrokenBeam* message. Finally, the *TailgatingAg* agent executes the *NotifyTailgating_a* action in order to notify the *Central_A* actor an unauthorized access. This checking is carried out until the *ContactSensorAg* agent notifies the *TailgatingAg* agent that the door is closed thanks to the *Access_IP* protocol. Tailgating detection is carried out in the same way for entrance and exit alternatives.

Fig. 2b shows a fragment of *system overview diagram* for our system design that includes the Tailgating Interaction Protocol. It can be observed that once roles have been grouped into agents, the information about percepts, actions, and data related to roles is automatically propagated and linked with the agents in the system overview diagram. Finally, the *agent acquaintance diagram* that contains the communication links between agents is automatically generated using the messages of information included in the interaction protocols. In short, every agent that perceives information captured by a device uses an intermediate agent (*CommunicatorAg*) to establish its communication. On the other hand, the agents responsible for detecting anomalous situations (*TailgatingAg* and *SabotageDoorAg*) can benefit from the knowledge offered by agents related to the contact and infrared sensors (*InfrarredSensorAg* and *ContactSensorAg*) to carry out their tasks.

3.3 Detailed Design

Now, we are in front of one of the major aspects and contributions of our proposal. The system overview diagram previously obtained (see Fig. 2b) describes the overall system architecture, that is, the agents identified, the perceptions received, the actions to be executed, the interaction protocols among agents, and the data read and/or written by agents. These entities are described using Prometheus concepts. However, the *VigilAgent* models of the detailed design phase are developed using INGENIAS concepts, which are different from the previous ones. Therefore, a transformation from Prometheus models to INGENIAS models must be carried out in order to perform the next modeling phase.

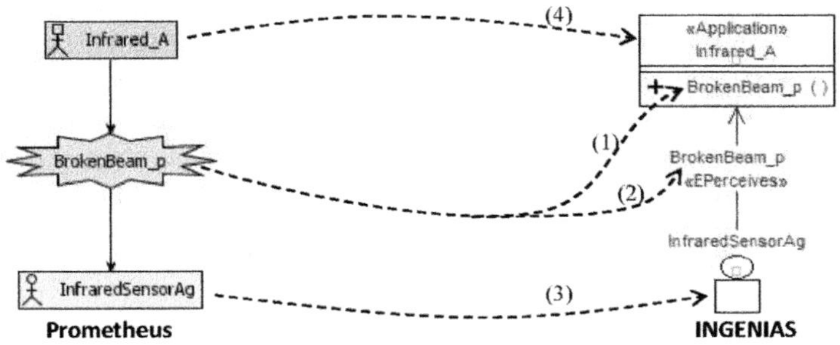

Fig. 3 Example of mapping information related with Prometheus percepts into INGENIAS.

We have developed four conceptual mappings to transform the structures that involve percepts, actions, messages and data related to agents. These mappings have been inferred considering both the definition of these concepts, and how each Prometheus structure can be modeled using an INGENIAS equivalent structure. For example, a *percept* is a piece of information from the environment received by means of a sensor. Percepts are sent by *actors* (Actor → Percept) by means of *interaction protocols*, these percepts are received by agents (Percept → Agent). The relations among actors, percepts and agents (Actor → Percept → Agent) are described in the *system overview diagram*. In INGENIAS, all the software and hardware that interacts with the system and cannot be designed as an agent is considered an *application*; and every agent that perceives changes in the environment must be in the *environment model* associated to an application. Therefore, as Fig. 3 shows (arrow 1), the percepts of a Prometheus agent can be triggered in INGENIAS by specifying a collection of *operations* in an *application*. A Prometheus percept has a field, *Information carried*, to specify the information it carries. As Fig. 3 depicts (arrow 2), in INGENIAS this information is described in a type of event named *ApplicationEventSlots* that is associated to the *EPerceives* relation established between the agent and the corresponding application. Notice that the Prometheus agent and actor concepts have been directly mapped to INGENIAS agent and application concepts (see arrows 3 and 4, respectively). Finally, let us highlight that action, message and data concepts in Prometheus are equivalents to operation defined in the applications, interaction unit and application concepts, respectively, in INGENIAS. The conceptual mappings have been automated specifying a Model-To-Model (M2M) transformation using the QVT language [4].

Once the transformation has been performed, three activities should be carried out to complete the modeling. Firstly, it is necessary to identify the tasks performed by each agent by examining the *interaction model* and the initial *environment model* which are automatically generated from the Prometheus model applying the transformation. Specifically, an agent should perform a task for every received perception or message. After that, a *state diagram* describing the behavior of each agent is specified using information about tasks, received percepts and messages. Secondly, the *Applications*, obtained after once the transformation is executed, only include methods to send perceptions to the agents and to specify actions to be executed by the agents. Therefore, these applications should be refined to include new methods depending on the specific needs of the system being developed. For example, in our case study, the *CheckUser* method was added to the *BDD* application to get information stored in a database, and the *GetId* method was added in *Reader_A* application to get its identifier. Finally, a model to describe the *application deployment* is specified attaching numbers in notes to represent the number of instances that there will be in execution for each type of entity.

3.4 Implementation and Deployment

Our approach considers that supporting tool for INGENIAS, Ingenias Development Kit (IDK) [7], is an exceptional agent tool to develop a Model-To-Text (M2T) transformation for generating code for any target language chosen, this is ICARO-T in *VigilAgent*, as it provides the necessary functionalities for developing new modules capable to carry out this task. These modules are developed following a general process based on both the definition of specific templates for each target platform, and procedures to extract information from INGENIAS models. In the literature there is no proposal for mapping INGENIAS models to ICARO-T code. So, our contribution to solve this gap has been to identify them and to develop modules to automatically carry out the Model-to-Text transformation from INGENIAS to ICARO-T.

The process for using our IDK modules in order to automatically generate ICARO-T code of the system being developed that can be updated with new code introduced by the user when necessary is carried out as follows. (1) The *INGENIAS ICARO-T Framework generator* module (*IIF*) is used to automatically generate code for the detailed design specification. IIF generates several XML files that describe the behavior of each agent, java classes for each agent and application, and the XML file describing the application deployment. (2) The developer manually inserts code in the protected regions of the generated java classes and implements those new classes he/she needs. (3) The developer uses the *ICAROTCodeUploader* module in order to update the model with the modifications introduced in the protected regions. Finally, the script file generated by IIF module will be executed by the deployment manager to launch the developed application.

4 Conclusions

The lack of consensus about which the best methodology to develop agent-oriented applications is has driven the main idea presented in this paper: not to propose a new methodology from scratch but to combine several methodologies, *VigilAgent*, to take advantage of each one of the analyzed methodologies. The learning curve of *VigilAgent* can be steep at first because users must get used to different terms that have the same meaning depending on the technology used in each phase (Prometheus and INGENIAS for modeling, and ICARO-T for implementation). However, this disadvantage is overcome thanks to the two transformations that are executed automatically. It is worth pointing out that the time spent learning how to develop and implement the INGENIAS ICARO-T Framework generator and the ICAROTCodeUploader modules described in section 3.4 was two months and fifteen days. This effort is rewarded as new applications are modeled and implemented. Our future work will be continue applying *VigilAgent* to face new cases study.

Acknowledgement. This work was partially supported by Spanish Ministerio de Ciencia e Innovación under TIN2010-20845-C03-01 and CENIT A-78423480 grants, and by Junta de Comunidades de Castilla-La Mancha under PII2I09-0069-0994 and PEII09-0054-9581 grants.

References

1. Bordini, R.H., Dastani, M., Dix, J., El Fallah Seghrouchni, A.: Multi-Agent Programming: Languages, Platforms and Applications. Springer, Heidelberg (2005)
2. Gascueña, J.M., Fernández-Caballero, A., Garijo, F.J.: Using ICARO-T framework for reactive agent-based mobile robots. In: Advances in Practical Applications of Agents and Multiagent Systems. Advances in Soft Computing, vol. 70, pp. 91–101 (2010)
3. Gascueña, J.M., Fernández-Caballero, A.: On the use of agent technology in intelligent, multi-sensory and distributed surveillance. The Knowledge Engineering Review (2011) (in press)
4. Object Management Group, Meta object facility (MOF) 2.0 query/view/transformation specification, version 1.0. OMG Document Number formal/2008-04-03 (2008)
5. Padgham, L., Winikoff, M.: Developing intelligent agents systems: A practical guide. John Wiley and Sons, Chichester (2004)
6. Patricio, M.A., Castanedo, F., Berlanga, A., Pérez, O., García, J., Molina, J.M.: Computational intelligence in visual sensor networks: Improving video processing systems. SCI, vol. 96, pp. 351–377. Springer, Heidelberg (2008)
7. Pavón, J., Gómez-Sanz, J., Fuentes, R.: The INGENIAS methodology and tools. In: Agent-Oriented Methodologies, pp. 236–276. Idea Group Publishing, USA (2005)
8. Pavón, J., Gómez-Sanz, J., Fernández-Caballero, A., Valencia-Jiménez, J.J.: Development of intelligent multi-sensor surveillance systems with agents. Robotics and Autonomous Systems 55(12), 892–903 (2007)
9. Räty, T.D.: Survey on contemporary remote surveillance systems for public safety. IEEE Transactions on Systems, Man, and Cybernetics - Part C: Applications and Reviews 40(5), 493–515 (2010)

Developing Secure Agent Infrastructures with Open Standards and Open-Source Technologies

Joan Bellver, Jose M. Such, Agustin Espinosa, and Ana Garcia-Fornes

Abstract. Security and privacy are key issues for open Multiagent Systems in which previously unknown parties can interact to each other. There are some Agent Platforms that provide support for both security and privacy. However, these Agent Platforms fail to provide openness and interoperability that are also crucial for developing secure and open Multiagent Systems. In this paper we present a security infrastructure that is based on open standards (AMQP, SSL, SASL, X.509 certificates, WS-Security, FIPA-ACL) and open-source technologies (Qpid, NSS, Axis2, Rampart). This infrastructure provides unique features regarding security and privacy on the one hand, and openness and interoperability on the other hand.

1 Introduction

As stated in [14], security related studies in the Multiagent Systems (MAS) research field have been increasing over the last few years due to the fact that an agent's incorrect or inappropriate behavior may cause non-desired effects such as money and data loss. Agent platforms (APs) provide all the basic infrastructure required to create a MAS [18]. There are many APs developed by the agent community – for a list of current APs and the features they provide refer to [1]. However, only a few of them currently take security concerns into account. For instance, Jade [7], Magentix [13], AgentScape [11], SECMAP [15], Tryllian ADK [19], Cougaar [10], and SeMoA [12] are security-concerned APs.

Privacy is also a great concern for users in the digital world. Indeed, just 10% of users are unconcerned about their privacy on-line [17]. Privacy requires security to control access and distribution of information [6]. From the above APs, Magentix,

Joan Bellver · Jose M. Such · Agustin Espinosa · Ana Garcia-Fornes
Departament de Sistemes Informàtics i Computació, Universitat Politècnica de València, Camí de Vera s/n, València, Spain
e-mail: {jbellver, jsuch, aespinos, agarcia}@dsic.upv.es

J.B. Pérez et al. (Eds.): Highlights in PAAMS, AISC 89, pp. 37–44.
springerlink.com

SECMAP, AgentScape, and Cougaar allow agent's principal identities to remain private by assigning a unique identity to each agent. Agents can then use their identity to interact to other agents preserving the privacy of their principals.

In order to avoid a lack of accountability that could cause a sense of impunity and encourage abuse, Magentix and AgentScape keep track of the association between principals and agents. This information is not known by the rest of the agents in the AP. Thus, principals' privacy is preserved while assuring accountability.

Agents developed in either Magentix or AgentScape cannot interoperate securely with agents developed using other APs. The security mechanisms for these APs only consider agents developed for running on top of them. As stated in [8], interoperability is crucial for the medium-term development of MAS. Moreover, Magentix only considers the implementation of secure but closed MAS, where an administrator must explicitly allow a principal to register as user of the AP. This is a major drawback for developing secure but also open MAS, in which, heterogeneous agents can enter and leave the MAS at any moment, must be able to interact to each other, and can span organizational boundaries [8].

In this paper, we present the security infrastructure we designed and implemented for the Magentix2 AP. This infrastructure takes into account security, privacy, interoperability, and openness. The remainder of the paper is organized as follows. Section 2 presents an overview of Magentix2. Section 3 presents the security infrastructure for Magentix2. Finally, section 4 presents some concluding remarks.

2 Magentix2 Overview

The Magentix2 AP [4] focusses on providing support for open MAS. The support that Magentix2 provides for agent interactions is composed of two main parts: agent communication and, on top of it, agent conversations. Magentix2 uses AMQP[1] [16] as a foundation for agent communication. This standard facilitates the interoperability between heterogeneous entities. Magentix2 allows heterogenous agents to interact to each other via messages represented following the FIPA-ACL [3] standard that are exchanged using the AMQP standard.

Several AMQP implementations are available[2]. We used the Apache Qpid[3] open-source implementation of AMQP for the implementation of the Magentix2 Agent Communication. Apache Qpid provides two AMQP servers, implemented in C++ (the one we use) and Java. Qpid also provides AMQP Client APIs supporting the following languages: C++, Java, C# .NET, Ruby and Python. Qpid allows distributed applications made up of different parts written in any of these languages to communicate with each other. What is more, any client developed using one of the Qpid Client APIs is able to communicate with any client developed using any other

[1] http://www.amqp.org/

[2] OpenAMQ http://www.openamq.org/,
RabbitMQ http://www.rabbitmq.com/ and Red Hat MRG
http://www.redhat.com/mrg/ are examples of such implementations.

[3] http://qpid.apache.org/

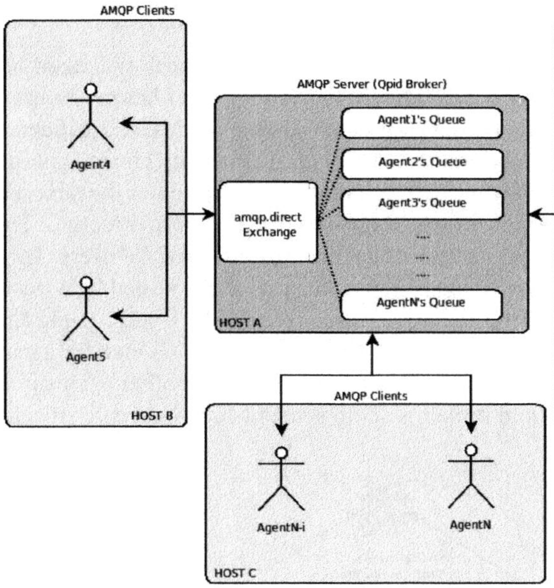

Fig. 1 Magentix2 Agent Comunication Architecture

AMQP-compliant API via any AMQP server implementation, as long as both server and clients implement the same version of the AMQP standard.

Figure 1 shows an overview of the Magentix2 agent communication architecture. Magentix2 is composed by one or more (in this case federated) AMQP Servers (QPid brokers). Magentix2 agents act as AMQP Clients (using Qpid Client APIs) which connect to the Qpid broker and, then, are able to communicate to each other. Magentix2 agents can be located in any Internet location, they only need to know the host on which the Qpid broker (or one of the federated Qpid brokers) is running.

Magentix2 provides a Java API to facilitate the development of agents. This API allows agent programmers to only focuss on creating FIPA-ACL messages and sending and receiving them, without dealing directly with the Qpid Client Java API. At this moment this API is only written in Java, but the existence of multiple QPid Client APIs for several programming languages enables the development of agents written in different programming languages. What is more, any proprietary implementation that follows both AMQP and FIPA-ACL standards would be interoperable with Magentix2 agents.

3 Magentix2 Security Infrastructure

In this section we explain the security infrastructure design and implementation for the Magentix2 AP. The building blocks of the Magentix2 security infrastructure are:

the Magentix2 PKI, the Magentix2 Management Service (MMS), the Magentix2 Agent Library (MAL), and the Qpid broker.

The Magentix2 PKI is set during installation time and is managed using the Mozilla Network Security Services[4] (NSS). During the installation process the Magentix2 Certificate Authority (MCA) is created. Using this MCA certificates for the MMS and the Qpid Broker are created. Agent certificates are created automatically and dynamically by the MMS at execution time (as described in the next section).

Figure 2 shows an overview of the Magentix2 security infrastructure. The MMS is in charge of controlling the creation of agent certificates for AP users. These users must authenticate themselves to the MMS with a certificate issued by a trusted third party certificate authority (the red one). Then, the MMS issues, using the MCA (the blue one) certificates for the agents (blue certificates). Once an agent gets a valid certificate issued by the MCA, it can communicate to other agents in a secure fashion. To this aim, the Qpid broker is set to accept MCA-issued certificates.

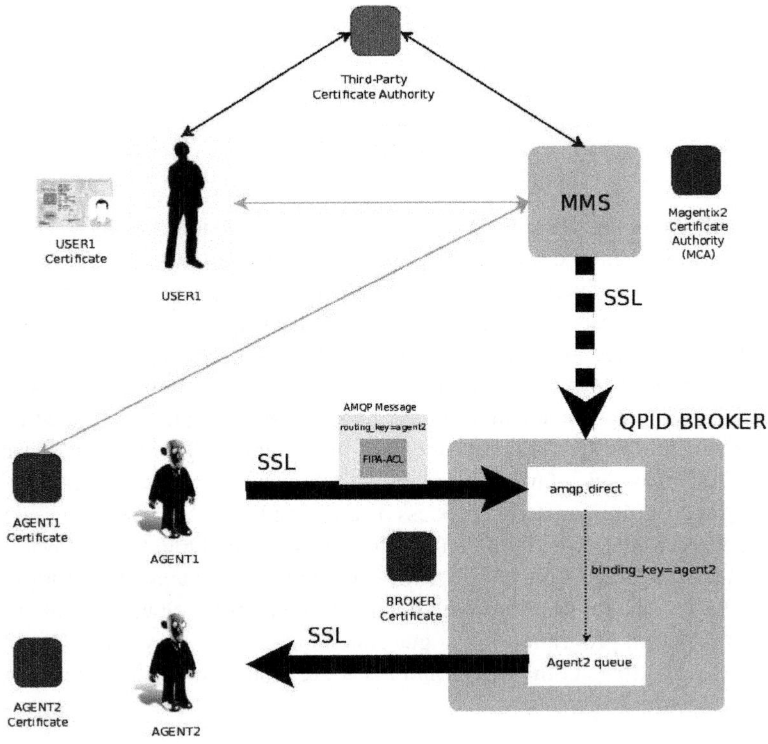

Fig. 2 Magentix2 Security Infrastructure

[4] http://www.mozilla.org/projects/security/pki/nss/

3.1 Magentix2 Management Service

The Magentix2 Management Service (MMS) is a WS-Security[5] compliant secure web service implemented using Apache Axis2[6] and its security module Apache Rampart[7]. WS-Security specifies how SOAP messages should be signed and encrypted. A secure service compliant with WS-Security is provided with basic security features, such as authentication, integrity, confidentiality and non-repudiation.

When an agent starts, if it does not have an agent certificate signed by the MCA, the MAL automatically calls the MMS to obtain one. This call is performed using WS-Security with X.509 certificates to obtain an identity and credentials for joining the AP. These X.509 certificates contains the identity of the principal behind the agent. Moreover, these certificates must be signed by a third party certificate authority accepted by the MMS. The administrator of a Magentix2 AP can, at any time, add and remove certificates of third-party certificate authorities in the MMS. If strong accountability is needed (e.g. law enforcement in case an agent misbehaves), the MMS can require certificates stating for the real identity of the agent's principal. For instance, in Spain, the MMS can require the X.509 certificates issued by either the Spanish Electronic Identification[8] (DNIe) or the *Fábrica Nacional de Moneda y Timbre*[9] (FNMT).

The MMS service needs two inputs, the name of the agent and its certificate; and produces one output, the agent certificate signed by the MCA. The name of the agent is its identity in the AP. The second input is a certificate for the agent containing the agent's public key. The agent key pair (private and public key) and this certificate is created by the MAL. The MMS is in charge of signing the certificate using the MCA.

The MMS keeps track of the association between the agent identity and the agent's principal identity. Moreover, the MMS is in charge of maintaining the Access Control List (ACL) file of the Qpid broker, to allow new agents to communicate with already existing agents (ACL files are explained in the next section).

The MMS provides the following key features regarding security, privacy, openness, and interoperability:

- **Security:** The MMS issues agent Certificates (using the MCA) proving their identity in the Magenix2 AP. The MMS also keeps track of the association between a principal and his/her agents. This is crucial for assuring accountability, e.g. in case of law enforcement.
- **Privacy:** The MMS does not make the association between users and agents public. Therefore, the rest of agents in a Magentix2 AP are not aware of this association.

---

[5] `http://www.oasis-open.org/committees/wss/`

[6] `http://ws.apache.org/axis2/`

[7] `http://ws.apache.org/rampart/`

[8] `http://www.dnielectronico.es/`

[9] `http://www.fnmt.es/`

- **Openness:** The MMS allows principals that are not previously known to launch agents that join the AP without requiring an administrator to perform the registration. It allows whatever principal holding a certificated from a trusted third party CA.
- **Interoperability:** The MMS is a Web Service with a well-defined interface and compliant with WS-Security. Thus, whatever web service client implementation following this interface and WS-Security is able to interoperate with the MMS. Moreover, the MMS is loosely coupled with the agent communication protocol (explained in the following section). The certificates it generates can be used for AMQP, HTTP or whatever other protocol allowing the use of certificates for securing communications.

3.2 Secure Agent Communication

Agent communication in Magentix2 is based on AMQP. The AMQP standard specifies secure communication by tunneling AMQP connections through SSL [5] (so-called amqps) and relies on the SASL standard [9] as the authentication framework. Qpid implements both SSL and SASL support.

In order to achieve a complete integration between SSL and SASL, we set the SASL mechanisms to EXTERNAL. The EXTERNAL mechanism specifies that SASL can use identities already authenticated using a mechanism *external* to SASL itself. In this way, agents are authenticated to the Qpid broker regarding their negotiated SSL security context, i.e., agents are authenticated considering their agent identities in their agent certificates (issued by the MMS using the MCA).

The AMQP connection of every agent to the Qpid broker is tunneled through SSL. Hence, the communication between two Magentix2 agents is provided with confidentiality, integrity and non-repudiation out of the box. To ensure authenticity of the sender identity in a FIPA-ACL message (recall that in Magentix2 FIPA-ACL messages are encapsulated into AMQP messages), an agent must verify that the identity of the sender in the AMQP sender message field is the same as the identity of the sender in the FIPA-ACL sender message field upon receiving a new message. This is performed automatically by the MAL, but must be considered for agents implemented in other APs when interoperating with Magentix2 agents.

The Qpid broker allows all authenticated AMQP clients to access all of its resources by default. This could cause, for instance, that an agent could be able to steal messages that are not directly addressed to itself. The Qpid broker allows to define access control permissions by means of an ACL file, that specifies for each authenticated entity its permissions. We set the default permissions to Magentix2 agents so that agents can only access resources for sending messages to other agents and read the messages directly addressed to themselves. In order to dynamically add these permissions for new agents, the MMS modifies this file when an agent calls the MMS to obtain its agent certificate and update the Qpid broker following the AMQP broker management protocol standard.

The agent communication mechanism provides the following key features regarding security, privacy, openness, and interoperability:

- **Security:** Agent communication is tunneled through SSL. Thus, agent communication is provided with confidentiality, integrity, and non-repudiation. Moreover, access to Qpid broker resources is controlled.
- **Privacy:** Agents can authenticate to each other using their agent certificate when they communicate to each other. Thus, they do not need to provide their principals' identities in interactions with other agents.
- **Openness:** The Qpid broker accepts connections from all the agents that hold an agent certificate issued by the MCA. Therefore, an agent that is not previously known can join the AP at any moment. This is a crucial feature for developing open MAS.
- **Interoperability:** Any agent implementation that is compliant with AMQP and FIPA-ACL will be able to communicate with Magentix2 agents. This is regardless the programming language they were developed in, and the AP and (remote) location they are running on.

4 Conclusions

In this paper we presented the Magentix2 security infrastructure. This infrastructure presents key features regarding security, privacy, openness and interoperability not provided by current APs. It is based on open standards (AMQP, SSL, SASL, X.509 certificates, WS-Security, FIPA-ACL) and open-source technologies (Qpid, NSS, Axis2, Rampart). It allows the development of not only secure but also open MAS in which previously unknown agents can enter and leave the MAS at any moment.

Further research is needed in order to improve privacy and interoperability in Magentix2. Agents are able to authenticate with their agent identity so that the identity of their principals remain private. However, the MMS keeps track of the agent's principal identity and its association to the agent identity. Thus, the MMS itself can be a threat for the privacy of the principals running agents on top of it. However, not knowing this association could result on a lack of accountability. We are currently exploring an extension of the MMS to use Privacy-Enhancing Identity Management Systems [2] that minimize data collection without losing accountability.

Acknowledgements. This work has been partially supported by CONSOLIDER-INGENIO 2010 under grant CSD2007-00022, and projects TIN2008-04446, TIN2009-13839-C03-01 and PROMETEO/2008/051.

References

1. Alberola, J.M., Such, J.M., Garcia-Fornes, A., Espinosa, A., Botti, V.: A performance evaluation of three multiagent platforms. Artificial Intelligence Review 34, 145–176 (2010)

2. Clauβ, S., Kesdogan, D., Kölsch, T.: Privacy enhancing identity management: protection against re-identification and profiling. In: DIM 2005: Proceedings of the 2005 Workshop on Digital Identity Management, pp. 84–93. ACM, New York (2005)
3. FIPA. FIPA ACL Message Structure Specification. FIPA (2001)
4. Fogués, R.L., Alberola, J.M., Such, J.M., Espinosa, A., Garcia-Fornes, A.: Towards Dynamic Agent Interaction Support in Open Multiagent Systems. In: Proceedings of the 13th International Conference of the Catalan Association for Artificial Intelligence, vol. 220, pp. 89–98. IOS Press, Amsterdam (2010)
5. Frier, A., Karlton, P., Kocher, P.: The secure socket layer. Technical Report MSU-CSE-00-2, Netscape Communications (1996)
6. Gangopadhyay, A. (ed.): Managing Business with Electronic Commerce: Issues and Trends. IGI Publishing, Hershey (2001)
7. JADE Board. Jade security guide, `http://jade.tilab.com`
8. Luck, M., McBurney, P., Shehory, O., Willmott, S.: Agent Technology: Computing as Interaction (A Roadmap for Agent Based Computing). AgentLink (2005)
9. Myers, J.: Simple authentication and security layer, sasl (1997)
10. Newman, A.E.: Cougaar developers guide (2004), `http://www.cougaar.org`
11. Quillinan, T.B., Warnier, M., Oey, M., Timmer, R., Brazier, F.: Enforcing security in the agentscape middleware. In: Proceedings of the 2008 Workshop on Middleware security, MidSec 2008, pp. 25–30. ACM, New York (2008)
12. Roth, V., Jalali-Sohi, M.: Concepts and architecture of a security-centric mobile agent server. In: ISADS (2001)
13. Such, J.M., Alberola, J.M., Espinosa, A., Garcia-Fornes, A.: A group-oriented secure multiagent platform. Software: Practice and Experience (2011) (in Press)
14. Such, J.M., Alberola, J.M., Garcia-Fornes, A., Espinosa, A., Botti, V.: Kerberos-based secure multiagent platform. In: Programming Multi-Agent Systems. LCNS, vol. 5442, pp. 197–210. Springer, Heidelberg (2009)
15. Ugurlu, S., Erdogan, N.: An overview of secmap secure mobile agent platform. In: Proceedings of Second International Workshop on Safety and Security in Multiagent Systems (2005)
16. Vinoski, S.: Advanced message queuing protocol. IEEE Internet Computing 10(6), 87–89 (2006)
17. Westin, A.: Social and political dimensions of privacy. Journal of Social Issues 59(2), 431–453 (2003)
18. Wooldridge, M.: An Introduction to MultiAgent Systems. Wiley, Chichester (2002)
19. Xu, H., Shatz, S.M.: Adk: An agent development kit based on a formal design model for multi-agent systems. Journal of Automated Software Engineering 10, 337–365 (2003)

A Fingerprint-Based Secure Communications Method

Angélica González, Rafael García Bermejo Giner, Eladio Sanz García, and Rosa María García

Abstract. This paper describes a session-based system for the exchange of information, the session key being based on a cypher key that is obtained from data acquired by scanning a finger print. The generation and use of the cypher key is carried out observing standards.

Keywords: Biometrics, Secure communications, cypher key, fingerprint.

1 Introduction

The main goal is to create an algorithm that makes it possible to generate cypher keys for symmetric cryptography from biometric features obtained from scanning a fingerprint. This mechanism offers various advantages when compared to methods currently used for the generation of symmetric cypher keys, since they are basically limited to randomly producing bytes for the key, or to the method known as PBE (Password-Based Encryption).

A second goal is to create a message-exchange mechanism that makes use of the previous algorithm, and which is to be secure, efficient, easy to implement and intuitive for users. These features stem from a requirement to provide secure exchanges of information between entities through an unsafe electronic channel, such as a public network. System security is based on the use of publicly available algorithms that follow international standards and have not been broken thus far, or at least not openly.

Not many works describe a procedure to generate a cypher key from a fingerprint. Most existing works use a fingerprint only for authentication, or they generate a static cypher key from it. Such is the case of [1] and [3].

Angélica González · Rafael García Bermejo Giner · Eladio Sanz García · Rosa María García
Computer Science Department, University of Salamanca, Spain
e-mail: {angelica, coti, esanz, rosa_mgs}@usal.es

J.B. Pérez et al. (Eds.): Highlights in PAAMS, AISC 89, pp. 45–52.
springerlink.com

2 System Description

The message exchange system is an application that makes use of an algorithm for the generation of cypher keys. The algorithm makes use of biometric information obtained from a person's fingerprint. The system is both secure and efficient, easy to implement and intuitive for users. The application makes use of the client-server architecture; server and client have been built separately. The server opens a port for client communication, which is based on the socket communication protocol. The server's functionalities are as follows:

- User registration within the system
- User authentication
- User database management
- Intermediation in the communication process between users, since no public key and no certification authority is used. The server must act as the certifying authority, and has to carry out the cyphering and de-cyphering of messages
- User authentication, which is done by comparing the stored fingerprint with the one received in the message[5]. This process is carried out by means of the API Fingerprint SDK 2009[7], developed by Griaule Biometrics

The client process' functionalities are as follows:

- To exchange messages with the server
- To capture the user's fingerprint, and to process said fingerprint by means of the Griaule library
- To deploy an interface for user interaction

2.1 Authentication

The system's authentication process comprises two stages: a registration phase and a proper authentication phase, which will start a session as a result[6].

The authentication stage involves sending a message from the user that wants to be authenticated to the server. This message includes the user's biometric information, as well as his or her name and email, which are the only items not cyphered. It must be remarked that it is very unlikely that two different readings of the same fingerprint will produce exactly the same results. Authentication algorithms perform their calculations by taking into account two groups of minutiae of both prints. This is done when the algorithm finds a major degree of correlation between them, according to an adjustable percentage. This percentage is not close to 100%, since this would make authentication practically impossible due to the large differences one finds between readings. The process makes use of the authentication functions found in the Griaule library, which is similar to other AFIS (Automated Fingerprint Identification System) programs [8] available on the market. Thus the program delegates responsibility for authorization to the library. The system makes use of the fingerprint stored in the database in order to compare it with the fingerprint received within the message, thus performing the authentication of the user. Once the user has been authenticated, the key that was received is used to generate the session's key, which is to be used just for that session, and will be discarded when the session concludes. This mechanism greatly strengthens the system's security, since it has been widely documented that attacks on cryptographic algorithms fail when disposable cypher keys are used, since the are employed just once.

2.2 *Message Cyphering*

The message cyphering mechanism is based on the Advanced Encryption Standard (AES) algorithm [9]. The AES algorithm is a symmetric block cypher that can use 128-, 192- or 256-bit keys in order to encode 128-bit blocks. The application makes use of 128-bit cypher keys in two combinations. The first one applies a CTR mode with no padding and the second one applies a CBC mode with padding according to the PKCS #7 [10] standard. The decision to use AES was taken due to its being a widely-accepted standard algorithm nowadays. In order to insure data integrity we have used the Secure Hash Standard SHA1 algorithm.

The system has been designed considering only the use of symmetric cryptography, since it is intended to be easy to implement and easy to use; nevertheless, all mechanisms have been implemented in a robust way, in order to insure authentication for the user, and also confidentiality, integrity and availability for information.

2.3 *Coding of Exchanged Messages*

Coding of messages is done by means of the ASN.1 (Abstract Syntax Notation) language, in its DER format. ASN.1 is both an ISO and ITU standard [11], based on coding messages in a certain specified format. This coding has been widely used in various fields of computing, electronics and communications. The DER coding has been used because it insures that given a data structure that must be digitally processed, it will produce a unique data sequence representation.

2.4 *Cypher Key Generation Algorithm*

Minutiae are also detected by the Griaule software, which processes the captured fingerprint and produces as a result a sequence of minutiae that characterize the print. Each of the minutiae consists of the following items:

- An integer value that represents the X coordinate of the minutia's location, and which is stored in a two-byte integer (short int).
- An integer value that represents the Y coordinate of the minutia's location, and which is stored in a two-byte integer (short int).
- An integer value that represents the orientation angle of the minutia. This value is measured in degrees and can take any value in the interval [1,360]. It is stored in an one-byte integer.

Fig. 1 Detection of minutiae in a fingerprint

- A one-byte integer value that represents the type of minutia. This depends on the software used for the identification of minutiae; in the case of Griaule, the API only detects the bifurcation and termination types.

The last byte of the minutia is the parity byte, which is calculated from the previous ones.

The algorithm has as a starting point the results obtained by the Griaule library, after processing the set of minutiae obtained by processing a fingerprint. The first step of the algorithm is to define an order relationship for all of the minutiae. This order is based on a geographic criterium, namely, the X coordinates of minutiae are taken into account first, and then the Y coordinates. Thus minutiae are sorted left to right first, and then top to bottom. The second step is a quantization of the number of minutiae. This is done by reducing the actual number of minutiae found to an integer value that belongs to the following range: $R = \{8, 16, 32, 64\}$, or more generally, $8 * 2^r$ where r belongs to the interval [0,3]. More formally, the quantized number of minutiae can be defined as follows:

1. Let n be the total number of minutiae found in a fingerprint $|n\varepsilon[8, 128]$. Less than 8 minutiae is taken to mean an erroneous reading of the print, and more than 64 would be a fairly strange result; the average number of minutiae in a fingerprint is normally about 35.

2. Let $\gamma \varepsilon R | \gamma = 8 * 2^r, | \gamma$ is the biggest allowable number that is not greater than n
3. Let $\Delta = n - \gamma$,

In order to quantify the number of minutiae, transforming it into one of the values available for γ, one must discard Δ minutiae, thus forming 2^r groups of 8 minutiae. In order to discard Δ minutiae, the procedure combines them, it does not simply discard some of them. This combination is done by taking the first $2 * \Delta$ minutiae, to which one applies the function **XORCrossed**. Thus they are reduced to just Δ minutia, and the number of minutiae has been quantified in such a way that it will be one of the values defined by δ.

Function XORCrossed intends to combine two minutiae by means of the XOR; to be precise, the function XORs the bytes of the minutiae, using an order of bytes that swaps their bytes. This operation reduces the number of minutiae.

Once the number of minutiae has been quantified, blocks of 8 minutiae are formed. The total number of blocks is 2^r and they are processed individually. The first part of processing is a cumulative sum of the angles of minutiae. This intends to establish a dependency on the order of minutiae, in such a way that the spectre of possible combinations is caused to grow, thus increasing the difficulty of a possible attack. More formally, the cumulative sum of the angles is defined as follows:

1. Let m_i be the ith minutiae within block B, $| i \varepsilon [1, 8]$
2. Let A_i be the two-byte segment within the minutia that represents the value of the angle $| A_i \varepsilon [1, 360], | i \varepsilon [1, 8]$

For all $m_i \varepsilon B, A_i = \sum_{j=1}^{i} A_j$, or in another way: $A_i = A_{i-1} + A_i$, for $i > 1$

The next step in the process carries displaces circularly (left to right or right to left) $(r+1)^2$ bits all of the minutia's bits, taking all 64 bits of the minutia as a single block, and according to the following rules:

1. displace $(r+1)^2$ bits left if the type of minutia m_i is equal to the type of minutia m_{i+1}. More formally: $m_i << (r+1)^2$, if $T_i == T_{i+1}, | i \varepsilon [1, 7]$.

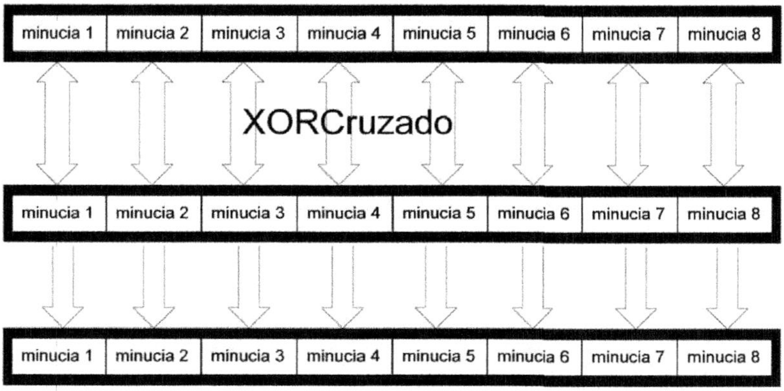

Fig. 2 Fusion of two blocks is carried out by means of function XORCrossed

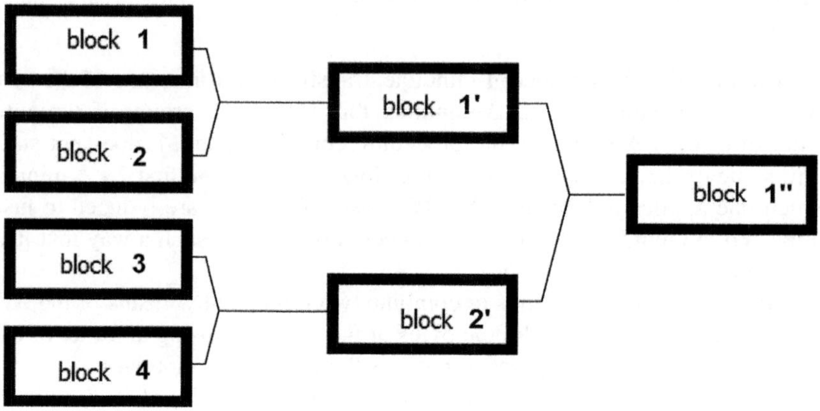

Fig. 3 Fusion of minutiae blocks

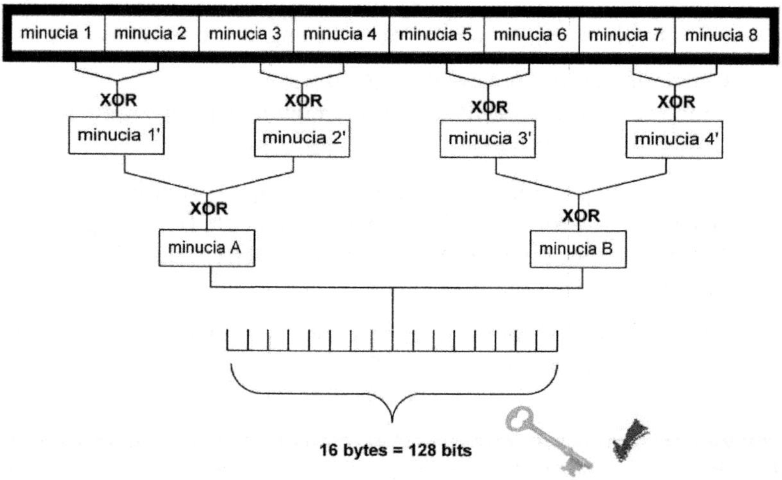

Fig. 4 Final process, by means of which a cypher key is created

2. displace $(r+1)^2$ bits right if minutia m_i is not the same as the type of minutia m_{i+1}. More formally:$m_i >> (r+1)^2, if T_i <> T_{i+1}, |i\varepsilon[1,7]$.
3. for the very last minutia, m_8, the same process is carried out, but comparison is made with the first minutia, m_1.

Once each block has been processed separately, the final process is a fusion of the various blocks in the way shown below. Thus a single block is obtained, from which the final cypher key will be generated.

Block fusion is carried out two by two, since one 2^r blocks. Fusion is performed by taking the block's minutiae and applying to them the XORCruzado function. More formally, block fusion is defined in the following way:

1. Let m_i be the ith minutia in block $B_1, |i\varepsilon[1,8]$.
2. Let n_i be the ith minutia in block $B_2, |i\varepsilon[1,8]$.

For each of the minutiae in blocks $B_1 and B_2$, one applies the XORCruzado function to $m_i and n_i$.

Block fusion is repeated until one obtains a final block When the final block has been obtained, there remains a last step in the process. The minutiae of the block are taken in pairs and XORed, taking each minutia as a block of 64 bits. The process finishes with just two minutiae, which are joined to form a 16-byte array, representing a 128-bit key.

3 Conclusions and Further Development

After checking various mechanisms, it was decided to use fingerprint's minutiae since they constitute a very secure authentication mechanism. All of the mechanisms employed follow standards, since the system was intended to completely follow international standards. The representation used for minutiae follows the NIST standard for the transfer and representation of fingerprints. The Griaule library takes care of this part of the process. Thus one can be sure that the system that has been created can be implemented independently of the libraries employed (Automated Fingerprint Identification Systems) and also independently of devices.

Once the devices and the libraries had been selected, the algorithm was designed by taking as a basis the most common symmetric cypher algorithms. Broadly speaking, the algorithm we propose makes use of the same operations that are implemented in the cypher algorithms. These include circular displacements, XOR and other algorithms used for the determination of parity bits. The concept that underlies the algorithm is the generation of a unique byte sequence. This byte sequence is to be created using as input information dependent of the minutiae obtained from reading a fingerprint. Calculations involve the operations mentioned above. The process involves firstly the definition of a unique representation of fingerprint data, and then processing that representation in order to eliminate any ambiguities. The security and robustness of the proposed algorithm is based on the fact that the various values that can be taken by the minutiae variables offer a very wide spectre of possible solutions. The algorithm is intended to be dynamic during its execution. It will not always follow the same execution path. This is due to the fact that the number of minutiae is variable, generally about 30. Besides, the coordinates and angles of the minutiae are not constant either, since the fingerprint can be displaced or rotated. The only values that should remain constant are the types of minutiae. Thus we get the advantage of changing session keys, and the communication system is more secure. To summarize, the application fulfils the requirements of information security and can be considered secure, robust, simple to implement and easy to use.

References

1. European Journal of Scientific Research 31(3), 372–387 (2009); ISSN 1450-216X ©EuroJournals Publishing, Inc. (2009), http://www.eurojournals.com/ejsr.htm
2. Irrevocable Cryptographic Key Generation from Cancelable Fingerprint Templates: An Enhanced and Effective Scheme. N. Lalithamani Assistant Professor, Department of Computer Science and Engineering Amrita School of Engineering, Amrita Vishwa Vidyapeetham Coimbatore, India. K.P. Soman- Professor and Head Centre for Excellence in Computational Engineering and Networking Amrita Vishwa idyapeetham Coimbatore, India
3. Secure and Private Fingerprint-based Authentication PhD Thesis submitted in fulfillment of the requirements for the Doctor of Philosophy in Mathematics by Arathi Arakala School of Mathematical and Geospatial Sciences Science, Engineering and Technology Portfolio RMIT University, Melbourne, Australia (November 10, 2008)
4. AFIS (Automated Fingerprint Identification Systems). A listing of web sites related to fingerprints for law enforcement. Other sites are government-related, and yet others deal with AFIS-manufacturing companies and related systems, http://onin.com/fp/afis/afis.html
5. Bajo, J., De Paz, J.F., Rodríguez, S., González, A.: Hierarchical neural network for clustering and classification. Logic Journal of the IGLP, http://bisite.usal.es/webisite/?q=en/node/160, ISSN 1368-9894 - Print ISSN 1367-0751
6. Pinzón, C.I., De Paz, J.F., Rodríguez, S., Corchado, J.M., Bajo, J.: A hybrid agent-based classification mechanism to detect denial of service attacks. Journal of Physical Agents (JoPHA) 3(3), 11–18 (2009), http://hdl.handle.net/10045/12528, ISSN: 1888-0258
7. http://www.griaulebiometrics.com/page/es/fingerprint_sdk
8. http://onin.com/fp/afis/afis.html
9. http://csrc.nist.gov/publications/fips/fips197/fips-197.pdf
10. ftp://ftp.rsasecurity.com/pub/pkcs/pkcs-7/pkcs-7v16.pdf
11. http://www.itu.int/ITU-T/studygroups/com17/languages/X.680-0207.pdf

Multiagent System for Indexing and Retrieving Learning Objects

Jonas Vian and Ricardo Azambuja Silveira

Abstract. This paper proposes a multiagent system application model for indexing and retrieving learning objects stored in different and heterogeneous repositories. The objects within these repositories are described by filled fields using different metadata standards. The searching mechanism covers several different learning object repositories and the same object can be described in these repositories by the use of different types of fields. Aiming to improve accuracy and coverage in terms of recovering a learning object we propose an information retrieval model based on the multiagent system approach and an ontological model to describe the knowledge domain covered.

Keywords: Recommender Systems, Learning Objects Repositories, Intelligent Learning Systems.

1 Introduction

The effort and investment needed by educators and educational institutions to produce qualified learning content are considerable. This is especially important in the production of content for distance education because almost all the information and knowledge must be completely and explicitly covered by the learning objects (Downes, 2001). Owing to the large cost of producing learning content, there is a strong advantage in improving the possibility of reusing it. Therefore, the development of mechanisms to facilitate the reuse of learning content has attracted the interest of several research groups, organizations and educational institutions around the world.

Nowadays the most well-known concept used to enable the organization and reuse of learning content is that of Learning Objects (LO) (McGreal, 2004). Research in this area has originated many different ideas, standards and specifications. This research produces different ideas that guide the politics of production and the storage of learning objects. At the moment it is not possible to highlight in

Jonas Vian · Ricardo Azambuja Silveira
Federal University of Santa Catarina (UFSC) Florianópolis - SC - Brazil
e-mail: jonas.vian@gmail.com, silveira@inf.ufsc.br

J.B. Pérez et al. (Eds.): Highlights in PAAMS, AISC 89, pp. 53–60.
springerlink.com

the current state of the art just one widely accepted standard or adopted model that could be used to guide these politics.

The recovery of learning objects is heterogeneous because of the different standards and specifications for production and storage, and the large distribution of the repositories across the web; it is dynamic as well because of the volume of research in this area and the magnitude of the field. Moreover, it is clear that the lack of effective and specialized LO search tools does not allow wide reuse of the learning objects produced metadata (Downes, 2001).

This paper proposes an LO intelligent search model capable of indexing and retrieving learning objects, regardless of the metadata standard used, located in different and heterogeneous repositories, using multiagent system technologies as well as domain ontologies. The existing tools are only usually based on a syntax search. This type of search is not efficient to recover learning objects as they used to be in retrieving ordinary documents in the web. Aditionaly the proposed model can find objects in a federated form of search according to CORDRA (2006) and FEB (2010), in different and heterogeneous repositories, using different sets of metadata and considering the semantic aspects of learning content.

2 Background

The theoretical framework of this research covers the basic concepts related to learning objects as well as formal specification of metadata for the LOs. Furthermore, the use of multiagent systems technology combined with retrieval techniques, usually used for Semantic Web to solve the problem of retrieving objects in distributed and heterogeneous environment using semantic aspects of learning object. The obtained results pointed to use representation of domain ontology to contextualize the domain specific concepts search problems and indexing and weighting information techniques to improve the accuracy, coverage and the performance of search tools.

Learning objects are educational resources that can be used in the learning process supported by technology (McGreal, 2004). The IEEE (IEEE-LTSC, 2005) defines learning objects as any material, digital or not, that can be used, reused and referenced in e-learning. The learning object is formally described by its metadata, stored in repositories, which can be combined with other learning objects to create larger objects such as lessons and courses (Nash, 2005).

Repositories of learning objects are mechanisms designed to store and organize learning objects so they can be located and retrieved. The learning object repository provides the interface for submitting, reviewing and recovering any object (Downes, 2001; Nash, 2005). The recovery is usually done through search engines, and the objects retrieved from a repository can be used directly by students, teachers and learning designers for authoring courses or used directly by any Learning Management Systems (LMS).

The learning objects repositories are usually based on database management technologies, LDAP, or XML, for ad hoc implementation but, recently have used specific specialized frameworks such as DSpace (DSPACE, 2010) and FEDORA (Flexible Extensible Digital Object Repository Architecture) (FEDORA, 2010).

As regards the theoretical background of agents we consider a multiagent system a loosely coupled network of problem solvers who work together to solve problems which go beyond their individual capability (Wooldridge, 2002). These problem solvers are essentially autonomous, distributed and heterogeneous in nature. We also adopt the FIPA reference model of agents (FIPA, 2010) and used JADE (Java Agent Development Framework) (JADE, 2010).For system modeling we adopt the O-Mase, meta-methodology (DeLoach et al. 2007).

Besides the theoretical background related to multiagent systems and learning objects we consider retrieval techniques used for finding (Manning et al. 2008) and techniques for knowledge representation to ontologies consisting of entities that can be classes, instances, relationships, and data types (Euzenat et al. 2007).

3 The Proposed Model

The objective of this model is to promote reusability of learning objects through the support of software agents capable of dealing with the heterogeneous scenario of learning object recovery environments. The proposed multiagent system model is designed to be able to index and retrieve learning objects in different repositories. The proposed multiagent system architecture for indexing and retrieval of learning objects is shown in Figure 1.

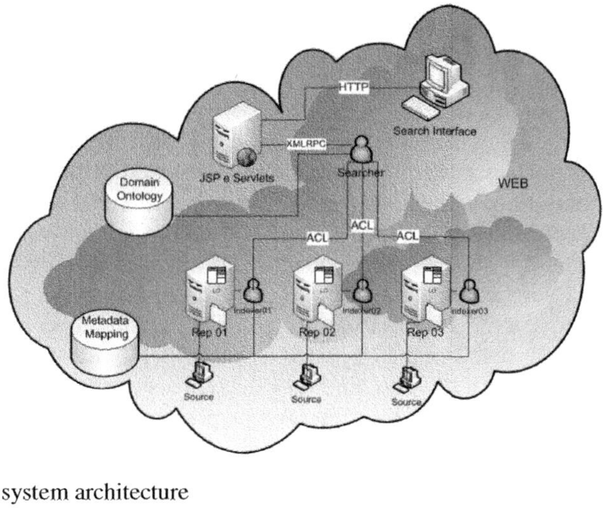

Fig. 1 Overall system architecture

According to Figure 1, the main components are: *Repositories*, which represent the set of repositories of learning objects on which the system performs the indexing and retrieval of objects; a *Multiagent System*, composed of two types of agents, Indexer Agent and Searcher Agent; *Metadata Mapping*, the database structure specially designed to correlate the different elements of each pattern of metadata processed by the system; *Domain Ontologies*, a set of ontologies used by the searcher agents which allows the expansion of the keywords used by the users to request searches by the agents; and *Search Service*, which provides the web server user interface where the users perform queries by asking the system with keywords to find learning objects in the set of repositories covered by the system and where the results of the queries are presented.

The multiagent system model was designed according to the O-MASE methodology, and the model diagrams were drawn with AgentTool III. This tool allows the construction of a series of model diagrams that give a formal view of the multiagent system. The diagram of agents is shown in Figure 2. In this diagram one can view the properties of the two types of agents proposed in this model: the Indexer Agent and the Searcher Agent.

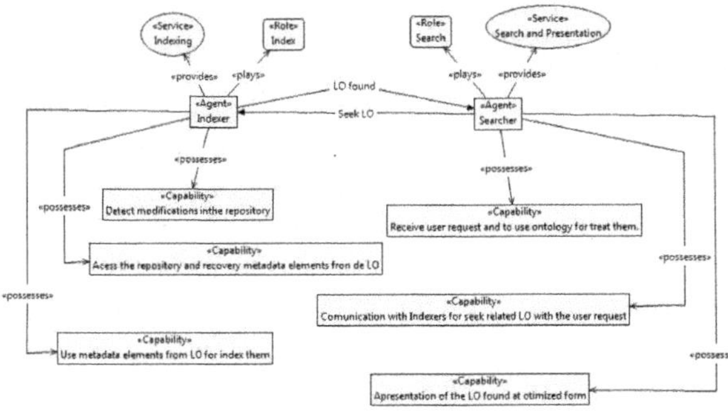

Fig. 2 Diagram agents

The Indexer Agent is able to get all the information about the learning objects from the repository using the metadata of a learning object stored in the repository according to the formal pattern of metadata used (LOM, DublinCore, CanCore, etc.) and according to how these metadata are registered in the specific repository. The Indexer Agent must know the physical location of the repository, and must be able to access the learning object metadata of all learning objects existing in the repository. The Searcher Agent is able to receive requests from the user, which types keywords to ask the system to find a group of learning objects related to the keywords. The agent receives the keyword terms from the user through the search interface as well as one of the domain knowledge areas available in the user interface selected by the user. The Searcher Agent performs query expansion of the keyword using the existent domain ontology to expand the query according to the knowledge domain area chosen by the user. For each repository there is an instance of the Indexer Agent responsible for indexing the repository, and there is an instance of the Searcher Agent responsible for expanding the user query and communicate with Indexers Agents.

4 The Prototype

The prototype described below was developed to assess the proposed model and validates the basic idea of using multiagent systems and ontology to find and recover learning objects from diverse and heterogeneous repositories. The prototype is based on Java programming language technology and uses the JADE Framework (JADE, 2010) for the agent's implementation, and JSP, Servlets and Tomcat Application Server for the user interface. Protégé (2010) was used to create and edit ontology in OWL format. Moreover, the framework Jena 6 (JENA, 2010) is used to implement the communication of the agents with the OWL ontologies. This tool is used by the agents to expand the queries by consulting the ontology indicated by the user.

The XMLRPC (XMLRPC, 2003) protocol is used for the communication between the agents and the search service. This protocol allows the exchange of XML content messages from calls to the remote processes, URL, JLDAP JDOM: Java Libraries is used to provide functionality for connecting agents to the learning object repositories to recover the metadata of the learning objects and the XML processing to extract specific information from the metadata elements set.

As the Tomcat web server and the application are started up, the system becomes available for users to search learning objects from all the registered repositories according to the existent ontologies by using the interface shown in Figure 3. The user uses this screen to perform a query for learning objects by typing keyword terms and pointing to the desired domain.

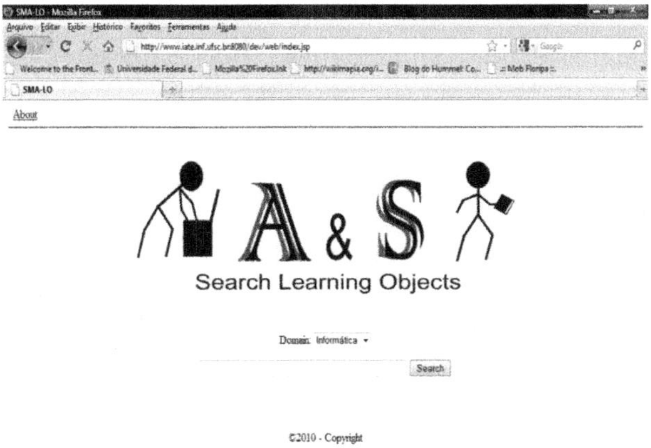

Fig. 3 Search interface

As the system agents are started by the JADE framework the Indexer Agent performs the indexing of the repositories by using the appropriate Java library to connect with the repositories. Further this agent retrieves metadata using the appropriate retrieval information technique for the specific metadata pattern used for each repository. The agents store the resulting index using XML. When the users ask the query the JSP page triggers the Servlet web server, which uses XMLRPC to perform communication with the multiagent system. The Searcher Agent receives the keyword from the Servlet and expands the query according to the ontology indicated by the user and consults the Indexer Agents using FIPA ACL and asking them about existent learning objects in each repository that match the expanded keyword provided by the user. The Searcher Agent receives the answers from Indexer Agents, organizes the list of recovered objects and sends back the outcome to the Servlet via XMLRPC which displays the object list in the user interface.

5 The Validation Test

The described prototype was evaluated using two real different repositories of learning objects and a very simple ontology describing the knowledge field of Information Security, shown in Figure 4. This ontology was, specially written for this test with Protégé (2010). This ontology consists of concept classes in the field of Information Security as well as several "*is-a*" and "*part-of*" relationships.

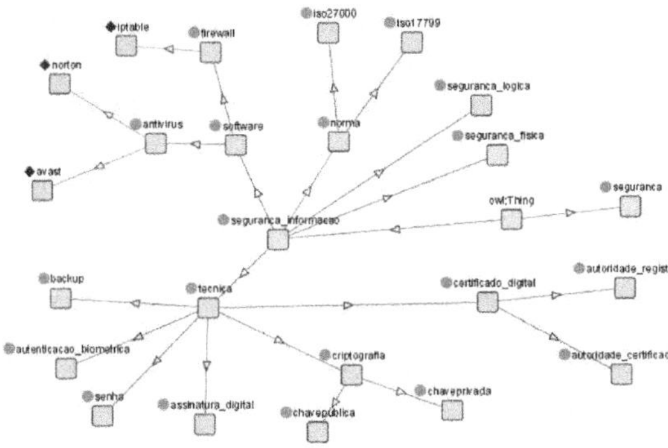

Fig. 4 Ontology is the domain of Information Security

Figure 5 shows the comparison of the answer of the queries submitted to the native search engine repositories and the answer of the queries submitted to the search engine of the prototype.

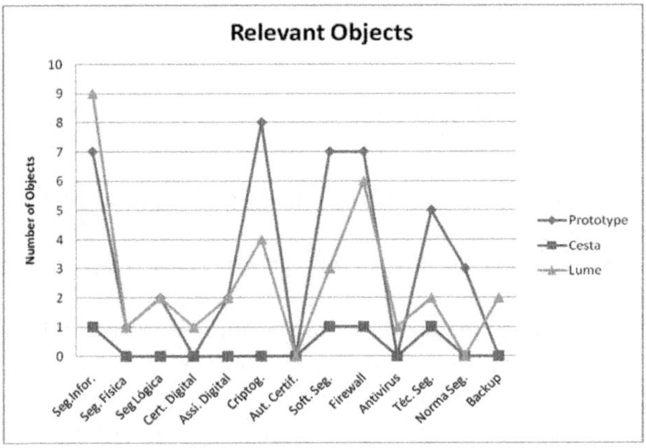

Fig. 5 The result of the queries

Two repositories were used to test the prototype: LUME (2010) and CESTA (2010). The first one was built with DSpace and uses Dublin Core metadata standard, and CESTA (2010) is based on an LDAP server and uses the LOM metadata standard. To carry out tests to validate the prototype, we first make several queries using the native repositories' search engines to compare the obtained results with the obtained by the agent system. The queries were made using some specific keyword about information security.

The number of object pointed in the answers of the queries submitted to the search engine of the prototype is usually higher than the number of the objects pointed by the specific search engine of each repository. We can see an improvement in the accuracy of recovered objects, considering the sampling results. These tests show that the proposed recovery model of learning objects can improve the reuse and sharing of learning objects from different repositories. More details of the evaluation test can be viewed in Vian (2010).

6 Final Remarks

The result of the evaluation tests of the prototype shows that the combination of multiagent system together with information retrieval techniques such as indexing and considering semantic aspects of the information by means of ontologies can improve the search tool adding precision, coverage and performance.

The proposed model allows the federated search of learning objects in different distributed and heterogeneous repositories. Recovery of learning objects in different and heterogeneous repositories encourages their reuse and sharing. The proposed model improves the reuse of learning objects by using a multiagent system to deal with differences between existing patterns of metadata, enables interoperability between heterogeneous repositories of learning objects and provides improves the accuracy and coverage of search mechanisms, using domain ontologies to expand queries. The indexing of the repository using relevant values medatada improves the performance of search engines. The proposed model can be applied in other scenarios integrated into a learning management system to be used by teachers or students to search learning content directly from the LMS during learning processes.

References

CESTA. Collection of Entities Support the use of Technology in Learning. Center for Interdisciplinary Studies in New Technologies in Education, CINTED (2010), `http://www.cinted.ufrgs.br/CESTA/` (accessed April 2010)

CORDRA. Content Object Repository Discovery and Registration / Resolution Architecture (March 2006), `http://cordra.net/introduction/` (accessed: January 2010)

Currie, S.: Metadata for Learning Resources: An Update on Standards Activity (2008)

DeLoach, S.A., Ojeda, J.C.G., Valenzuela, J., Oyenan, W.H.: Organization-based Multiagent System Engineering (O-MASE) - Description Framework. Kansas State University, Manhattan (2007)

Downes, S.: Learning Objects: Resources for distance education worldwide. The International Review of Research in Open and Distance Learning 2(1) (2001)

DSpace (2010), http://www.dspace.org/ (accessed April 2010)

Euzenat, J., Shvaiko, P.: Ontology Matching. Springer, Heidelberg (2007)

FEB. Federation of Educa Brazil (2010), http://feb.ufrgs.br/ (accessed January 2010)

Fedora. Fedora Commons Repository Software (2010), http://www.fedora-commons.org/ (accessed May 2010)

FIPA. Foundation for Intelligent Physical Agents (2010), http://www.fipa.org/ (accessed December 2009)

IEEE-LTSC. WG12: Learning Object Metadata (2010), IEEE Learning Technology Standards Committee (2005), http://ltsc.ieee.org/wg12/ (accessed June 2010)

JADE Java Agent Development Framework (2010), http://jade.tilab.com/ (accessed January 2010)

JENA. Jena - A Semantic Web Framework for Java (2010), http://jena.sourceforge.net/ (accessed Tue 2010)

LUME. Lume - Repository of the Federal University of Rio Grande do Sul (2010), http://www.lume.ufrgs.br/ accessed February 2010)

Manning, C.D., Raghavan, P., Schutz, H.: Introduction to Information Retrieval. Cambridge University Press, England (2008)

McGreal, R.: Online Education Using Learning Objects. Routledge, London (2004)

Nash, S.S.: Learning Objects, Learning Object Repositories, and Learning Theory: Preliminary Best Practices for Online Courses. Interdisciplinary Journal of Knowledge and Learning Objects (2005)

Protégé. Protégé Ontology Editor (2010), http://protege.stanford.edu/ (accessed January 2010)

Vian, J.: Multiagent System for Indexing and Retrieving Applied to Learning Objects. Dissertation, Federal University of Santa Catarina (2010) (in Portuguese)

Wooldridge, M.: An Introduction to Multiagent Systems. John Wiley, England (2002)

XMLRPC. XML-RPC Home Page (2003), http://www.xmlrpc.com/ (accessed April 2010)

Data Mining Techniques for Web Page Classification

Gabriel Fiol-Roig, Margaret Miró-Julià, and Eduardo Herraiz

Abstract. Nowadays, the Web is an essential tool for most people. Internet provides millions of web pages for each and every search term. The Internet is a powerful medium for communication between computers and accessing online documents but it is not a tool for locating or organizing information. Tools like search engines assist users in locating information. The amount of daily searches on the web is broad and the task of getting interesting and required results quickly becomes very difficult. The use of an automatic web page classifier can simplify the process by assisting the search engine in getting relevant results. The web pages can present different and varied information depending on the characteristics of its content. The uncontrolled nature of web content presents additional challenges to web page classification as compared to traditional text classification, but the interconnected nature of hypertext also provides features that can assist the process. This paper analyses the feasibility of an automatic web page classifier, proposes several classifiers and studies their precision. In this sense, Data Mining techniques are of great importance and will be used to construct the classifiers.

Keywords: Data Mining, Artificial Intelligence, Decision Trees, Web page classification.

1 Introduction to the Problem

The growth of information sources available on the World Wide Web has made it necessary for users to utilize automated tools in finding the desired information resources. There is a necessity of creating intelligent systems for servers and clients that can effectively mine for knowledge. Web mining can be broadly defined as the discovery and analysis of useful information from the World Wide Web by the extraction of interesting and potentially useful patterns and implicit information

Gabriel Fiol-Roig · Margaret Miró-Julià · Eduardo Herraiz
Math and Computer Science Department
Universitat de les Illes Balears
Ctra. de Valldemossa km. 7,5, 07122 Palma de Mallorca, Spain
e-mail: {biel.fiol,margaret.miro}@uib.es

J.B. Pérez et al. (Eds.): Highlights in PAAMS, AISC 89, pp. 61–68.
springerlink.com

from artifacts or activity related to the World Wide Web. There are roughly three knowledge discovery domains that pertain to web mining: Web Content Mining, Web Structure Mining, and Web Usage Mining. Web content mining is the process of extracting knowledge from the content of documents or their descriptions. Web structure mining is the process of inferring knowledge from the World Wide Web organization and links between references and referents in the Web. Finally, web usage mining is the process of extracting interesting patterns in web access logs [1].

Web content mining is an automatic process that goes beyond keyword extraction. Since the content of a text document presents no machine readable semantic, some approaches have suggested restructuring the document content in a representation that could be exploited by machines. The usual approach to exploit known structure in documents is to use techniques to map documents to some data model. There are two web content mining strategies: those that directly mine the content of documents and those that improve on the content search of other tools. From a Data Mining point of view, Web mining, has three areas of interest: clustering (finding natural groupings of users, pages etc.), associations (which URLs tend to be requested together), and classification (characterization of documents).

Web page classification is the process of assigning a Web page to one or more predefined category labels [2]. Classification can be understood as a supervised learning problem in which a set of labeled data is used to train a classifier which can be applied to label future examples. The general problem of Web page classification can be divided into more specific problems. Subject classification is concerned about the subject or topic of a Web page. Functional classification cares about the role that the Web page plays. Sentiment classification focuses on the opinion that is presented in a Web page, that is, the author's attitude about some particular topic. Other types of classification include genre classification, search engine spam classification. This paper focuses on subject and functional classification.

The objective of this paper is to design an automatic classification system for web pages. The decision system designed has three main stages as indicated in Figure 1: the data processing phase, the data mining phase and the evaluation phase.

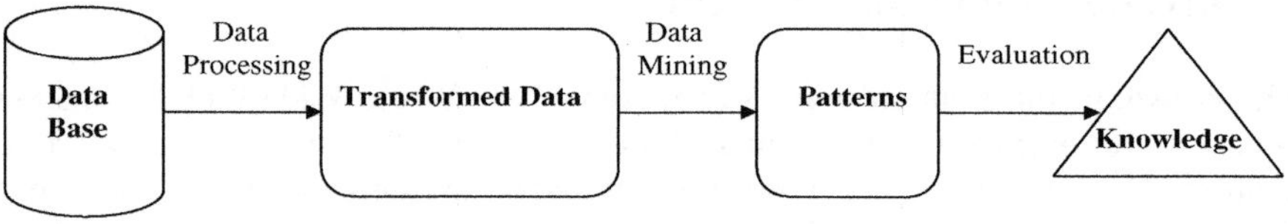

Fig. 1 The Decision System.

The data processing selects from the raw data base a data set that focuses on a subset of attributes or variables on which knowledge discovery is to be performed. It also removes outliers and redundant information, and uses HTML code to represent the processed data by means of an Object Attribute Table (OAT) [3]. The data mining phase converts the data contained in the OAT into useful patterns in particular decision trees are found [4]. The evaluation phase proves the consistency of pattern by means of a testing set. The positively evaluated decision system can then be used in real world situations that will allow for its validation.

2 The Data: Features of Web Pages

Web pages are what make up the World Wide Web. A Web page is a document or information resource written usually in HTML (hypertext markup language) and translated by your Web browser. Web pages are formed by a variety of information, such as: images, videos or other digital assets that are addressed by a common URL (Uniform Resource Locator).

These pages are typically written in scripting languages such as PHP, Perl, ASP, or JSP. The scripts in the pages run functions on the server that return things like the date and time, and database information. All the information is returned as HTML code, so when the page gets to your browser, all the browser has to do is translate the HTML, interpreting it and displaying it on the computer screen. Since all pages share the same language and elements it is possible to characterize each of them accurately in an automatic way.

All web pages are different, in order to classify them, data extracted from the HTML code will be used. Pages can then be classified according to multiple categories. Throughout this document, the following labels have been used: blog, video, images and news.

- Blog: short for weblog, is a personal online journal with reflections, comments provided by the writer. Blogs are frequently updated and intended for general public consumption. Blogs generally represent the personality of the author or reflect the purpose of the Web site that hosts the blog. Blogs can be distinguished by their structure: a series of entries posted to a single page in reverse-chronological order. Blogs contents are basically text, occasionally images and videos are included.
- Video: these web pages provide a venue for sharing videos among friends and family as well as a showcase for new and experienced videographers. Videos are streamed to users on the web site or via blogs and other Web sites. Specific codes for playing each video are embedded on the Web page.
- Image: a photo gallery on a website is collection of images or photos that is uploaded to a website and available for website visitors to view. These web pages hardly have any text and generally carry a navigator that allows the visitor to move around the images.
- News: an online newspaper is a web site which provides news on a basis which is close to real time. A good news site updates its content every few minutes.

The retrieval of the data available in web pages is not immediate and presents drawbacks. The extraction of useful data is a delicate process that requires careful handling. The use of an automatic program will simplify and improve the data acquisition process.

3 Transformed Data: Generation of the Object Attribute Table

The original data base is formed by a list of web pages. In the data processing phase an automatic program, written in Python, that processes the HTML code of the different web pages has been used. In particular, the program gathers information and generates the OAT. The attributes of the OAT, the columns, are the following:

- Page's text length (***TL***): number of text characters in the web page.
- External links (***EL***): number of links to external web pages.
- Internal links (***IL***): number of links to internal web pages.
- Image (***Im***): number of <img> elements in the web page.
- External Images (***EI***): number of external links with images.
- Internal Images (***II***): number of internal links with images.
- Multimedia Objects (***MO***): number of <object> elements in the web page, such as videos or flash player.
- Word Flash (***WF***): number of appearances of the word "flash" or similar in the text of the page.
- Word Video (***WV***): number of appearances of the word "video" or similar in the text of the page.
- Word Image (***WI***): number of appearances of the word "image" or similar in the text of the page.
- Word Blog (***WB***): number of appearances of the word "blog" or similar in the text of the page.
- Word News (***WN***): number of appearances of the word "news" or similar in the text of the page.

Table 1 illustrates the format of the OAT. Each row of the OAT describes the characteristics of a web page using the above defined attributes. There are a total of 12 numerical attributes and 271 rows. In the following step, an expert assigns classes to each of the rows according with the four categories mentioned above. The last column of Table 1 corresponds to the class.

Table 1 Object Attribute Table.

TL	*EL*	*IL*	*Im*	*EI*	*II*	*MO*	*WF*	*WV*	*WI*	*WB*	*WN*	*CLASS*
44299	96	38	20	20	0	0	0	3	166	2	2	IMAGE
54795	36	260	36	5	20	0	0	2	10	7	24	NEWS
235056	588	319	110	66	38	0	1	12	274	282	0	BLOG
84131	120	26	30	15	11	1	5	125	70	4	0	VIDEO
....	....	....	....	....	....	....	....	....	....	....	....	....

Some significant properties are connected to the use of OAT. Among these properties we single out the possibility to deal with binary or multivalued attributes [5], and to represent incomplete and vague knowledge [6, 7].

The OAT also facilitates the incorporation of an inferencial mechanism. This mechanism is based on the abstraction principle and considers the evolving characteristic of the environment. Moreover, it achieves a more efficient agent's decision stage.

4 Classification Methods

Classification methods allow the construction of a predictive model that is later used to assign labels to web pages. In order to create this model a training set must be used. The entries corresponding to 271 web pages are used to generate the model and evaluation is carried out using 20-fold cross validation. Figure 2 illustrates this process.

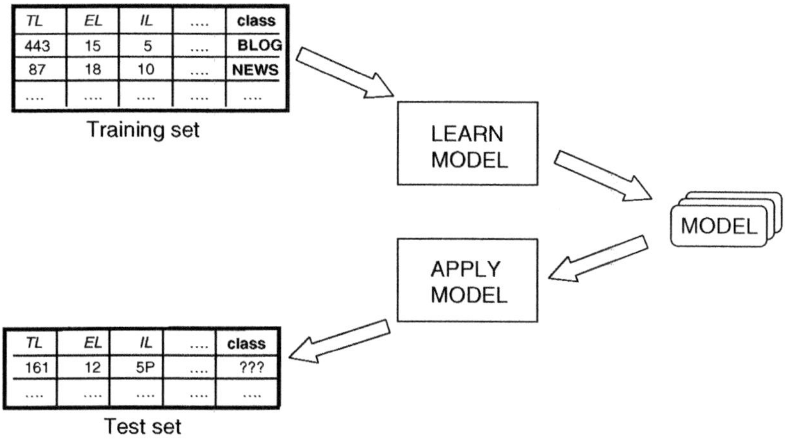

Fig. 2 The Classification Process.

5 Decision Trees

Among available classification methods, decision trees were selected for their simplicity and intuitiveness. Decision trees were developed using WEKA [8], (Waikato Environment for Knowledge Analysis) a collection of machine learning algorithms for data mining tasks.

Different classification algorithms were tried out, among them Best First Decision Tree (BFT), Logistic Model Tree (LMT), J48 Graft Tree (J48GT) and J48 Pruned Tree (J48PT). Obviously, the trees obtained by the different methods differ. Essentially, this difference becomes apparent in the tree's complexity and its precision.

The classification methods were applied to the set of 271 instances and 12 variables. The results of some of the most relevant trees, using 20-fold cross validation, are shown in Table 2.

Table 2 Classification Results.

METHOD	Correct	Incorrect	Correct (%)	Incorrect (%)
BFT	246	25	90.78	9.22
LMT	245	26	90.41	9.59
J48GT	249	22	91.88	8.12
J48PT	245	26	90.41	9.59

The confusion matrices, shown in Table 3, indicate the number of classified instances according to class. The notation used for the classes is the following **a** = IMAGE, **b** = NEWS, **c** = BLOG and **d** = VIDEO.

Table 3 Confusion Matrices.

		BFT					**LMT**		
	a	**b**	**c**	**d**		**a**	**b**	**c**	**d**
a	61	2	3		**a**	60	2	2	2
b	4	83	3	0	**b**	5	83	2	0
c	3	2	78	1	**c**	3	0	77	4
d	3	2	2	24	**d**	2	1	3	25

		J48GT					**J48PT**		
	a	**b**	**c**	**d**		**a**	**b**	**c**	**d**
a	60	3	2	1	**a**	61	3	2	0
b	1	86	3	0	**b**	5	82	3	0
c	3	1	79	1	**c**	4	1	78	1
d	3	2	2	24	**d**	3	2	2	24

The results of some of the most relevant trees are shown below in Figure 3 and Figure 4.

The LMT tree has a total of 11 nodes, 5 internal nodes corresponding to the 4 attributes used and 6 leaf nodes or branches representing the classes. The average branch length is 3 and the maximum depth is 4. The attributes representing the internal nodes are: ***WN***, ***EL, WV*** and ***TL***.

The J48 pruned tree has a total of 19 nodes, 9 internal nodes corresponding to the 6 attributes used and 10 leaf nodes or branches representing the classes. The average branch length is 3.5 and the maximum depth is 5. The attributes representing the internal nodes are: ***WN***, ***EL, TL, WV, Im*** and ***WI***.

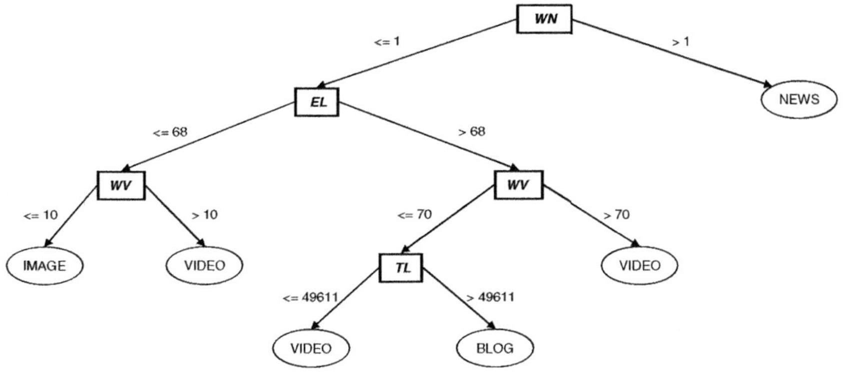

Fig. 3 The Logistic Model Tree.

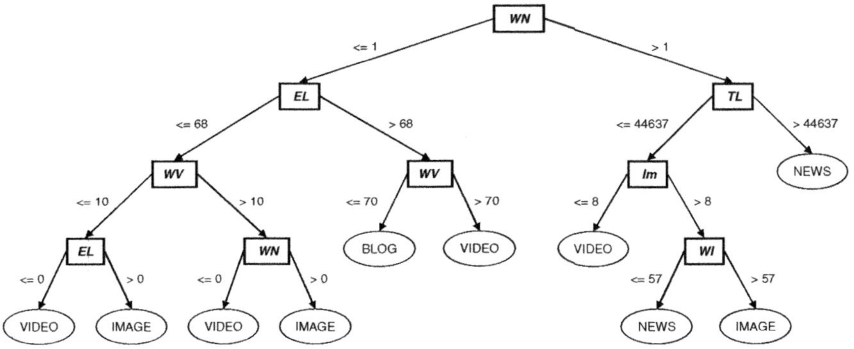

Fig. 4 The J48 pruned Tree.

If both trees are compared, it can be seen that attributes ***WN*** and ***EL*** appear at the upper levels of the LMT tree and the J48 pruned tree suggesting their importance. The number of appearances of the word "news" (***WN***) is a deciding factor for the classification of web pages. None of the pages with 1 or less appearances of the word "news" is a news page. On the other hand, a high number of external links and a reduced number of appearances of the word "video" correspond with a blog page.

6 Conclusions and Future Work

Classification methods based on data mining techniques have been applied to the area of web mining in order to construct efficient trees that classify web pages depending on their features.

In the development of the method, several steps must be considered. First of all, raw data must be collected and analyzed. The selection of data is an arduous and necessary task and requires an in depth analysis of all the problem requirements. Secondly, a variable analysis process takes place. Those unnecessary variables for data classification are removed and, if necessary, new variables are considered. Once the data is consolidated, the data's final format in terms of an Object Attribute Table (OAT) is introduced. Third, data mining methods are applied to the OAT, in order to find valid and efficient models that classify the instances. The resulting trees are analyzed in terms of complexity and precision. Generally speaking, classification accuracy requires complex trees.

The results show that a web page classification based on web page features available in the page's HTML code is possible. The success rate is acceptable even though this paper offers a limited vision of one of the many solutions available. The following aspects can be considered as future work: other different variables, such as page style or text style may offer better results and should be studied. Also other decision trees can be obtained using different learning algorithms and should be evaluated.

References

1. Dunham, M.H.: Data Mining. Introductory and Advanced Topics. Prentice Hall, New Jersey (2003)
2. Qi, X., Davison, B.D.: Web Page Classification: Features and Algorithms. ACM Computing Surveys 41(2), Article 12 (2009)
3. Miró-Julià, M., Fiol-Riog, G., Vaquer-Ferrer, D.: Classification using Intelligent Approaches: an Example in Social Assistance. Frontiers in Artificial Intelligence and Applications 202, 138–146 (2009)
4. Fiol-Roig, G.: UIB-IK: A Computer System for Decision Trees Induction. In: Raś, Z.W., Skowron, A. (eds.) ISMIS 1999. LNCS, vol. 1609, pp. 601–611. Springer, Heidelberg (1999)
5. Miró-Julià, M., Fiol-Roig, G.: An Algebra for the Treatment of Multivalued Information Systems. In: Perales, F.J., Campilho, A.C., Pérez, N., Sanfeliu, A. (eds.) IbPRIA 2003. LNCS, vol. 2652, pp. 556–563. Springer, Heidelberg (2003)
6. Fiol-Roig, G.: Learning from Incompletely Specified Object Attribute Tables with Continuous Attributes. Frontiers in Artificial Intelligence and Applications 113, 145–152 (2004)
7. Miró-Julià, M.: Degenerate Arrays: a Framework using Uncertain Data Tables. In: Moreno Díaz, R., Pichler, F., Quesada Arencibia, A. (eds.) EUROCAST 2005. LNCS, vol. 3643, pp. 21–26. Springer, Heidelberg (2005)
8. Witten, I.H., Frank, E.: Data Mining: Practical machine learning tools and techniques, 2nd edn. Morgan Kaufmann, San Francisco (2005)

Analizying Factors to Increase the Influence of a Twitter User

José del Campo-Ávila, Nathalie Moreno-Vergara, and Mónica Trella-López

Abstract. Twitter is a free social media service that allows anyone to say almost anything to anybody in 140 characters or less. Over the past few months, Twitter has experienced an explosive growth which has aroused the interest of many developers. In consequence, there have appeared many analytic tools which, besides other characteristics, calculate how influential a user is. This meaningful value can be estimated using different metrics and tools. In this paper, we study the reliability of them and show how data mining techniques can help: (a) to identify the actions which can increase the influence of a user (depending on the concrete tool), (b) to discover if those actions are related to different tools (and whether we can increase influence in more than one way), and (c) to advise people (or companies) about how they can get a greater impact.

Keywords: data mining, social networks, Twitter, influence.

1 Introduction

In just four years, Twitter [10] has become one of the most popular social-media networking tools. Its more than 145 million users have made it a public platform for the discussion and dissemination of news related to business, politics, education and gossip. In Twitter, the users are characterised by a basic profile that includes personal or corporate information (bio, location, name, web) and details about their contact network (for example, the set of users who they follow and the set of users followed by them).

Twitter users – or twitterers– are not only people, but also organisations and companies that use Twitter as a broadcast medium. They could be interested in extending their influence to reach a bigger audience and "influencers" to work with them. In

José del Campo-Ávila · Nathalie Moreno-Vergara · Mónica Trella-López
Dpto. Lenguajes y Ciencias de la Computación, University of Málaga (Spain)
e-mail: {jcampo,moreno,trella}@lcc.uma.es

J.B. Pérez et al. (Eds.): Highlights in PAAMS, AISC 89, pp. 69–76.
springerlink.com

order to do that, users should know what parameters are the most relevant ones when it comes to calculating influence, and which actions will lead to increase it.

Mining social networks is becoming a very interesting task because of its promising impact in the real world [4] and Twitter is one of the most relevant platforms [2, 7]. Currently, there are few studies analysing statistics to derive measures of influence in social networks, that is, trying to calculate what the impact of different users' actions is or how influential they are [3]. New tools are being developed to fill the existing niche. They use a broad range of variables to obtain influence scores (number of followers, followers/following ratio, frequency at which the user is mentioned, and so on) by applying unknown algorithms to different sets of parameters (usually almost disjoint sets). Their criteria differ from each other so much that ranking values calculated for the same user can differ significantly from one tool to another.

This paper presents a preliminary study that aims to define the concept of influence and to identify dominant parameters that could help to reach a greater "influence" in social networks (Section 2). We analyse the reliability of different influence Twitter metrics and tools (Sections 3 and 4) and, finally, we infer some interesting relations between them using data mining techniques (Section 5). Last but not least, some conclusions and future research activities are offered (Section 6).

2 Influence Concept on the Twittersphere

There seems to be no consensus about what exactly *influence* means in the Twitter universe and there are several definitions which try to determine which aspects are relevant when estimating a twitterer's influence. Nevertheless, all of them focus on explaining the impact (power, authority, reach) of a user in the community. So, generally speaking, Twitter's influence can be defined as *a measure of the communication effectiveness and capability to engage the audience*.

The attributes used by existing analytic tools to measure the influence vary from one to another. The criteria to establish the degree of influence of the users can range from quantitative data about the number of their connections in the social network (personal network) to more qualitative aspects related to the content of their tweets. In this section, we present the most common parameters used to measure the influence. Most of them are reported directly by Twitter through a public API that allows to explore a user's network and their tweets.

Parameters obtained from the personal network:

- *Followers*: number of people following the user. The more followers they have, the bigger the audience for their messages is.
- *Following*: number of people the user is following. They would form one of the sources for the user to get information. If someone is following a user that publishes relevant information, they can retweet it.
- *Followers/Following ratio*: it provides a measure of the type of interaction the user has in Twitter by examining the number of people referenced by the user

(following) and the number of people referencing them (followers). As the ratio increases, the audience for their messages could appear to be more suggestive.
- *Friends*: number of people that follow each other. It can give you an idea of how big the user's influence is in the sector which you are interested in.

Parameters based on tweets content:

- *Retweets*: number of times that the user has been retweeted by other people. It indicates interesting and informative tweets.
- *Replies*: number of times that people answered a user back. Like retweets, this is a measure of the writer's capacity to engage other users.
- *Mentions*: number of times that other twitterers have mentioned the user in their tweets. They are less usual than retweets but a high number of mentions signals attractive content.
- *Attributions*: number of times that people cite a previously published content. They are rarer than retweets but their final purpose is similar: passing some content catalogued as rather interesting to the third party.
- *Follow Friday Recommendations*: the users that are "Follow Friday" recommended, have been recognised as valuable people to follow because of their content and/or reputation.

3 Analytic Tools to Calculate the Twitter Influence

There is a considerable number of analytic tools that calculate a twitterer's influence. Simple counters are insufficient to determine the power of a user in the network. Influence should be measured on the basis of how people interact. So, these tools combine some of the parameters mentioned above with others to calculate the influence degree. The three tools used in this paper to obtain and compare data are Twinfluence [9], TwitterGrader [11] and Klout [5].

Twinfluence. It is a simple tool for measuring the combined influence of twitterers and their followers. Twinfluence calculates two types of ranks:

- *Absolute ranking*: it represents the user's reach (number of first-order plus second-order followers, with a maximum of 30,000) considering all the other twitterers that have been analysed by Twinfluence. The measure is given in two formats: position in the list and percentage of users that have lower score.
- *Relative scores*: they are obtained comparing users with similar number of followers. They are:
  1. Velocity: average number of first and second-order followers attracted per day.
  2. Social capital: measure of how influential the followers of a user are.
  3. Centralisation: measure of how high the influence of a user is invested in a small number of followers.

TwitterGrader. It only provides an absolute ranking that compares a user with the whole population graded. As Twinfluence, the score is given either as the position in the ranking, or the percentage of users that have lower score. Apart from quantitative measures (as number of followers or follower/following ratio), TwitterGrader infers some metrics that are used to calculate the influence score:

- *Power of followers*: it counts the TwitterGrader grade of the people following you (it is similar to the "social capital" of Twinfluence).
- *Updates frequency*: more and more recent updates generally leads to a higher grade.
- *Engagement*: mainly based on retweets.

Klout. It gives a score meaning *how successful a person is at engaging their audience and how big impact their messages have on people* as a combination of the following measures:

- *True reach*: it takes into account how active a user's network of followers actually is. It represents the number of followers minus the spam followers, the inactive accounts and the people who have little influence.
- *Amplification ability*: the likelihood that your message will generate retweets or cause a conversation.
- *Network score*: a measurement of how influential the people who retweet, mention, list or follow you are.

4 Comparing Tools and Rankings

To analyse the reliability of previous tools we have compiled a list of Twitter users, trying to make it as heterogeneous as possible (domain, popularity, participation, etc.). To do this task, firstly, we manually created an initial subset of users selecting different kinds of users: the most influential users according to the studied tools (they have some lists with the top twitterers), lists of well-known companies, categories in Twitter, etc. Thus, once we have a preliminary subset and we have assured that we would have a minimum variety of users, we gathered the rest of the users automatically by adding people following users in that preliminary subset. With this process we composed a list of 2559 users.

Later, we needed to collect information about these users so we consulted in the three tools websites for them. Parsing the results we automated the collecting of information, but different tools showed problems to calculate the influence of different users (fail whale error, cancellation because of time delay, etc.), so the final dataset that contained all the information was composed by 431 users.

The scripts used to automate this tasks, the final dataset and the list of twitterers can be downloaded from [1].

Once the data have been collected, we have compared scores computed using the three systems. Twinfluence and TwitterGrader give the same ranking to a significant percentage of users (their calculations match in a 16%) as is shown in Figure 1, despite the fact that their algorithms take into account different parameters. Indeed, relaxing the match criterion, both tools return values in the same decile in as many as 31.01% of cases, and in the same quartile in up to 33.79% of them. However, they

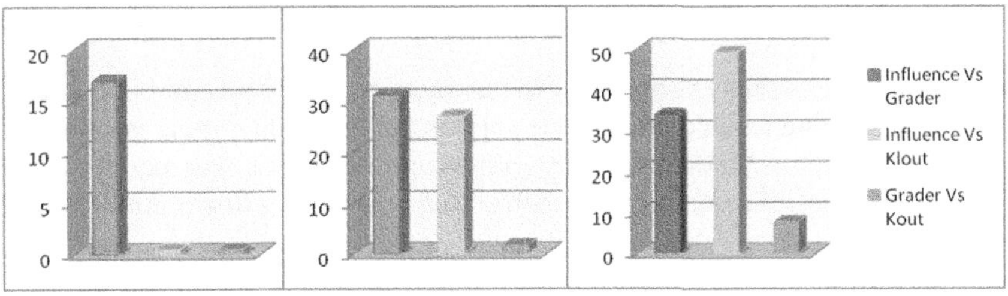

Fig. 1 Percentage of users whose influence matches in the same quantile. From left to right, percentiles, deciles and quartiles

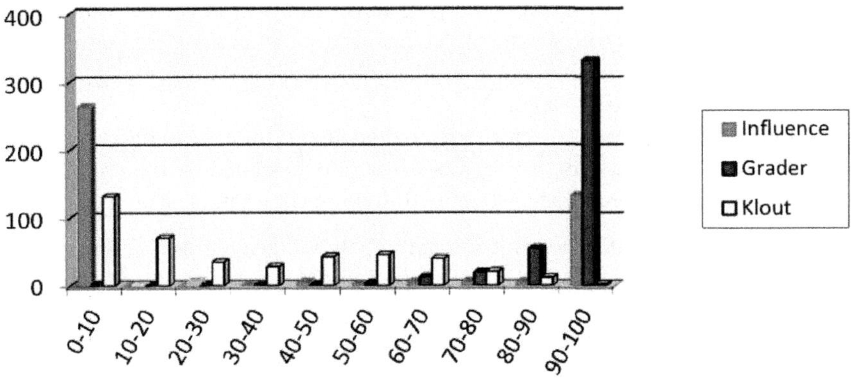

Fig. 2 Frequency of the users' rankings for the different tools

differ considerably from Klout. In particular, while a user can have a rank value of 96.7% using TwitterGrader, it is not higher than 6% following the Klout criteria.

Bearing this in mind, we observe that no ranking distribution follows a normal frequency (see Figure 2). While TwitterGrader concentrates scores around the last decile (aprox. 250 users are in 90%-100%), Klout does it around the first quartile (aprox. 230 users are in 0%-25%). Furthermore, if we look at Twinfluence results, twitterers' ranking occupy two extremes (aprox. 260 users are in 0%-20% and 140 are in 80%-100%).

Our analysis suggests that these tools will become far more useful and relevant once they agree on how the influence must be calculated. Actually, they support their decisions on sets of parameters which are too disjointed. The end user cannot spend time and effort on carrying out an exhaustive analysis of the influence of each tool (or their algorithms). Without that knowledge, the value computed by these systems is not significant because, on the one hand, no guidelines are available as to which tool must be used in each case and, on the other hand, the score does not tell us anything about how to improve the ranking. So, in this context, could we identify which parameters are more relevant for each tool and its corresponding metrics?

5 Applying Data Mining to Identify Dominant Parameters

When the amount of data is high and the extraction of new relevant knowledge is needed, data mining techniques become very useful [6]. Data mining is an area that uses statistics and machine learning algorithms in order to discover models, patterns or relations between the data that are being studied. Bearing this in mind, we want to know if there is some relation between the simple parameters, the derived metrics and the influence calculated by tools presented in Section 4.

For that purpose, we have used the previously described dataset (431 users). Besides the location, homepage, joining date, number of followers, followings and friends (data that can be accessed using the Twitter API), we have also taken into account data offered by Twinfluence, TwitterGrader and Klout on their corresponding web sites ([9, 11, 5]). In particular, we have used: (a) *From Twinfluence*, the number of second order followers, the velocity, the social capital, the centralisation and the ranking; (b) *From TwitterGrader*, the grade score and the ranking value; and (c) *From Klout*: the ranking value.

We have applied regression algorithms to this dataset in order to find useful models because the attribute class is a number (the rank assigned by the tool). The induction of models has been done using the following automatic learning algorithms:

- REPTree: it is a fast decision tree learner which builds a regression tree using information gain as the splitting criterion, and prunes it using reduced-error pruning. It only sorts values for numeric attributes once.
- M5P: it implements the routines for generating M5 model trees [8, 12]. They combine a conventional decision tree with the incorporation of linear regression functions at the leaves. The advantages are a clear structure and simple regression functions that usually involves few variables.
- DecisionStumps: it is an algorithm that basically induces a one-level regression tree where the split at the root level is based on a specific attribute/value pair.

All of them are implemented in the weka framework [13] that supports several standard data mining tasks and it is the tool that we have used to induce the models.

Once we have generated the different models with all the algorithms, we have detected that those induced by REPTree are the most accurate attending to the relative absolute error (mean absolute error divided by the corresponding error of the ZeroR classifier – classifier predicting the prior probabilities of the classes observed in the data –). After having used the induced models, some of them shown in Figure 3, we can conclude that:

1. *Twinfluence*: models generated by M5P and REPTree use only two parameters to expand the nodes: *velocity* and *number of second order followers* (DecisionStumps uses *velocity* too). So it seems clear that both parameters are relevant if a user wants to increase their influence in Twitter. The model is simple (it uses few nodes) and accurate (its relative absolute error is 2.93%, tested using 10-fold cross validation). In Figure 3 (left) we show the REPTree model of the influence calculated by this tool.

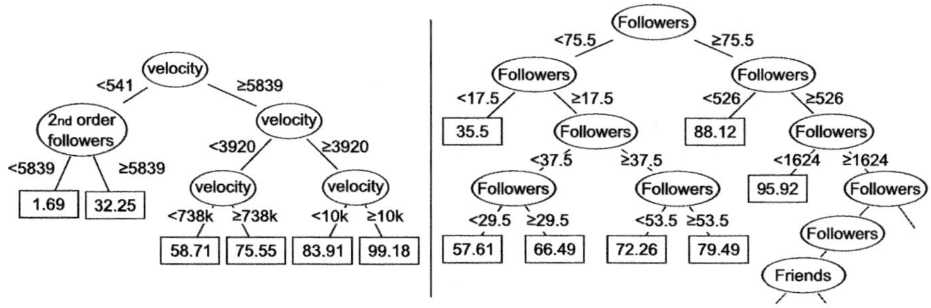

Fig. 3 REPTree models for Twinfluence (left) and TwitterGrader (right).

2. *TwitterGrader*: models induced by M5P and REPTree use more parameters (*number of followers, friends, followings, etc.*) and they are more complex. In addition, the accuracy is not so high (relative absolute error is 25.68% for REPTree). We could say that this tool considers some aspects from the topology of the social network, mainly the number of followers (the higher the number of followers, the greater the influence), but this assertion is not as confident as in the case of Twinfluence. In Figure 3 (right) we show the upper part of the REPTree model of the influence calculated by TwitterGrader.
3. *Klout*: for this tool we cannot induced any accurate model, so it seems that the relations between parameters are not as direct as in the previous tools. As it was explained in Section 2, 25 parameters are used to calculate the influence, but none of them is provided on the Web and, therefore, we could not include them in the dataset.

6 Conclusions and Future Works

Our study demonstrates that current tools may help to measure how influential a twitter user is. However, none of them provides an accurate measure of a standardised reach or scope by themselves. In that sense, we have shown how data mining techniques can help: (a) to identify the actions which can increase the influence of a user (depending on the concrete tool), (b) to discover if those actions are related to different tools (and if we can improve influence in more than one way), and (c) to advise people (or companies) about how they can get a greater impact.

In particular, we have established some rules to increase the influence of a twitterer (at least for Twinfluence and TwitterGrader tools). So, once we have identified dominant parameters in Twitter influence measures, the next step is to study the cost associated with increasing popularity ranking by means of modifying particular variables and the behaviour of a user.

Now we are trying to collect more data in several dimensions: number of tools, number of parameters (metrics) and number of users. Our intention is to carry out a more intensive data mining process, taking in consideration that useful conclusions have been reached in this first step. We would also like to propose a new metric that takes into account the conclusions extracted in future studies.

Acknowledgements. This work has been partially supported by the projects SESAAME (with code number TIN2008-06582-C03-03), MDD-MERTS (with code number TIN2008-03107) and ARTEMISA (with code number TIN2007-67515), all of them granted by the MEC, Spain.

References

1. ATIM, `http://albireo.lcc.uma.es/~jcampo/software/ATIM/`
2. Cha, M., Haddadi, H., Benevenuto, F., Gummadi, K.P.: Measuring user influence in twitter: The million follower fallacy. In: Proceedings of the Fourth International AAAI Conference on Weblogs and Social Media, ICWSM (2010)
3. Kempe, D., Kleinberg, J., Tardos, E.: Maximizing the spread of influence through a social network. In: Procc. of the 9th ACM SIGKDD International Conference on Knowledge Discovery and Data Mining, pp. 137–146 (2003)
4. Kimura, M., Saito, K., Nakano, R., Motoda, H.: Extracting influential nodes on a social network for information diffusion. Data Mining and Knowledge Discovery 20, 70–97 (2010)
5. Klout, `http://klout.com/`
6. Kononenko, I., Kukar, M.: Machine learning and data mining: Introduction to principles and algorithms. Horwood Publishing, England (2007)
7. Leavitt, A.: The influentials: New approaches for analyzing influence on twitter (2009), `http://webecologyproject.org`
8. Quinlan, J.R.: Learning with continuous classes. In: Proceedings of the Australian Joint Conference on Artificial Intelligence, pp. 343–348 (1992)
9. Twinfluence, `http://twinfluence.com/`
10. Twitter, `http://twitter.com/`
11. Twitter-Grader, `http://twitter.grader.com/`
12. Wang, Y., Witten, I.H.: Induction of model trees for predicting continuous classes. In: van Someren, M., Widmer, G. (eds.) ECML 1997. LNCS, vol. 1224. Springer, Heidelberg (1997)
13. Witten, I.H., Frank, E.: Data Mining: Practical machine learning tools and techniques. Morgan Kaufmann, San Francisco (2005)

Lightweight User Modeling – A Case Study

Nuno Luz, Ricardo Anacleto, Constantino Martins, and Ana Almeida

Abstract. In the context of previous publications, we propose a new lightweight UM process, intended to work as a tourism recommender system in a commercial environment. The new process tackles issues like cold start, gray sheep and over specialization through a rich user model and the application of a gradual forgetting function to the collected user action history. Also, significant performance improvements were achieved regarding the previously proposed UM process.

Keywords: User Modeling, Tourism, Recommendation.

1 Introduction

Tourism brings quite a challenge regarding (semi–)automated advising and planning systems, which must rely in very complex decision support techniques. The heterogeneity of places, their specificities, and the endless amount of variables involved in traveling adds a considerable amount of complexity to the problem, thus making it difficult to solve without assistance. User interests and preferences should also affect the decision process, and how to efficiently model, retrieve and evolve the system beliefs about the user is, in fact, a complex problem by itself [2].

Despite recent evolutions in User Modeling (UM) and recommendation systems (RS), it is still hard to create a well balanced user model regarding complexity and performance [7]. Lately, a lot of on-line tourism social networks have emerged or evolved to be the successors of the previous tourism information systems. Some examples are TripAdvisor, TripWolf, Dopplr, Wayn, TripSay, Driftr, Sair+, Real Travel, TravBuddy, Exploroo and TripConnect. Although we didn't find published scientific work about most of these tourism information systems, empirically, it can

Nuno Luz · Ricardo Anacleto · Constantino Martins · Ana Almeida
GECAD – Knowledge Engineering and Decision Support Group, R. Dr. António Bernardino de Almeida, 431. 4200-072 Porto, Portugal
e-mail: `{nmal,rmao,acm,amn}@isep.ipp.pt`

J.B. Pérez et al. (Eds.): Highlights in PAAMS, AISC 89, pp. 77–84.
springerlink.com

be stated that they have been moving from an approach where data about individual users is collected explicitly, to an approach where social (collective) data is collected implicitly [3]. Although reducing the amount of explicitly collected information is important to reduce friction [9], we believe both the user as an individual and as a social being must be considered when designing the user model. Even so, some of these systems have been neglecting the user as an individual. This happens possibly because of the computationally intensive decision support techniques that must be used in order to infer new knowledge from all these data.

In previous publications [1, 4, 5], a hybrid tourism UM process using a complex and rich model was proposed and implemented in the TOURSPLAN decision support system. In this paper, a lightweight version of this UM process, intended to work as a tourism recommender system in a commercial environment, is presented along with significant improvements. Our goal, besides improving and removing flaws in the current process, is to find a balance between two important aspects:

- Maintaining the richness of the user model;
- Improving performance in knowledge acquisition and adaptation processes.

This paper is organized as follows: section 2 introduces the original TOURSPLAN UM process as described in [1]. Section 3 presents the new lightweight process and its improvements. Finally, in section 4 and 5, the process evaluation is presented along with some conclusions and future work.

2 TOURSPLAN User Modeling

UM can be defined as an iterative process with three steps, which begins with the retrieval of user feedback. From user feedback, new knowledge can be inferred and added to the user model (knowledge discovery). Finally, both the initial user representation and the inferred knowledge can be used in an adaptation step that enhances the system suitability as perceived by the user [5]. This approach is followed by TOURSPLAN.

In TOURSPLAN, the user model contains several elements classified either as domain independent data, which includes personal, demographic (*e.g.*, age, country), psychological (*e.g.*, creativity, liveliness) and handicap (*e.g.*, sight, mobility) information, or domain dependent data, which includes a point of interest (POI) likelihood matrix, tags (or keywords), user stereotypes (*e.g.*, partying, cultural) and trips.

Different implicit and explicit feedback retrieval techniques are used, such as forms, tests and click analysis (*e.g.*, over tags, taxonomy classes, POIs). When feedback is retrieved, the knowledge discovery step infers new knowledge about the user and updates several components of the user model. These components are:

- **Likelihood Matrix:** links the user with classes of the POI taxonomy;
- **Psychological Model:** contains a degree of association with several psychological attributes;

- **Tag (or Keyword) Model:** associates the user with specific tags, which describe POIs;
- **Stereotype Model:** links the user with the POI classes through an abstraction that can be seen as a conceptualization of the taxonomy;
- **Jennings Model [6]:** provides related and popular POIs.

While the richness of the user model tackles common recommender system issues like *cold start*, *gray sheep* and *over-specialization*, it also brings new challenges. Providing real-time system adaptation using such a complex model can become computationally intensive, specially when collaborative filtering methods are used without the aid of a clustering or classification process. Also, the synchronous execution of the knowledge discovery step each time user feedback is given, heavily contributes to the deterioration of the system response time.

A common disadvantage to the likelihood matrix and the tag model is that, over time, new interactions tend to not affect the model, thus leading to *over-specialization*. This happens because the model values are updated according to an absolute value that tends to be infinite. When the user profile contains high absolute values, the interaction weights will provide almost no impact over the resulting relative value. Although the Jennings model component and the employed collaborative filter [5] try to fill this gap, since both the psychological and the stereotype models rely on the likelihood matrix, most of the user model will become affected by this issue.

3 Lightweight User Modeling

In order to enhance performance, the lightweight approach provides a way to perform the knowledge discovery step asynchronously, along with a parameterized execution frequency (an ideal frequency would be running the adaptation task for every retrieved feedback). This way, response time is reduced when retrieving user feedback, since the system only has to store information (in the action history) for later processing.

Users are associated with stereotypes defined according to psychological attributes and tags (see fig. 1). Since the taxonomy is also defined according to them, a classification of both users and classes can be achieved over time. The likelihood matrix and tags associate the user with the taxonomy and recommendable items through values ranging from 0 (unlikely) to 1 (likely). Handicap values range from 0 (less) to 1 (most) and can be set for different types of handicaps (*e.g.*, sight, mobility). An action history keeps a record of all relevant user-system interactions (*e.g.*, clicking a POI), allowing the application of a gradual forgetting function that gives more credit to recent user actions.

The main purpose of the Jennings model component was to provide fresh results according to the ratings of the overall user community and to tackle *over-specialization*. Since these can be achieved through both the application of the gradual forgetting function to the action history, and by using global recommendable item ratings, this component has been removed from the model. Also, stereotypes

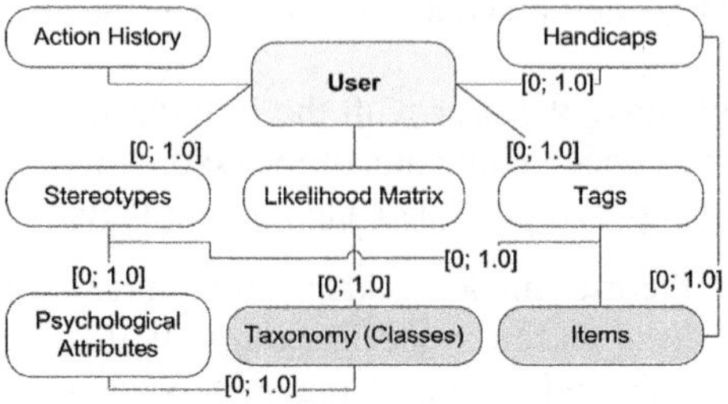

Fig. 1 New Simplified User Model

and the psychological model have been merged in order to allow initialization of the user model and a fuzzy classification of users and item classes.

3.1 Feedback Retrieval

User feedback is given through user-system interactions. In that sense, we have identified a small number of interaction types (see table 1).

All actions are kept in two sets: one for recent actions and another for forgotten actions. The final action weight is measured using an adaptation of the gradual forgetting function proposed in [8], which is then used by the knowledge discovery step to penalize action influence (as given in table 1). For each observed user action i from an action set of length equal to n, starting from the most recent, the gradual forgetting function weight is given by:

$$w_i = 1 - \frac{2k_i(i-1)}{n-1} \tag{1}$$

$$k_i = \frac{t_i}{2s} + \alpha \tag{2}$$

Where k_i represents the overall *forgetfulness* for action i, which is influenced by the α factor, by a ratio between the action timespan (time since the action was performed), given by t_i, and the total action set timespan, given by s. When the resulting weight value drops below 0, the action is forgotten, *i.e.*, it stops affecting the user model.

Table 1 Action types in the Lightweight RS. *Effect* goes from 0 (negative) to 1.0 (positive), and *influence* from 0 (none) to 5.0 (heavy).

Action Types	Affects	Effect	Influence
Item Click	Overall POI Rate, Likelihood Matrix, Tags	1.0	1.0
Class Click	Likelihood Matrix	1.0	1.0
Tag Click	Overall Tag Rate, Tags	1.0	1.0
TourBasket Add	Overall POI Rate, Likelihood Matrix, Tags	1.0	2.0
TourBasket Remove	Overall POI Rate, Likelihood Matrix, Tags	0.0	2.0
Like Rate	Overall POI Rate, Likelihood Matrix, Tags	1.0	4.0
Dislike Rate	Overall POI Rate, Likelihood Matrix, Tags	0.0	4.0

3.2 Knowledge Discovery

The knowledge discovery step is asynchronous and its execution frequency can be parameterized. For immediate system adaptation, this step must be executed after feedback retrieval. The complexity of the implemented algorithm mainly depends on the amount of actions performed by the user, stereotypes and psychological attributes (see algorithm 1).

Algorithm 1. Knowledge discovery step.

```
1: if userLastActionDate ≤ userLastKDDate ∨ userActionCount < 2 then
2:     return
3: end if
4: updateProfileTags (userId, α, minSetTimespan)
5: updateProfileItemClasses (userId, α, minSetTimespan)
6: updatePsychologicalAttributes (userId, β)
7: updateStereotypes(userId)
8: userLastKDDate ← CurrentTimestamp
```

The algorithm starts by checking if the user performed new actions since the last knowledge discovery step. If so, and if there is a minimum amount of two actions, both the tag and likelihood matrix components are updated. New association values for the likelihood matrix and tag components are calculated according to equation 3.

$$assocVal_j = \frac{\sum_{i=0}^{n} effect_i \times influence_i \times w_i}{\sum_{i=0}^{n} influence_i \times w_i} \tag{3}$$

Where i is the index of the user action and j is the index of the concept (tag or item class) involved in the action. Notice that the action set must be ordered in a way such that most recent actions appear first. Also, depending on the component, only certain types of actions can be evaluated, *i.e.*, actions of type *Class Click* must be excluded from the evaluation action set when updating the tag component (in *updateProfileTags*). Overall rates are immediately updated when the action occurs.

Afterwards, the algorithm updates the psychological model of the user with new values based on the likelihood matrix. For an user u, the value of the psychological attribute p is given by:

$$\begin{aligned} paValue_{u,p} = {} & avg(userLMVal(u,i) \times classPAVal(i,p)) \times \beta \\ & + paValue_{u,p} \times (1-\beta), i \in likelihoodMatrixClassesOf(u) \end{aligned} \tag{4}$$

To avoid drastic changes in the psychological attribute values, an update factor β, ranging from 0 to 1.0, is used to define the weight of the new value relatively to the old one. Stereotype association values are also updated according to the new psychological attribute values. The association value between user u and stereotype s is given by:

$$stereotypeValue_{u,s} = 1 - avg(|userPAVal(u,i) - stereotypePAVal(s,i)|),$$
$$i \in psychologicalAttributesOf(u) \quad (5)$$

3.3 System Adaptation

Our focus regarding profile adaptation (*e.g.*, according to personal interests) is in providing recommendations of items (*e.g.*, POIs, events). Providing recommendations is a straightforward step since the model provides almost direct valued associations between users and items. The implemented recommendation algorithm allows item filtering through one or more components of the user model (see algorithm 2).

Algorithm 2. System adaptation step.

```
mixedWeight ← 0
if useLikelihoodMatrix then
    mixedResults ← getLikelihoodMatrixResults(user_id, LMThreshold)
    mixedWeight ← LikelihoodMatrixWeight
end if
if usePsychologicalModel then
    mixedResults ← mixedResults∪ getPsychologicalModelResults(user_id, PMThreshold)
    mixedWeight ← mixedWeight+PsychologicalModelWeight
end if
if useTags then
    mixedResults ← mixedResults∪ getTagResults(user_id, TThreshold)
    mixedWeight ← mixedWeight+TagWeight
end if
finalResults ←mergeResults(mixedResults)
finalResults ←filterWithHandicaps(finalResults)
```

The resulting recommendation value for an item i to an user u is given by $lmRating_{u,i}$ for the likelihood matrix component, $pmRating_{u,i}$ for the psychological model component, and $tagRating_{u,i}$ for the tag component.

$$lmRating_{u,i} = userLMVal(u, classOf(i)), u \in Users, i \in Items \quad (6)$$

$$pmRating_{u,i} = 1 - avg(|userPAVal(u,j) - classPAVal(classOf(i), j)|),$$
$$j \in psychologicalAttributesOf(u) \quad (7)$$

$$tagRating_{u,i} = avg(userTagValue(u,j)), j \in tagsOf(i) \wedge j \in tagsOf(u) \quad (8)$$

Each of the components are allowed to have a different impact in the resulting recommendation. In that sense, the final recommendation values are the result of a weighted average between the results of each component. The recommendation algorithm ends after filtering the results using the explicitly declared user handicap values. If the user handicap value is higher than the item handicap value, the item is excluded.

4 Evaluation

In order to evaluate and compare the Toursplan and Lightweight RS, three scenarios were created: a system startup (to check for system cold start problems), new user registration (to check for user cold start problems) and active user (with given feedback) recommendation request. The dataset contains a total of 300 generated users, 469 POIs, 56 POI classes, 126 tags with an average of 3 tags per POI, 6 stereotypes and 4 psychological attributes.

To evaluate performance ten samples of the execution time of recommendation requests for six different users were taken. The results for all three scenarios are presented in fig. 2. Notice that both RS are implemented in T–SQL.

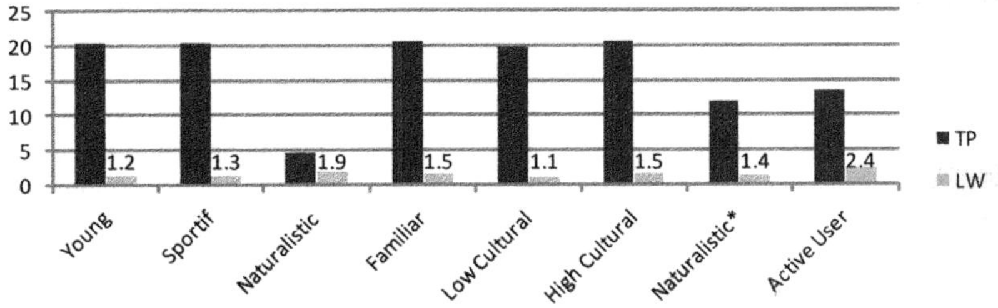

Fig. 2 Average recommendation execution time (in ms) for new users with different stereotypes at system startup (first 6), and for a new user (Naturalistic*) and an active user in an active system.

Performance improvements were also verified in the feedback routine, with the Lightweight RS performing an average of 3.6 ms, and the Toursplan RS an average of 15.2ms. In the first evaluation scenario (system cold start) both systems provided interesting results. However, the ability to discriminate between different

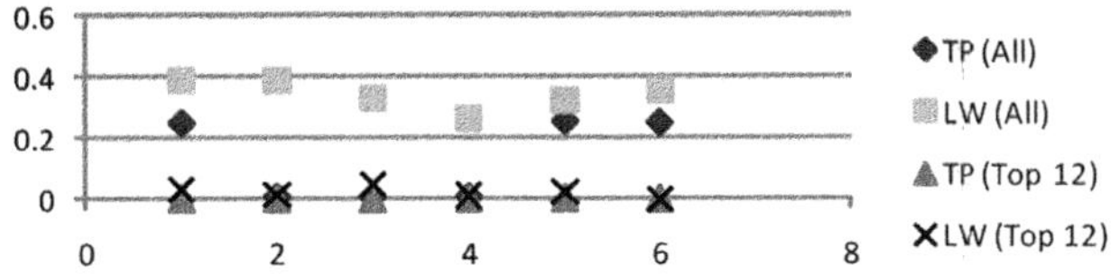

Fig. 3 Difference between maximum and minimum rating values given to POIs by both RS in the system cold start scenario.

results was better in the Lightweight RS. Fig. 3 depicts the range of ratings, after scale normalization, given by both RS. In the third scenario, a user with actions spanning over two years was chosen. When faced with new, more recent feedback, the Lightweight RS adapted well providing results that reflect a gradual forgetfulness of past feedback. On the other hand, the Toursplan RS reflects a gradual increase in over specialization in most of its recommendation modules, specially for heavily active users.

5 Conclusions and Future Work

The lightweight UM process contains significant performance improvements in critical routines, namely the recommendation and feedback submission routines. Although some modules were excluded in the Lightweight process, most common UM issues are well tackled mainly due to the application of the gradual forgetting function [8]. Along with a record of the user action history it allows quick and efficient adaptation of the system and recommendation results.

Future work includes the analysis of collaborative filtering methods that exploit clustering and classification techniques in order to reduce the evaluation set and evaluation under a commercial environment.

References

1. Almeida, A., Coelho, B., Martins, C.: Intelligent hybrid architecture for tourism services. In: Bramer, M. (ed.) IFIP AI 2010. IFIP Advances in Information and Communication Technology, vol. 331, pp. 205–214. Springer, Heidelberg (2010)
2. Brusilovsky, P., Maybury, M.T.: From adaptive hypermedia to the adaptive web. Communications of the ACM 45(5) (2002)
3. Chi, E.H., Pirolli, P., Chen, K., Pitkow, J.: Using information scent to model user information needs and actions and the web. In: Proc. of SIGCHI on Human Factors in Comp. Systems (2001)
4. Coelho, B., Figueiredo, A., Martins, C.: Tours planning decision support, Portugal (2009)
5. Coelho, B., Martins, C., Almeida, A.: Adaptive tourism modeling and socialization system. In: 2009 Int. Conf. on Comp. Sc. and Eng (CSE), Vancouver, Canada, pp. 645–652 (2009)
6. Jennings, A., Higuchi, H.: A personal news service based on a user model neural network. IEICE Trans. on Inf. and Systems E75-D(2), 198–209 (1992)
7. Kobsa, A.: Generic user modeling systems. In: Brusilovsky, P., Kobsa, A., Nejdl, W. (eds.) Adaptive Web 2007. LNCS, vol. 4321, pp. 136–154. Springer, Heidelberg (2007)
8. Koychev, I., Schwab, I.: Adaptation to drifting user's interests. In: Proc. of ECML 2000 Workshop: Machine Learn. in New Info. Age (2000)
9. Porter, J.: Designing for the Social Web. Peachpit Press (2008)

Semantic Zoom: A Details on Demand Visualisation Technique for Modelling OWL Ontologies

Juan Garcia, Roberto Theron, and Francisco Garcia

Abstract. The ontological design and reuse processes sometimes become difficult, specially for medium or large ontologies, so visualisation techniques help to deal with them. Diverse tools such as ontology editors and modelling tools have been proposed to develop ontological engineering, most of them using graphs or UML-like diagrams. These techniques are good enough for small ontologies, nevertheless, they fail for large ontologies due to they display too much information, resulting in cluttered graphs or overcrowded visualisations. The use of the semantic zoom technique represents a good option to deal with this problem, resulting in visualisations more clear and understandable. This paper analyses the currently most used tools for modelling or visualise ontologies, and proposes a visualisation making use of semantic zoom with our tool OWL-VisMod to develop a details on demand visualisation technique.

Keywords: Semantic Zoom, OWL ontologies, OWL-VisMod.

1 Introduction

Ontologies are explicit representations of domain concepts, they provide the basic structure or armature around which knowledge bases can be built. Each ontology is a system of concepts and their relationships, in which all concepts are defined and interpreted in a declarative way. Typically, an ontology consists of a finite list of terms and the relationships between these terms. The terms denote important con-

Juan Garcia · Roberto Theron
Computer Science Department, University of Salamanca, Spain
e-mail: {ganajuan, theron}@usal.es

Francisco Garcia
Computer Science Department, Science Education Research Institute (IUCE), GRIAL Research Group, University of Salamanca, Spain
e-mail: fgarcia@usal.es

J.B. Pérez et al. (Eds.): Highlights in PAAMS, AISC 89, pp. 85–92.
springerlink.com

cepts (classes of objects) of the domain with a hierarchy. Apart from subclass relationships, ontologies may include information such as properties, value restrictions, disjoint statements, specifications of logical relationships between objects. There is no a specific number of elements to determine if an ontology is large, due to this is relative to the ontology's field. For instance, an ontology representing biological elements can be considered large for some hundreds of thousands of elements, in contrast to an ontology representing academic concepts that can be considered large for some hundreds of elements. In terms of visualisation, a large ontology is one that has enough elements to make the visualisation to seem overcrowded and less understandable.

Diverse tools to modelling ontologies have been proposed; they use the most common visualisation techniques such as directed graphs, hierarchical trees, treemaps or representations based on conceptual maps. Most of visualisations become overcrowded with some hundreds of elements, so that would be the a parameter to identify a large ontology in terms of visualisation, no matter the subject. These tools share the same problem dealing with medium or large size ontologies. Representing all the information in a unique visualisation results in a less clear and understandable way to represent ontologies. Our approach is based on the principle of "divide and conquer", using diverse visualisations to represent diverse aspects and applying the concept of semantic zoom. It has been defined as: "A graphical technique to balance detail and context is known as semantic zooming or multiscale interfaces. A physical zoom, on the one hand, changes the size and visible detail of objects. A semantic zoom, in the other hand, changes the type and meaning of information displayed by the object. Semantic zooming avoids the physical distortions of fisheye views, by using a semantic transition between detailed and general views of information" [7]. In contrast to ordinary graphical zoom, semantic zoom does not only change parameters of a graphical representation, but modifies the selection and structure of data to be displayed. We are interested in semantic zoom as a technique to improve the ontologies visualisation and modelling.

This paper is organised as follows: we have provided a brief introduction, then we discuss some related work. In the third section we analyse the current tools and whether or not they implement semantic zoom or any other visualisation techniques. In the fourth section we describe our proposal for modelling using semantic zoom approach in OWL-VisMod and finally we discuss the conclusions and the future work.

2 Related Work

Diverse commercial tools have been proposed to modelling ontologies, being currently the most important, SemanticWorks, TopBraidComposer, IODT and IODE. Some of these tools offer a free version with reduced funcionality; on the other hand, Protege is currently the most widely used tool to edit ontologies, with the advantage that is a project completely free.

SemanticWorks[1] is a commercial tool designed to edit RDF documents in a GUI and check its sintaxis, as well as design RDF schema and OWL ontologies using a graphical design view. It is also able to check the syntax and semantics of OWL Lite, OWL DL ontologies and export ontologies in the RDF/XML and N-Triples formats. The modeller uses a representation based on squarified and expandable boxes for properties and classes, where the hierarchy is represented with a line above the box. This representation is based on the conceptual maps approach, intended to convey complex conceptual knowledge bases is a clear, understandable way. The properties associated with classes are also expanded and linked with the classes. Classes are represented using a squared box with left-side corners rounded, while properties are represented using a normal squared box. Diverse concepts are represented using specific self-designed symbols. This is an efficient way to show or hide information according to the user interests. The main disadvantage of this modelling tool, is that it duplicates nodes. Duplicating elements result in a less efficient and redundant model; this redundancy causes that the user easily becomes confused navigating the model. This strategy of redundancy make extremely difficult to navigate a large ontology, and to be conscious where are we going through this full of possibilities labyrinth of paths.

TopBraid Composer[2] is an enterprise-class modelling environment for developing Semantic Web ontologies and building semantic applications. There are three available versions: a Free Edition, Standard Edition and Maestro Edition. TopBraid Composer is a UML-based modelling plug-in eclipse, part of the TopBraid Suite. We tested using TopBraid Composer Free Edition version 3.3.0 that does not support the UML representation, it is provided just with the paid versions. TopBraid Composer is a fully Protege-based tool that performs the most common operations over ontologies, such as: inference, consistency checking as well as the inclusion of SPARQL query engine. It is composed by diverse visual components, it uses a simple tree like view visualisation to show the hierarchy of the classes and properties. The look and feel is similar to Protege ontology editor, with the possibility to display also a UML-like view. There is no a properly global visualisation, instead the tool just visualise a UML-like class-diagram, to help to understand the relationships among classes and properties in the ontological model.

IODT (IBM Integrated Ontology Development Toolkit)[3] is a toolkit for ontology-driven development, including EMF Ontology Definition Metamodel (EODM), EODM workbench, an Scalable Ontology Repository (SOR, before version 1.5 it is named Minerva) and some extenders to the core components. EODM is derived from the OMG's Ontology Definition Metamodel (ODM) and implemented in Eclipse Modeling Framework (EMF). EODM includes RDFS/OWL parsing and serialization, reasoning, and transformation between RDFS/OWL and other data-modelling languages. EODM is also an open source project of Eclipse.org[4]. EODM Workbench is an integrated, ontology-engineering environment that supports ontol-

[1] http://www.altova.com/semanticworks/owl-editor.html
[2] http://www.topquadrant.com/products/TB-Composer.html
[3] http://www.alphaworks.ibm.com/tech/semanticstk
[4] http://www.eclipse.org/emft/projects/eodm/

ogy building, management, and visualisation. It has UML-like graphic notions to represent OWL class, restrictions and properties in a visual way. It can also have multiple views to support visualisation of an ontology; these views are independent but synchronized so changes made in one visualisation affect all of them. The Visual Workbench represents the model in a UML-like view where classes and properties are differentiated by the assigned colour. Classes are yellow-coloured and its individuals and datatype properties are included inside the class representation. This modelling tool is completely based on the IBM Rational for developing software. Advantages of this modelling tool include the widely known UML standard representation of classes, properties and the hierarchy; the facility to create diagrams and the comprehensibility of them. The main disadvantage of this tool is the scalability to model a large ontology. Large ontologies become difficult to clearly being modelled and understood due to the surplus quantity of information displayed. Visualising all the information of an ontology in only one visualisation, results in an overcrowded view of the knowledge model.

Protege[5] [5],[6] is a free, open source ontology editor and knowledge-base framework. Protege includes diverse plugins developed and maintained by the Protege community. One of these plugins is OWLViz, a graph-based visualisation that represent classes, properties, hierarchy, and the classical hierarchy tree views. Classes are represented as nodes in the graph, while properties are represented as edges connecting nodes, where the edges represent is-a relations (hierarchy). Jambalaya [1][2] is another plugin intended to visualise OWL ontologies with Protege. Currently is no longer supported to newer versions of Protege than 3.3.1, we have had to use an older version of Protege (3.3.1), to use it. Jambalaya project can be found in its official site[6].

The first view represents the self-contained model and is based on a treemap used to show the hierarchy and the structure. Classes are represented as squared boxes and the hierarchy is represented by nesting subclasses inside their superclass. The second view is based on traditional graphs where relationships are represented in the same manner that the self-contained model. This graph connects classes with classes (ISA relationships and object properties representing coupling relationships among classes) as well as classes with their instances. This tool offers a great variety of configuration options, to hide components, to change colours and shapes, to filter data. Although Jambalaya represents a very good tool to visualise an ontology, scalability represents the main disadvantage due to large graph visualisations are well known to become cluttered. This can be clearly depicted in figure 1; where a graph with some hundreds of elements can become a web of nodes and edges where first, the hierarchy is not clear, and second, representing all the relationships at the same time distinguished by colours make the user difficult to remember the colour codifications. Eventhough Jambalaya lets the user to apply filters to hide or show relationships, the use of graphs is well known to be not the best way to deal with large datasets.

---

[5] http://protege.stanford.edu/

[6] http://webhome.cs.uvic.ca/ chisel/projects/shrimp/demo/jambalaya-applet.html

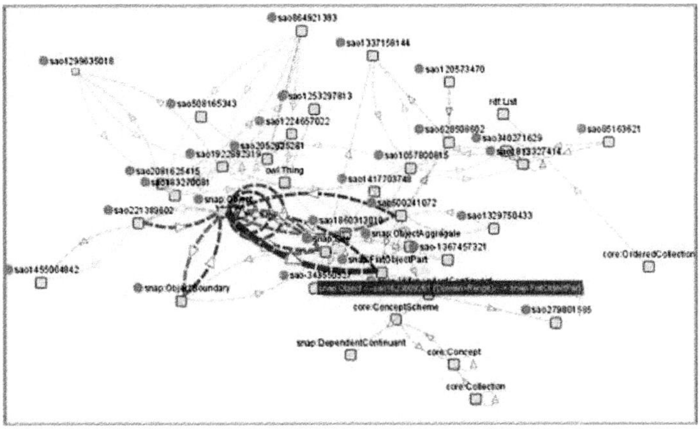

Fig. 1 Jambalaya represents relationships as coloured arrows, and the arrowhead indicating the direction of the property, from the domain to the range. It has implemented six different layouts: grid, radial, spring, sugiyama, tree and treemap[3], to analyse the ontology from diverse perspectives.

3 Semantic Zoom with OWL-VisMod

OWL-VisMod aims to provide a helpful tool for modelling ontologies; it is based on the use of visual modelling for building or modifying OWL ontologies. According to the analysis of the diverse current tools, we have identified that the main problem is the saturation of information in the visual models. For instance, all the tools using graphs to represent the ontology, have the same problem dealing with large ontologies. As consequence, these graphs become completely cluttered full of edges and nodes and it makes more difficult to comprehend the knowledge base, even when most of them let the user to apply filters on the visualised data. On the other hand, the tools using UML-like views have to face with the same problem, they represent both properties and hierarchy relationships using edges as well as graphs do. Our proposal tries to exploit the advantages of diverse visualisations but using a different approach. We use diverse visualisations in order to separate all the information to avoid cluttered and overcrowded models. The navigation model proposed in OWL-VisMod starts with a specific visualisation (tree or treemap) showing the hierarchy of concepts to look for a class to be analysed. Due to the user is navigating the hierarchy, is not needed to show at this time the relationships. Once the user has navigated the hierarchical model and is interested in a specific class, then the semantic zoom is performed. It starts by querying the information of a class in the ontological model, according to the visualisation is going to be displayed. Our proposal illustred in figure 2 is oriented to display the individuals of a class and their values in the properties. It shows the selected class at the center as the most important element, then the individuals are located surrounding the class in a radial layout and purple-coloured to be clearly distinguished. It establishes a link among

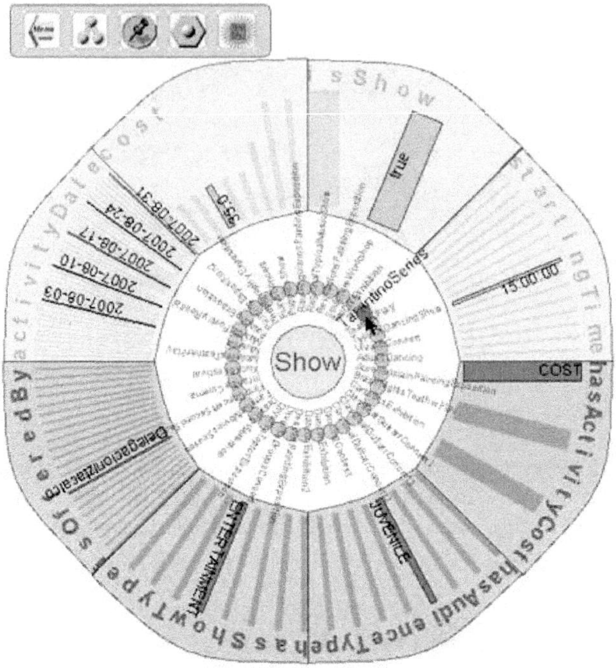

Fig. 2 A semantic zoom over a class called *Show* located at the center of the visualisation. Individuals are located surrounding with a radial layout the selected class, and properties surrounding the individuals. Properties have histograms representing the internal values related with the individuals.

them and the class without defining edges; in the same manner, the properties are located surrounding the individuals in a radial layout, and they are provided with a space in order to put inside histograms representing their internal values. This visualisation organises more efficiently the information, even when the displayed information has some hundreds of elements (properties, individuals and property values). In contrast, a graph with the same number of elements visualised, would be cluttered and less clear to be understood.

The navigation over this visualisation is illustrated in figure 3. The navigation is relatively easy to perform, and in contrast to Jambalaya, SemanticWorks or TopBraid Composer, that the visualisation grows up when the user is selecting elements, our proposal just provide with more space to the selected element by incrementing its angle. This represents another advantage of our visualisation, because those ones that are always growing up and demanding the use of scrolling to be viewed, make the user lost the whole perception, having to deal with scrolling up and down or left and right. Contrasting both figures 2 and 3, they would represent two opposite cases: the first one would answer the question: which values for each property have this specific individual?; while the second case would answer the question:

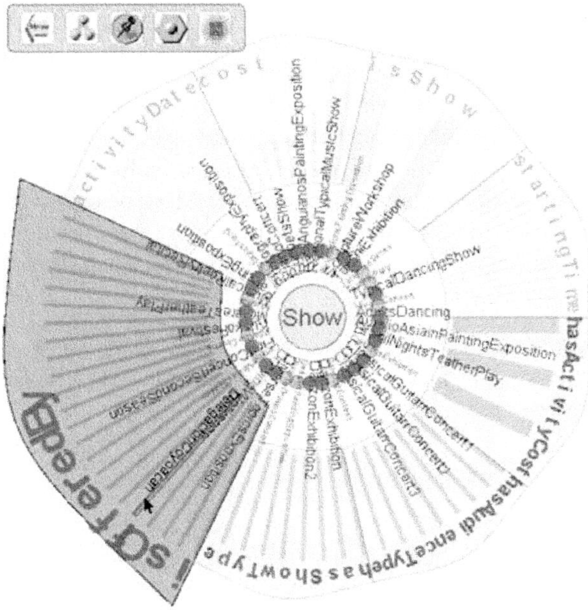

Fig. 3 Navigating the visualisation, once a property is selected and an internal value is highlighted, all the individuals having this value for this property are also highlighted. Another possibility is to highlight an individual, in this case all the internal values in the properties are also highlighted, so the navigation is in both directions.

which individuals have this specific value for this specific property?. It is important to highlight the previous query process developed before the semantic zoom is done.

As we have previously mentioned, this visualisation is just one part of a set of visualisations from our tool; it is displayed when the user is demanding the internal components from a class. We take advantage of the radial layouts that efficiently organise elements in a circular view, surrounding the focused element located at the centre. Moreover, the visualisation is intuitive and the navigation simple. Another visualisation of our modelling tool that uses semantic zoom has been defined in [8].

4 Conclusions and Future Work

We have analysed the most widely used tools to model ontologies and we have pointed out their main disadvantages, that include redundancy in the case of SemanticWorks, or cluttering in the case of the tools that use graphs or UML diagrams. Even though the analysed tools represent good proposals to visualise or modelling small ontologies, they become less efficient when dealing with large ontologies. Moreover, these disadvantages have a negative effect during the visualising or modelling phases, because they make the user get confused and sometimes even lost in the visualisation. We have proposed a visualisation based on the use of the semantic zoom technique over a selected class, to represent individuals, properties and the internal values of each property for a selected class in an OWL ontology. Our proposal uses simple and known visual elements such as a radial layout and histograms to represent elements in a visualisation, but organised in a compact representation that improves the visual navigation. This visualisation is included as a part of a whole modelling tool we are developing, called OWL-VisMod, where diverse linked visualisations interact each other to analyse or modelling OWL ontologies. The future work includes the evaluation of the usability of this visualisation, together with the other visualisations of the tool.

Acknowledgements. This work was supported by Spanish Government project TIN2010-21695-C02-01 and by the Castile and Lion Regional Government through GR47 and the project FFI2010-16234.

References

1. Storey, M.A., Musen, M., Silva, J., Best, C., Ernst, N., Fergerson, R., Noy, N.: Jambalaya: interactive visualization to enhance ontology authoring and knowledge acquisition in Protege. In: Workshop on Interactive Tools for Knowledge Capture (K-CAP 2001), Victoria, British Columbia, Canada (2001)
2. Storey, M.A., Noy, N., Musen, M., Best, C., Fergerson, R.: Jambalaya: an interactive environment for exploring ontologies. In: Proceedings of the International Conference on Intelligent User Interfaces, San Francisco, California, United States, p. 239 (2002)
3. Johnson, B., Shneiderman, B.: Treemaps: a space filling approach to the visualization of hierarchical information structures. In: Proceedings of the 2nd. IEEE Visualization Conference, pp. 284–291 (1991)
4. Keim, D., Mansmann, F., Schneidewind, J., Schreck, T.: Monitoring Network Traffic with Radial Traffic Analyzer. In: IEEE Symposium on Visual Analytics Science and Technology, pp. 123–128 (2006)
5. Gennari, J., Musen, M., Fergerson, R., Grosso, W., Crubezy, M., Eriksson, H., Noy, N.: The Evolution of Protege: An Environment for Knowledge-Based Systems Development. Stanford Medical Informatics (2002)
6. Knublauch, H., Fergerson, R., Noy, N., Musen, M.: The protégé OWL plugin: An open development environment for semantic web applications. In: McIlraith, S.A., Plexousakis, D., van Harmelen, F. (eds.) ISWC 2004. LNCS, vol. 3298, pp. 229–243. Springer, Heidelberg (2004)
7. Modjeska, D.: Navigation in Electronic Worlds: A Research Review, Technical Report. Computer Systems Research Group, University of Toronto (1997)
8. Garcia, J., Garcia, F., Theron, R.: Visualising semantic coupling among entities in an OWL ontology. In: Sicilia, M.-A., Kop, C., Sartori, F. (eds.) ONTOSE 2010. LNBIP, vol. 62, pp. 90–106. Springer, Heidelberg (2010)

PCMAT – Mathematics Collaborative Learning Platform

Constantino Martins, Paulo Couto, Marta Fernandes, Cristina Bastos,
Cristina Lobo, Luiz Faria, and Eurico Carrapatoso

Abstract. The aim of this paper is to present an Educational Adaptive Hypermedia Tool, PCMAT, Based on Progressive Assessment and adapted to the student model and learning style. The adaptation of the application is based on progressive self-assessment (exercises, tasks, etc.). The learning platform was already implemented, tested and evaluated in learning processes in Basic Schools. Also, the paper defines and evaluates the characteristic of the User Model to be used in the Student Modeling.

Keywords: Adaptive Hypermedia, Cooperative Learning, Student Modeling.

1 Introduction

The main objective of Adaptive Systems is to adequate its relation with the user (content presentation, navigation, interface, etc.) according to a predefined but updatable model of the user that reflects his objectives, preferences, knowledge and competences [2, 5]. For Educational Adaptive Systems, the emphasis is placed on the student knowledge in the domain application and learning style, to allow him to reach the learning objectives proposed in his training [4]. Although numerous research and already developed systems provided good results, more development,

Constantino Martins · Paulo Couto · Marta Fernandes · Cristina Bastos
Cristina Lobo · Luiz Faria
GECAD – Knowledge Engineering and Decision Support Group, R. Dr. António Bernardino de Almeida, 431. 4200-072 Porto, Portugal
e-mail: {acm,pjco,mmaf}@isep.ipp.pt, cristinabastos70@gmail.com, {mcgm,lef}@isep.ipp.pt

Eurico Carrapatoso
Faculty of Engineering – University of Porto, R. Dr. Roberto Frias, s/n. 4200-465 Porto, Portugal
e-mail: emc@fe.up.pt

J.B. Pérez et al. (Eds.): Highlights in PAAMS, AISC 89, pp. 93–100.
springerlink.com

experimentation and implementation are still necessary to conclude about the adequate features and effectiveness of these systems [11]. The OCDE PISA 2006 study [13].concerning Scientific Competence of Portuguese students concluded that the knowledge level about mathematics did not show any improvements when compared with a similar study undertaken in 2003. The results of this study reveal that students of 7 and 8 grades achieved a low behavior in mathematics [13]. The results show that mathematic knowledge of Portuguese students is lower when compared with the majority of OCDE countries [13]. The aim of this paper is to present the project, PCMAT (Mathematics Collaborative Learning Platform). PCMAT is an Collaborative learning platform with a constructivist approach, assessing the user knowledge and presenting contents and activities adapted to the characteristics and learning style of the student of mathematics in basic schools. This paper is organized as follows. Section 2 provides a general approach to Adaptive Hypermedia Systems (AHS). The section 3 defines Student Model and section 4 Learning Styles concept. Platform Development and Some Results are presented in section 5 and 6.

2 Hypermedia Systems

Adaptive Hypermedia (AH) is generally referred as a crossroad in the research of Hypermedia and User Modeling (UM) [2, 3, 5]. An AHS builds a model of the objectives, preferences and knowledge of each user and uses it, dynamically, through the Domain Model and the Interaction Model, to adapt its contents, navigation and interface to the user needs. De Bra in 2004 [4] indicates that these systems must present the functionality to change content presentation, links structure or links annotation The global architecture proposed by Benyon [1] and De Bra [5], indicates that AHS must have three essential parts: The User Model, the Domain Model and the Interaction Model. In Educational Adaptive Hypermedia Systems (EAHS), the emphasis is placed on students knowledge in the domain application and learning style, in order to allow them to reach the learning objectives proposed in their training [12].

3 Student Model

In generic AHS, the User Model allows changing several aspects of the system, in reply to certain characteristics (given or inferred) of the user [2]. These characteristics represent the knowledge and preferences that the system assumes that the user (individual, group of users or no human user) has [8, 11, 12]. In EAHS, the UM (or Student Model) has increased relevance: when the student reaches the objectives of the course, the system must be able to re-adapt, for example, to his knowledge [2, 11, 12]. A Student Model (SM) includes information referring to the specific knowledge that the system judges that the user possesses on the domain, known as the Domain Dependent Data (DDD). The components of the Domain Dependent Data correspond to the Domain Model with three-level functionality [1, 10, 11]: Task level, Logical Level and Physical Level. The Domain Independent Data (DID) are composed of two elements: the Psychological Model and the Generic Model of the Student Profile [1, 8, 16]. The data related to the user interests, common knowledge and background are kept in the Generic Model of the Student Profile. For each AHS, it will be necessary to define which are the characteristics and relevant parameters of the user to be kept [2, 3, 7, 11, 12].

4 Learning Styles

The key of constructivism theory is that student must be actively involved in the learning process. It is important that teachers understands that the construction of knowledge acquisition occurs from knowledge that student already possess and differs from Student to Student. The role of the Teachers is now to be a guide for the student [6, 11]. Students learn in different ways and depend upon many different and personal factors [15]. The emphasis in student individual differences is also important in a context to recognize, design and support students activities (tasks). In constructivism learning theory, Students have different Learning Styles (LS) [6, 11]. Generally, Learning Styles is understood as something that intent to define models of how a person learns. Generally it is understood that each person has a Learning Style different and preferred with the goal to achieve better results. Some case studies have been proposed that teachers should assess the learning styles of their students and adapt their classroom and methods to best fit each student's learning style [9, 14]. There are different Learning Styles models (based on different psychological theories) such as for example [15]: Models based on personality (Witkins and Myers-Briggs Type Indicator); Models based on information processing approach (Schmecks and Kolbs); Models based on Social Interaction (Reichmann and Grasha); Models based on multidimensional factors (Keefe and Dun & Dun). The model proposed by Kolb is the most commonly used inventory and is based on Pigets model on cognitive and learning development [15].

5 Platform Development

The platform, PCMAT (Mathematics Collaborative Learning Platform), is based on AHA! (Adaptive Hypermedia Architecture) [17]. The learning platform developed is based on a constructivist approach, assessing the user knowledge and presenting contents and activities adapted to the characteristics and learning style of the student. Also, the platform allows the students and teachers to autonomously create and consolidate knowledge, with permanent automatic feedback and support, through instructional methodologies and educational activities explored in a constructivist manner [11]. The adaptation of the application is based on progressive self-assessment (exercises, tasks, etc.). The scheme is set by the teacher but is individualized to each students level of knowledge, competences, abilities and learning path. The platform is also connected to tutorials than are contextually accessed by the students when they fail a progression step. One of the goals of the project is to define and to evaluate the characteristic of the User Model to be used in the Student

Model. With the objective to consolidate the knowledge of the student, our system is able to make permanent automatic feedback and support, through instructional methodologies and educational activities explored in a constructivist manner. In addition the platform can adapt the use of learning objects in accordance to the constructivist analysis of the student and his performance. The constructivist approach is also used to suggest some references to the student according to the response of the progressive self-assessment (exercises, tasks, etc.) [11]. In order to evaluate the system, one Basic School was used. The course chosen was Direct Proportionality (mathematics) [11].

5.1 System Architecture Definition

The system architecture is based on some strategies already used for the implementation of Adaptive Hypermedia System (AHS), like for example the [17]: Dexter Model; Amsterdam Hypermedia Model (AHM); Adaptive Hypermedia Application Model (AHAM) or Munich Reference Model. The architecture used for the implementation of the system is almost the same referred by De Bra and Benyon [1, 5] (Fig. 1).

Therefore, in our system, the user requests an assessment by clicking on a link in a Web page. Every assessment/page corresponds to a "concept" in the domain model and the user model (which is an overlay model). The system checks the suitability

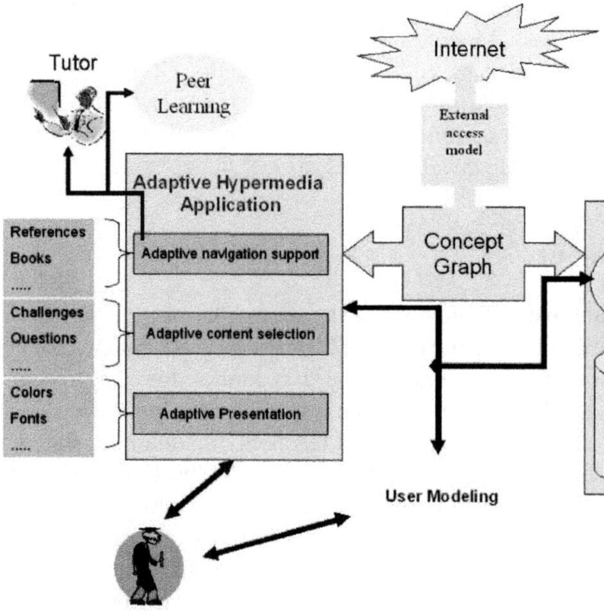

Fig. 1 Architecture of the System.

of the requested page for the current user. The adaptation rules used to check if the page is suitable are defined in the adaptation model. Updates to the User Model are inferred from the interaction between the user and the application. The correct or wrong answers of the user allow the system to estimate the users knowledge level about the concept related with the requested content.

5.2 Student Model Implementation

Two different types of techniques can be used to implement the Student Model: Knowledge and Behavioral based [8, 11]. The Knowledge-Based adaptation typically results for data collected through questionnaires and studies of the user, with the purpose to produce a set of initial heuristics. The Behavioral adaptation results from monitoring of the user during his activity [12]. The use of stereotypes allows to classify users in groups and generalizes student characteristics to that group [11, 12]. The Behavioral adaptation can be implemented in two forms: the Overlay and the Perturbation methods [12]. These methods relate the level of the student knowledge with the learning objectives/competences that the student is supposed/intended to reach [12]. The approach used to build the User Model (UM) is the Stereotype Model with the overlay model for the knowledge representation of the student. The representation of the stereotype is hierarchical. Stereotypes for user with different knowledge have been used to adapt information, interface, scenario, goals and plans. The user modeling process starts with the identification of the user subgroup (using for example questionnaires and Learning Styles), then the identification of key characteristics (which one to identify the members of a user-subgroup), and finally the representation in hierarchical ordered stereotypes with inheritance. The user plan is a sequence of user actions that allows him to achieve a certain goal. The System observes the user actions ant try to infer all possible user plans. This goal is possible because our system has a library of all possible user actions and the preconditions of those actions. The definition of the characteristics of the student will took into account the Domain Model and the constructivist approach of the application. For example, the generic profile includes the name, age, knowledge, deficiencies and the domain of the application. In the psychological profile, the learning style, cognitive capacity and traces of the personality of the student will be stored. In the Domain Dependent Data, the objectives, result of assessments and aptitude of the user will be followed. The tools used to collect data are: Questionnaires, Learning Styles, Psychological tests and exams. For the definition of the Learning Styles of the student we are using the Kolb Learning Styles Matrix [11].

5.3 Domain and Adaptation Models Development

The Domain Model represents concept hierarchies and the related structure for the representation of the user knowledge level (quantitative value). The Domain and Adaptation Model use the student characteristics from the User Model (UM). The knowledge about the user, represented in the User Model, is used by the Adaptation

Model to define a specific domain concept graph, adapted from the Domain Model, in order to address the current user needs. The path used in the graph is defined by: The interaction with the student using a progressive assessment; the student knowledge representation defined by the Overlay Model and the user characteristics in the UM. The system adaptation (adaptation to content or links) to the user can produce user model updates as well. The results of Domain and Adaptation Models achieve are: The development of the concept graph by each user to use in the Adaptation rules and the Definition of the Adaptation Model using the characteristics of the student in the User Model.

5.4 Interaction Model

The Interaction Model represents and defines the interaction between the user and the application [11]. In the Interaction Model, the system presents the functionalities to change the content presentation, the structure of the links or the links annotation with the objective to allow the student to reach the learning goals proposed in their training [5]. To guide the user to the relevant information and keep him away from the irrelevant information or pages that he still would not be able to understand, it is used the technique generally known by link adaptation (Hiding, disabling, removal, etc.). Also, the platform supplies, in the content (page), additional or alternative information to certify that the most relevant information is shown. The technique that is used for this task it is generally known by content adaptation. The constructivist approach is also in the sense to suggest references and activities to the student according with the response of the progressive self-assessment exercises, tasks, etc. [11].

6 Results and Conclusion

The first version of the framework presented in previous section, was already implemented, tested and evaluated in learning processes in mathematics basic schools. The collected evaluation data has showed a very high degree of interest and motivation from students and teachers alike, resulting from its use. Students also perceive this tool as very relevant for their learning, as a self-operating application to be integrated in a more global learning strategy that includes also tutoring (direct contact with the teacher) and peer learning. Teachers agree with these definitions of the platform, as well [11]. Another result is the definition of a new strategy and architecture for the implementation of an Educational Adaptive Hypermedia platform in basic schools in Portugal [11] . The main result of the present development is the validation of a reference picture for the user model that will support new adaptive functionalities based on the use of learning objects to truly support a constructivist learning and cognitive path [11]. The definition of the characteristics of the student to be stored and the selection of the techniques of the Overlay Model and stereotype for the representation of the user knowledges in the UM and the Adaptation Model were defined. The number and type of characteristics to use depend on the finality of each system, but some relevance is in the cognitive part, learning styles and student knowledge [12]. At present, our research of the Student Model and AHS, goes in the direction to make possible the reuse of each student model in different systems. The standards are more and more relevant for this effect, allowing the systems to communicate and to share data, components and structures, at syntax and semantic level [4], even if most of them still only allow syntax integration [3, 5, 11].

Acknowledgements

The authors would like to acknowledge COMPETE Program, and the Portuguese Science and Technology Foundation for their support to GECAD unit, and the project PCMAT (PTDS/CPE-CED/108339/2008).

References

1. Benyon, D.: Adaptive systems: A solution to Usability Problems. Journal of User Modeling and User Adapted Interaction 3(1), 1–22 (1993)
2. Brusilovsky, P., Millán, E.: User models for adaptive hypermedia and adaptive educational systems. In: Brusilovsky, P., Kobsa, A., Nejdl, W. (eds.) Adaptive Web 2007. LNCS, vol. 4321, pp. 3–5. Springer, Heidelberg (2007)
3. Brusilovsky, P., Nejdl, W.: Adaptive Hypermedia and Adaptive Web. In: Singh, M.P. (ed.) Practical Handbook of Internet Computing, pp. 1.1–1.14. Chapman Hall & CRC Press, Baton Rouge (2005)
4. Chepegin, V., Aroyo, L., De Bra, P., Heckman, D.: User Modeling for Modular Adaptive Hypermedia. In: SWEL 2004 Workshop at the AH 2004 Conference, TU/e Computing Science Report 04-19, pp. 366–371 (2004)
5. De Bra, P.: Web-based educational hypermedia. In: Romero, C., Ventura, S. (eds.) Data Mining in E-Learning, Universidad de Cordoba, Spain, pp. 3–17. WIT Press (2006), ISBN 1-84564-152-3
6. Jonassen, D.: Objectivism versus constructivism: Do we need a new philosophical paradigm? Educational Technology, Research and Development 39(3), 5–14 (1991)
7. Kavcic, A.: The Role of User Models in Adaptive Hypermedia Systems. In: Proceedings of the 10th Mediterranean Electrotechnical Conference MEleCon 2000, Lemesos, Cyprus (May 2000)
8. Kobsa, A.: User Modeling: Recent Work, Prospects and Hazards. Adaptive User Interfaces: Principles and Practice. North Holland Elsevier, Amsterdam (1993)
9. Kolb, A., et al.: Learning styles and learning spaces: Enhancing experiential learning in higher education. Academy of Management Learning and Education 4(2), 193–212 (2005)
10. Kules, B.: User Modeling for Adaptive and Adaptable Software Systems (2000)
11. Martins, C., Faria, L., Carrapatoso, E.: Constructivist Approach for an Educational Adaptive Hypermedia Tool. In: The 8th IEEE International Conference on Advanced Learning Technologies (ICALT 2008), University of Cantabria, Santander on July 1st to 5th (2008)
12. Martins, C., Faria, L., Carrapatoso, E.: User Modeling In Adaptive Hypermedia Educational Systems. Educational Technology & Society, [Journal indexed in Social Sciences Citation Index (listed in Web of Science)] (2007), ISSN 1436-4522

13. Pinto-Ferreira, C., Serrão, A., Padinha, L.: OCDE PISA 2006 Competências Científicas dos Alunos Portugueses, Gabinete de Avaliação Educacional Ministério da Educação (2006)
14. Stash, N., Cristea, A., De Bra, P.: Explicit Intelligence in Adaptive Hypermedia: Generic Adaptation Languages for Learning Preferences and Styles. In: Workshop CIAH 2005, Combining Intelligent and Adaptive Hypermedia Methods/Techniques in Web Based Education Systems, in conjunction with HT 2005, pp. 75–84 (2005)
15. Ritu, D., Sugata, M.: Learning Styles and Perceptions of Self. International Education Journal 1(1) (1999)
16. Vassileva, J.: A task-centred approach for user modeling in a hypermedia office documentation system. In: [Brusilovsky et al, 1998], pp. 209–247 (1998)
17. Wu, H., Houben, G.J., De Bra, P.: User Modeling in Adaptive Hypermedia Applications. In: Proceedings of the Interdisciplinary Conferentie Informatiewetenschap, Amsterdam, pp. 10–21 (1999)

Applying Recommender Methodologies in Tourism Sector

Joel Pinho Lucas, Bruno E. da Silva Coelho, María N. Moreno García, Ana Maria de Almeida Figueiredo, and Constantino Lopes Martins

Abstract. Nowadays, there is a constant need for personalization in recommender systems. Thus, they try to bring it by making suggestion and providing information about items available. There are numerous options of methods to be employed in recommender systems. However, they still suffer from critical limitations and drawbacks. Therefore, current recommender techniques try to minimize the affects of such drawbacks. In this work we describe two different recommender methodologies proposed. To do so, we implemented such methodologies in a real recommender system for tourism. Afterwards, we analyzed and compared the recommendation given by both methodologies in order to find out if they are effective and able to deal with common drawbacks.

Keywords: Recommender Systems, Tourism, User Modelling.

1 Introduction

Nowadays, e-commerce systems present loads of products available for sale. This way, users would probably have difficulty in choosing the products they prefer and, consequently, have difficulty purchasing them. Due to such facts and to a more and more competitive industry, these systems need to personalize the presentation of their products to the consumers.

A way to reach such personalization is by means of the "recommender systems", which are used in e-commerce to make suggestion of products and to provide them information in order to aid them in the selection of products [1]. In this

Joel Pinho Lucas · María N. Moreno García
Departamento de Informática y Automática, Universidad de Salamanca, Plaza de la Merced s/n, 37008, Salamanca, Spain
e-mail: {joelpl,mmg}@usal.es
Bruno E. da Silva Coelho · Ana Maria de Almeida Figueiredo · Constantino Lopes Martins
Knowledge Engineering and Decision Support Research Center (GECAD,
Instituto Superior de Engenharia do Porto, R. Dr. António de Almeida 431,
4200073, Porto, Portugal
e-mail: {ana,const}@dei.isep.ipp.pt

J.B. Pérez et al. (Eds.): Highlights in PAAMS, AISC 89, pp. 101–108.
springerlink.com

sense, recommender systems aim at enabling the creation of a new store personally designed for each consumer [1].

Taking into account that data mining techniques are applied for identifying patterns within datasets, according to Cheung et al. [2] these techniques can be successfully applied for recommender systems. In this work we analyze the implementation of two distinct recommender methodologies that we inserted in a recommender system for tourism. Through this case study, we compare both methodologies and highlight the strengths and weakness of each.

In the next section we describe general features and concepts related to recommender systems. Subsequently, in section 3, we describe the main features and concepts of the two recommender methodologies and, in section 4, we analyze the performance of both in a real recommendation scenario.

2 Recommender Systems

According to Cheung et al. [2] and Lee et al. [3], the methods implemented in recommender systems can be divided into two main classes: collaborative filtering and content-based methods. Content-based methods compare text documents to user profiles, where web objects are recommended to a user based on those he has been interested in the past [3]. Hence, recommender systems that use such type of methods do not take into account information acquired by other users. On the other hand, in collaborative filtering methods the recommendation process is based on products' opinions collected from other users [4].

Breese et al. [5] classified collaborative filtering methods into two groups: memory-based and model-based methods. In the first, the nearest neighbors of a target user are found by matching the opinions of such user to the opinions of all system's users. On the other hand, model-based methods build a predictive model by means of a training set which comprises opinions acquired from a small portion of the users. Such methods have been developed more recently in order to avoid the sparsity problem, which usually arises when memory-based methods are employed, because e-commerce systems generally offer millions of products for sale [6]. As a result, recommender systems typically do not employ merely memory-based methods. Likewise, content-based recommender methods usually are not employed. They are not effective due to the lack of mechanisms to extract Web objects features. Therefore, both methods are commonly employed together.

2.1 Drawbacks

The most critical drawback such methods presents is probably associated to data sparsity, due to the large number of items that current recommender systems usually present. According to Sarwar et. al. [7], users of e-commerce systems are able to purchase, in general, only 1% of the products available by the system. This constraint is more problematic for memory-based collaborative methods, because it may not be feasible to obtain enough ratings from users of a system. Model-based collaborative filtering methods can minimize effects of sparsity.

Another drawback originated from the large number of items available in recommender systems is scalability. Such drawback may turn into the major concern to the system performance depending on the volume of items available. The performance is a key feature in recommender systems, because these systems need to provide their users fast feedback. Generally, scalability is not a drawback for model-based methods, because in such methods, differently from others, the induction of the predictive model is rarely performed at run time. On the other hand, the first rater problem [4] [8] is a drawback that may occur in all type of collaborative filtering methods. It is related to the restraint of having few opinions on which to base the predictions.

Conversely, there are drawbacks, such as the gray sheep problem [4], that occur only in collaborative filtering methods. The gray sheep problem refers to the users who have opinions that do not consistently agree or disagree with any group of users. However, such problem does not occur in content-based methods, because they do not consider opinions acquired from other system's users in order to make recommendations. According to Condliff et. al. [8], since a content-based system does not consider the social background of its users, the system is limited to recommend just items that are similar to those that a user has liked in the past.

2.2 Tourism Recommender Systems

Given its complexity, tourism is a privileged area for the application of artificial intelligence [9], and, in particular, Decision Support Systems [10], which are a generalization of Recommender Systems. According to the Travel Industry Association of America (www.tia.org), in 2003, 30% of the United States adult population (64 million) used the Internet to look for information about destinations or to check prices and schedules. In addition, 66% of them booked travel via the Internet. Moreover, according to the World Travel & Tourism Council, travel and tourism represents approximately 11% of the worldwide GDP (gross domestic product). In this way, the tourism domain consists in a crucial and strategical application field in recommender systems.

Tourism Recommender Systems can be classified as an intermediary between customer and travel agency [11]. They are not restricted by the human factor like travel agents, whose knowledge about destinations and places of interest are mostly limited and the recommendations he/she offered agent may be restricted to personal opinions and tastes.

The tourism context is especially interesting because, according to Werthner and Ricci [12], recommendations may refer to a variety of products, such as locations, attractions, accommodations, and flights, in order to provide a meaningful picture of the proposed travel. Current recommender systems for tourism are generally designed for accomplishing one of the two (or even both) following purposes: to aid the user in planning his trip (i.e. choosing one or more destinations) and to aid the user in planning what to do in a certain place (usually a city). In this work, we focus on the second recommendation purpose.

3 Case Study on Recommendation

In this section we analyze the behavior and results provided by two recommender methodologies developed by different authors. Such methodologies were implemented in a real recommender system. The PSiS (Personalized Sightseeing Planning System) was developed by Coelho et al. [13] to aid tourists to plan their stay in the city of Porto, Portugal. The items offered by this system consist of tourist attractions, named here as Points of Interest. Within PSiS each visiting plan tries to select the most adequate points of interest items according to the user profile. The user may access the detailed information of every point of interest in the system, including their description, address, average cost, duration, etc. It is also possible to view, through a map, where it is located in the city and to watch multimedia contents available for it (like videos and news).

Currently, the PSiS database contains information provided by the municipality of Oporto, from where it was gathered data related to the 241 points of interest currently available on the system. It is important to highlight that the current version of PSiS is a prototype.

In this way, the recommender methodologies implemented in PSiS allows it to suggest a list of points of interest to active user when he/she requests it. The first methodology was proposed in [14] and provides the user a general list of points of interest to visit. The second methodology we implemented was proposed by Coelho et al. [13] and provides the user a list of points of interest grouped by categories. In the next subsections we describe both methodologies.

3.1 The First Methodology

The main contribution and novelty of this methodology is the employment of classification based on association for recommender systems and also the use of fuzzy sets concepts. In this sense, a fuzzy associative classifier (CBA-Fuzzy) was developed based on the standard CBA algorithm [15]. The CBA-Fuzzy algorithm is the basis of this methodology, since it is responsible for generating the rules composing the model employed to make recommendations.

Basically, it is composed by three components: conception of groups of users, rules' set generation and recommendation. The first two components are built offline and are responsible for building the recommender model. The third component is responsible for classifying the active user at runtime and to provide him recommendation.

Firstly, in the first component, we find out groups of users with similar preferences and characteristics to the active user. To do so, it employs an adapted version of the K-Means algorithm within users' demographic data. In this sense, this process may be considered as a collaborative filtering approach, because the system makes use of the information gathered from other users (instead of only from the active user). After that, an ordered list of items (or products) P = {p1, p2, p3, …, pm} is assigned to each group gi, where i ∈ {1, 2, 3, …, N}. The top items in each list will be the ones who better represent each group.

The second step in the construction of the recommender model is the conception of the classification rules (by means of CBA-Fuzzy). Such rules are used for classifying the active user, where data gathered from active user's last interaction with the system is considered, which is represented by a transaction "y". As it consists of past information from the active user, in this context this methodology may be considered as a content-based approach.

At this moment, the system compares "y" to the classification rules generated before. Then, a set Rc = {r1, r2, r3, ..., rN} containing "N" rules satisfying such condition is obtained. Subsequently, the possible groups to which the active user owns to is obtained. At this point, it calculates the membership function to every group of users found on Rc's rules consequent terms. To do so, a discriminator function "g" is defined in order to calculate the membership degree of the active user to every class found in Rc. Thus, more than one class is assigned to the active user. The classes satisfying a given membership threshold will be the ones taken into consideration. Afterwards, it considers the sets of items (first component) assigned to the groups of users. The recommendation provided will be a suggestion containing the "n" best ranked items in each group.

3.2 The Second Methodology

Basically, this methodology differs from the first one in the way the recommender model is built, because the second is more focused in the user model and the other focus mainly in the user interaction with the items. In this way, this methodology focuses its implementation in building knowledge representation formalisms that co-exist in order to constitute the system's view of the user.

Nevertheless, the methodology needs to extract information from points of interest as well. It which identifies two groups of factors that influence destination choice: personal features and travel features [16]. In the context of this work, personal features refer to user data and travel features to points of interest data. Hence, PSiS needs to evaluate both points of interest information and user information. To do so, a modeling architecture for structuring user information was defined, which was based on the user modeling architecture proposed by Benyon [17]. Such author proposed a "Student User Modeling Architecture" that separates user information elements in an information hierarchy. It may be perfectly usable in a variety of other situations, such as tourism [13]. It defines that user data can be divided into two main modules: Domain Dependent Data (DDD) and Domain Independent Data (DID).

The first consists of user information which is particularly related to the system domain (related to items), whereas the second consists of user information that is not related to the system. The user modeling process performed in this methodology is mostly not domain dependent, because it employs just one type of domain information for user modeling, which consists of the use of keywords to relate items' features to users. To do so, a set of keywords was created for each point of interest and then the user, when he/she enrolls the system, inserts personal data that are intrinsically related to those keywords. This is clearly a content-based approach. On the other hand, it was included in PSiS some pre-defined (by means of a statiscal analysis of tourist data in the city of Porto) user stereotypes and basic psychological models. Moreover, a Likelihood Matrix was created in order to users express their feedback about point of interest they have visited. Afterwards, the system returns a set of points of interest that owns to the best classes that match his/her likelihood matrix, given the optimal threshold. In this sense, these domain independent methods are clearly a collaborative filtering based strategy.

4 Applying and Comparing the Methodologies

In this section we simulate a critical user (in terms of recommendation) asking for recommendation. Thus, we are able to evaluate if it can provide reasonable suggestions of points of interest. In order to simulate the first-rater and the gray sheep problem, we created a user U1 whose data is composed by the less frequent values of users' database in PSiS. Hence, U1 encompasses non-frequent data, for which it would be certainly hard to find similar users and correlated items.

It is important to highlight that the user may ask for two types of recommendation: "Just recommend me points of interest" (using the first methodology) and "Give me recommendations by category" (using the second methodology). In figure 1 and 2 we show a capture of PSiS's screen when the user U1 asked for the first and the second, respectively, option of recommendation.

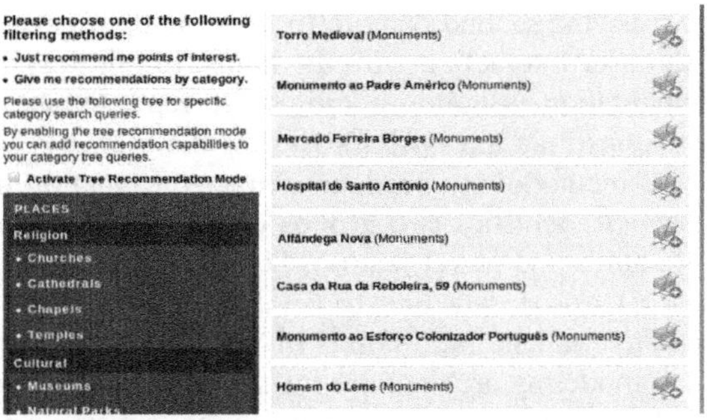

Fig. 1 General points of interest being recommended.

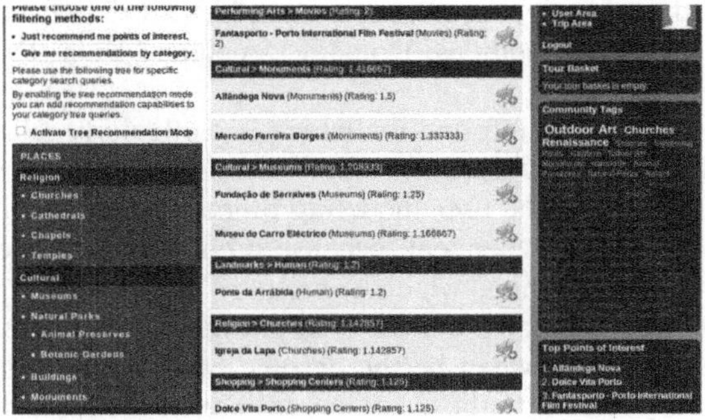

Fig. 2 Points of interest by categories.

Through figure 1 and 2, we can affirm that U1 received recommendations, in both cases, in the same way of a regular scenario. In both cases the system recommended eight items, but the ones recommended by the first methodology were limited to just one category. On the other hand, items recommended by the second methodology were diversified in many categories. Thus, we may say that the two methodologies presented complementary results, what enhances the overall recommendation quality. Besides, the second methodology recommended the same points of interest in the "Monument" category, which were also recommended by the first methodology.

Hence, we may conclude that effects of the gray-sheep and first-rater problems did not affect significantly the recommendation quality. Moreover, scalability would not be a critical issue and they are neither affected by limitations caused by sparsity, because they build their recommender model off-line.

5 Concluding Remarks

Through the case study described before, we may say that both methodologies provide more value and efficiency for PSiS, because both methodologies develop a hybrid approach, taking advantage from the strengths of both collaborative filtering and content-based approaches. Both methodologies were able to provide meaningful recommendations even for a critical user. Thus, both are able to enhance system personalization and make able to explore even more resources. Besides, both methodologies keep recommendations updated with the user profile.

Hence, this work proved that both methodologies are effective. Since the first one is more concerned on user interactions with items, it tends to avoid the gray sheep problem, whereas the second tends to avoid the first-rater problem because it associates keywords to every new item and user. Therefore, the two methodologies are complementary and may be employed together with no loss.

References

1. Schafer, J., Konstan, J.: E-commerce recommendation applications. In: Data Mining and Knowledge Discovery (2001)
2. Cheung, K.-W., Kwok, J.T., Law, M.H., Tsui, K.-C.: Mining customer product ratings for personalized marketing. Decision Support Systems 35(2), 231–243 (2003)
3. Lee, C.-H., Kim, Y.-H., Rhee, P.-K.: Web personalization expert with combining collaborative filtering and association rule mining technique. Expert Systems and Applications 21(3), 131–137 (2001)
4. Claypool, M., Gokhale, A., Miranda, T., Murnikov, P., Netes, D., Sartin, M.: Combining content-based and collaborative filters in an online newspaper (1999)
5. Breese, J.S., Heckerman, D., Kadie, C.: Empirical analysis of predictive algorithms for collaborative filtering, pp. 43–52 (1998)
6. Sarwar, B.M., Karypis, G., Konstan, J.A., Reidl, J.: Analysis of recommendation algorithms for e-commerce. In: ACM Conference on Electronic Commerce (2000)

7. Sarwar, B.M., Karypis, G., Konstan, J.A., Reidl, J.: Item-based collaborative filtering recommendation algorithms. In: World Wide Web, pp. 285–295 (2001)
8. Condliff, M.: Bayesian mixed-effects models for recommender systems (1999)
9. Stock, O., Zancanaro, M.: Intelligent interactive information presentation for cultural tourism. In: International CLASS Workshop on Natural Intelligent and Effective Interaction in Multimodal Dialogue Systems (2002)
10. Felfernig, A., Gordea, S., Jannach, D., Teppan, E., Zanker, M.: A short survey of recommendation technologies in travel and tourism. OEGAI Journal 25(7), 17–22 (2007)
11. Loh, S., Lorenzi, F., Saldaña, R., Licthnow, D.: A tourism recommender system based on collaboration and text analysis. Information Technology and Tourism (2004)
12. Werthner, H., Ricci, F.: E-commerce and tourism. Com. ACM 47(12), 101–105 (2004)
13. Coelho, B., Martins, C., Almeida, A.: Adaptive tourism modeling and socialization system. In: 2009 International Conference on Computational Science and Engineering (2009)
14. Lucas, J., Laurent, A., Moreno, M., Teisseire, M.: Fuzz-CBA: Classification à base de règles d'association floues. In: Rencontres Francophones sur la Logique Floue et ses Applications (2009)
15. Liu, B., Hsu, W., Ma, Y.: Integrating classification and association rule mining. In: Knowledge Discovery and Data Mining, pp. 80–86 (1998)
16. Ricci, F.: Travel recommender systems. Technical report, eCommerce and Tourism Research Laboratory, ITC-Irst, Trento, Italy (2002)
17. Benyon, D.: Adaptive systems: a solution to usability problems. Journal of User Modelling and User-Adapted Interaction, 1–22 (1993)

Automatic Learning Object Extraction and Classification in Heterogeneous Environments

A.B. Gil, F. De la Prieta, and S. Rodríguez

Abstract. This paper proposes the use of federated databases techniques in searching for educational resources by using a learning object paradigm that describes these resources based on metadata. Combining a complete agent-based architecture that implements the concept of federated search along with IR technologies may help organizing and sorting search results in a meaningful way for educational content. The paper presents also the ground for an approach for semantic-aware learning content retrieval based on abstraction layers between the repositories and the search clients.

Keywords: Multi-agent systems, Distributed Computing, e-learning, learning objects, repositories, Simple Query Interface, learning technology standards, web services.

1 Introduction

The field of distance education evolves by the incorporation of standards and technologies in a meaningful way to facilitate reusability and interoperability of the educational resources [1]. The mainstay of this development is based on the learning object paradigm [2]. These learning objects (LOs) incorporate both a self-contained modular resource and a set of educational descriptive metadata. There are different standards to labeled educational metadata like Dublin Core [3] and especially Learning Object Metadata, LOM [4]. LOs are typically stored in widespread digital repositories, called Learning Objects Repositories (LOR). These repositories mainly belong to academic institutions and universities. They generally provide some kind of search interface that makes it possible to retrieve the LOs. Any interaction for retrieving LOs can be done manually or be automated through different software systems. The main standards of interfaces for LORs are CORDRA, IMS-DRI, OKI-OSID or SQI [5]. Based on a database model in which

A.B. Gil · F. De la Prieta · S. Rodríguez
Computer and Automation Dept., University of Salamanca, Plaza de la Merced s/n, 37008, Salamanca, Spain
e-mail: {abg,fer,srg}@usal.es

J.B. Pérez et al. (Eds.): Highlights in PAAMS, AISC 89, pp. 109–116.
springerlink.com

labeling plays an important role, there are approaches which use special query languages for retrieving educational material in standardized repositories. One of the most important, and currently the most used, is the SQI (Simple Query Interface) [6] which provides the repositories with an abstraction level between the internal and external heterogeneity that includes a specification with a wide range of possibilities (different service-based communication protocols, synchronous and asynchronous queries, stateless and stateful repositories, etc.).

A big effort has been focused on the creation of sustainable repositories that are highly heterogeneous due to their different storing systems, access to objects, query methods, etc. The problem of heterogeneity affects not only to LOR systems but also to the educational content creation. The learning object creation involves authors in labeling each one of the content. This work is a high cost and non automatic activity. Consequently, it generates semantic mismatch like lack of important labeled fields, erroneous fields, etc. that makes more difficult the retrieval processes. It is thus imperative to be able to rely on an efficient and interoperable search system that can guarantee the recovery of quality learning objects from the repositories.

The remainder of this paper is organized as follows: section 2 describes relevant works related to learning object retrieval and selection, section 3 present the federated search system for LOs in digital repositories developed. Section 4 gives the conclusions of this study and several works in progress.

2 Related Works

The current increase in resources for online education calls for the efficient management of those services and the elements involved. The greatest efforts within the context of e-learning are nowadays focused on achieving the interoperability by adding labeling techniques in the educational content creation in a way to facilitate the storage, search, retrieval, etc. Many successful works have been proposed to meet the user's required learning objects.

Duval et al. have been actively working on learning object selection [7] [8]. They use the called Contextualized Attention Metadata (CAM) to capture information about actions throughout learning object lifecycle including creation, labeling, offering, selecting, using, and retaining. Four metrics using LOM and CAM are proposed for ranking and recommendation: Link Analysis Ranking, Similarity Recommendation, Personalized Ranking, and Contextual Recommendation. These metrics calculate various categorized rankings for learning objects, such as popularity ranking, object similarity based on number of downloads, etc. How these different rankings contribute to the learning object selection and how to combine them together are still questions faced. Based on semantic aspects by considering the contextual information as the learner's cognitive activities and the learning object content structure, Qiyan Han et al. [9] propose a learning object recommenda-tion framework to adapt to the learner's cognitive activities through an ontology-based method.

Researchers and developers in e-learning have begun attempts to apply recommender technologies, especially collaborative filtering, in learning object recommendation. Among the most popular characteristics from information filtering techniques applied to learning object recommendation are de information item (content-based or item-based approach) and user's social environment (collaborative approach).

McCalla [10] proposed an enhanced collaborative filtering approach, called the ecological approach, for designing e-learning systems. The key aspects of his approach involve gradually accumulating information and focusing on end users. Recker et al. [11] [12]are developing and evaluating their Internet-accessible system called Altered Vista where collaborative filtering techniques are applied within an ad hoc designed metadata structure. Manouselis et al. performed a case study on data collected from users of European Schoolnet's CELEBRATE portal to determine an appropriate collaborative filtering algorithm [13] [14]. Lemire's group proposed the RACOFI, Rule-Applying Collaborative Filtering, architecture to customize learning object selection [15]. Their recommendation is narrowed down and personalized by combining the collaborative filtering algorithm with an inference rule system.

In all these collaborative learning object recommendation systems, the key problem of cold start has not been addressed. Some more recent work has been done towards solving this problem. Tang et al. practice collaborative filtering in their evolving research paper recommender system [16] [17]. They emphasize the importance of pedagogical characteristics and try to use artificial learners to overcome the cold-start problem. The domain of their system, however, is limited to research papers thus the factors that influence the paper selection are much less complicated than those affecting learning object selection.

Tsai et al. take the hybrid approach [18] [19]. Similar to collaborative learning object recommendation systems, correlation-based algorithms are used to calculate helpfulness score via analyzing similar learners' feedback. In addition, preference-based algorithms enhance the selection with learners' preference. A Learner Preference Pattern is kept for each learner to record the preference history, which is generated and updated according to the learner's preference feedback. If a learning object is selected, and is given by a positive feedback; an increment is made to preference scores of all features of the learning object. The combination of scores of a learning object determined by the two algorithms decides its rank in the recommendation result. Their preference-based algorithm helps with the cold-start problem. However, all features of the selected learning object are treated equally.

Recently, there is an interesting work based on the use of ant colony optimization (ACO) algorithm, proposed several attributes-based ant colony system (AACS) to help learners find an adaptive learning object more effectively. On this method, Yang[20], based on using learning activities and education items predict optimal paths associated with the ACO algorithm to recommend Learning object sequence.

Real E-learning is mainly web-based supported by means of Internet connect learners, teachers, courses and all technologies implies. Most of the related works we have revised have been applied to a local case of study, for small and usually local parameters. Some of the recommender contents are not real learning objects and just research papers, or by simulating learning objects and users, etc. We focus our work on learning object retrieval in a real environment by using federated search in several public operative Learning Objects Repositories. It is needed bring the user with an unify framework that search and retrieval learning object process easier; that filters and catalogs the retrieved learning object according some rules to give useful content based on the educational metadata provided; to be browsed in order to view multiple perspectives in an open and free environment for learners, teachers, etc. We propose then an architecture that facilitates the searching and cataloging of learning object in different LOR.

3 Federated Search System for Learning Object

A semantic-based knowledge representation is used for content analysis and learning object recognition. At this work the semantic specification used for learning objects is LOM (Learning Object Metadata) standard but could be any other. LOM specifies a conceptual data schema that defines the structure of a metadata instance for a learning object. A metadata instance for a learning object describes relevant characteristics for the learning object to which it applies. Such characteristics may be grouped in general, life cycle, meta-metadata, educational, technical, educational, rights, relation, annotation, and classification categories. These metadata documents record significant information to be used in reasoning processes and for enabling user-friendly and intelligent applications for searching and retrieved relevant educational content for any user.

AIREH (Architecture for Intelligent Recovery of Educational content in Heterogeneous Environments), see Fig 1, is an agent-based system developed by using an organizational perspective to avoid the heterogeneous environment depicted in earlier sections. It uses the SQI interface to perform federated searches in distributed repositories and sort the retrieved learning objects according to the different needs of the application users [21]. This application allows integrate relevant context information from user, learning objects and for the heterogeneous and dynamical environment by the organizational agents. The search process takes place in a totally transparent way to the user along user-friendly interfaces. AIREH goes from the information extraction from repositories to data integration and relationship between LOs.

By empirically studying learning object across LORs we find strong heuristics to quickly prune in our system most of irrelevant objects and then filter by quality parameters [22]. A more effective record detection algorithm should take into account the semantics of the learning object and technical aspects applied to make search and catalog more accurate and intelligent. Some aspects in the learning object recovery are reliability, completeness and ranking accuracy.

- Completeness. Learning object retrieved should be as complete as possible to provide trust worthy answers. Omitted metadata are an important hole in the precision record detection. LOM has irrelevant labeled fields whereas others are absolutely indispensable.

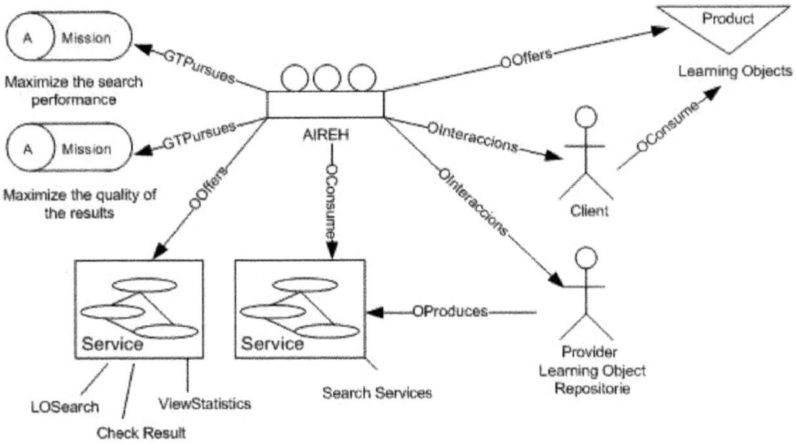

Fig. 1 Overall system architecture of the e-learning retrieval system

- Reliability. High quality structured data is needed for recovery LOs to gen-erate direct answers rely also on confidence LORs. Any learning object can have syntactic or semantic faults. The syntactic faults reveal a lack of con-formity with the specified in the standard. In case of vocabularies, it is supposed that one must use someone of the available terms and not any other. Other problems are inconsistent attribute values, spelling mistakes, and so on. Semantic faults only can be detected by the inspection of an ex-pert, so that a field of metadata agrees with the information deducible of the above mentioned content. Some attributes are ambiguous or inadequate so certain degree of subjectivity exists then with this information.
- Ranking accuracy. The set of potential learning objects to a query can be very large so an optimal ranking mechanism is critical for location relevant object information to conduct the relevance to the query. Individualized learning object selection or recommendation with domain attributes is needed.

Besides the search of learning objects, the system makes a ranking of the ob-jects returned after a first filter process (by eliminating duplicates and non reusable learning objects). This facilitates a homogeneous organization of learning object from heterogeneous sources facilitating the final selection for the users of the application by cataloguing the learning objects based on two variables. The first one does an estimation of the quality of the objects depending on such different parameters as size, incorporation of key characteristics, metadata measures, completeness, etc. The second one of the variables in the classification of the recovered objects imposes the use of collective aspects observed by final users of the application. This second one allows providing the classification by adding collaborative recommendation. At this moment the user in the platform AIREH, could vote the learning objects recovery and then system makes feedback to the ranking algorithm. Although is a light use of social recommendation attending to the LO context is an open branch we are working more widely to improve.

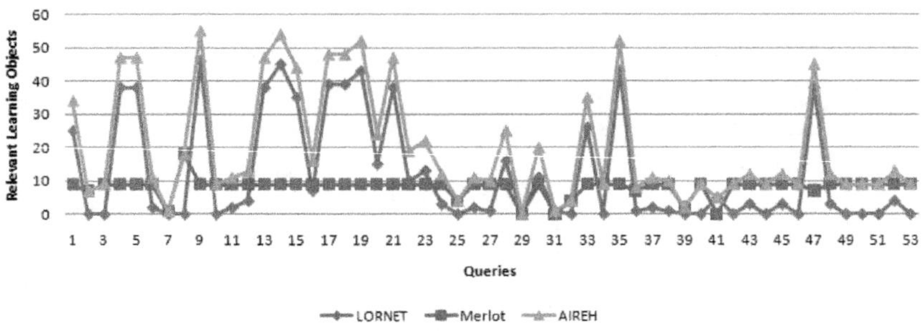

Fig. 2 Evaluation of relevant learning objects recovery

AIREH has been evaluated with a battery of 53 queries to validate its effectiveness in a real environment. The tests are performed on two real repositories, each one with a different session management. MERLOT repository is chosen (stateful) and LORNET (stateless). The agent translator is implemented for VSQL query language. The recovery performance for the architecture is based on the relevance for the metadata documents retrieved. This work fits the relevance related to profit or potential use of the materials recovered based on [23]. In the context of this paper, the metadata retrieved were categorized by a criterion of relevance binary: when learning object can't be accessed (information about the educational content source not labeled), is described as irrelevant and qualify with value 0. Otherwise learning object is relevant with value 1. Figure 2 shows the increment for the relevant learning objects by comparing with isolated repositories evaluated.

4 Conclusions and Work in Progress

The innovation of the architecture presented, AIREH, is the organizational model agent based proposed to accomplish the goals to search and location user relevant educational contents, facing highly dynamic environments such as the deep Web presented. The search process is integrated into the organization agents in a completely transparent way to the user, addressing the problems of distribution and integration of different repositories, abstraction of the internal logic of each one and classification, storage and search for LOs.

Taking into account the results obtained from the construction of the AIREH application, it is possible to conclude that the agent-based architecture is efficient to solve the problem of federated searches in heterogeneous repositories, due to the adaptation and learning facilities involved. The above mentioned architecture allows the adjustment of the system in a changeable environment depending on the load of the system and of the condition that is observed in the repositories depending on the on fly performance and includes several aspects related to learning object context and user preferences as language or repositories selection.

We realize that recommender systems in e-learning differ from other domains because at this environment recommender system must consider the learning context and also the learning sequence to be efficient. But also learning object recommendation problem involves almost three actors; the author's learning object content, the learner and the learning objects. At present the application is improving for stronger methods of recommendation to precise the personalized search and learning object's cataloguing by including these latest aspects.

References

1. Hatala, M., Richards, G., Eap, T., Willms, J.: The Interoperability of Learning Object Repositories and Services: Standards, Implementations and Lessons Learned. In: 13th World Wide Web Conference, Educational Track, New York, May 2004, pp. 19–27 (2004)
2. Lujara, S.K., Kissaka, M.M., Bhalaluseca, E.P., Trojer, L.: Learning Objects: A new paradigm for e-learning resource development for secondary schools in Tanzania. World Academy or Science, Engineering and Technology, 102–106 (2007)
3. DCMI Specifications (2008), `http://dublincore.org/specifications/`
4. IEEE 1484.12.1-2002, Draft Standard for Learning Object Metadata. The Institute of Electrical and Electronics Engineers, Inc.
5. De la Prieta, F., Gil, A.B.: A multi-agent system that searches for learning objects in heterogeneous repositories. In: Demazeau, Y., Dignum, F., Corchado, J.M., Bajo, J., Corchuelo, R., Corchado, E., Fernández-Riverola, F., Julián, V.J., Pawlewski, P., Campbell, A. (eds.) Trends in PAAMS. AISC, vol. 71, pp. 355–362. Springer, Heidelberg (2010)
6. Simon, B., Massart, D., Van Assche, F., Ternier, S., Duval, E., Brantner, S., Olmedilla, D., Miklos, Z.: A Simple Query Interface for Interoperable Learning Repositories. In: 1st Workshop on Interoperability of Web-based Educational Systems, Chiba, Japan (2005)
7. Ochoa, X., Duval, E.: Use of Contextualized Attention Metadata for Ranking and Recommending Learning Objects. In: Proceedings of 1st International Workshop on Contextualized Attention Metadata: Collecting, Managing and Exploiting of Rich Usage Information, pp. 9–16 (2006)
8. Wolpers, M., Najjar, J., Duval, E.: Tracking Actual Usage: the Attention Metadata Approach. Educational Technology & Society 10(3), 106–121 (2007)
9. Han, Q., Gao, F., Wang, H.: Ontology-based learning object recommendation for cognitive considerations. In: 8th World Congress on Intelligent Control and Automation (WCICA), July 7-9, pp. 2746–2750 (2010)
10. McCalla, G.: The Ecological Approach to the Design of E-Learning Environments: Purpose-based Capture and Use of Information about Learners. Journal of Interactive Media in Education 1, 18 (2004); Special Issue on the Educational Semantic Web
11. Recker, M., Walker, A., Lawless, K.: What do you recommend? Implementation and analyses of collaborative information filtering of web resources for education. Instructional Science 31(4-5), 299–316
12. Recker, M., Wiley, D.: A non-authoritative educational metadata ontology for filtering and recommending learning objects. Journal of Interactive Learning Environments. Special issue on metadata, 1–17 (2001)

13. Manouselis, N., Costopoulou, C.: Experimental Analysis of Design Choices in Multi-Attribute Utility Collaborative Filtering. International Journal of Pattern Recognition and Artificial Intelligence 21(2), 311–331 (2007)
14. Manouselis, N., Vuorikari, R., Van Assche, F.: Simulated Analysis of Collaborative Filtering for Learning Object Recommendation. In: SIRTEL Workshop, EC-TEL 2007 (2007)
15. Lemire, D., Boley, H., McGrath, S., Ball, M.: Collaborative Filtering and Inference Rules for Context-Aware Learning Object Recommendation. Technology and Smart Education 2(3), 179–188 (2005)
16. Tang, T.Y., McCalla, G.: Utilizing artificial learners to help overcome the cold-start problem in a pedagogically-oriented paper recommendation system. In: De Bra, P.M.E., Nejdl, W. (eds.) AH 2004. LNCS, vol. 3137, pp. 245–254. Springer, Heidelberg (2004)
17. Tang, T.Y., McCalla, G.I.: Smart Recommendation for an Evolving E-Learning System. In: AIED 2003 Workshop on Technologies for Electronic Documents for Supporting Learning (2003)
18. Tsai, K.H., Chiu, T.K., Lee, M.C., Wang, T.I.: A learning Object Recommendation Model based on the Preference and Ontological Approaches. In: Proceeding of the Sixth International Conference on Advanced Learning Technologies, ICALT 2006 (2006)
19. Wang, T.I., Tsai, K.H., Lee, M.C., Chiu, T.K.: Personalized Learning Objects Recommendation based on the Semantic Aware Discovery and the Learner Preference Pattern. Educational Technology and Society 10(3), 84–105 (2007)
20. Yang, Y.J., Wu, C.: An attribute-based ant colony system for adaptive learning object recommendation. Expert Systems with Applications 36, 3034–3047 (2009)
21. Gil, A., Prieta, F., López, V.F.: Hybrid Multiagent System for Automatic Object Learning Classification. In: Corchado, E., Graña Romay, M., Manhaes Savio, A. (eds.) HAIS 2010. LNCS (LNAI), vol. 6077, pp. 61–68. Springer, Heidelberg (2010)
22. Prieta, F., Gil, A., Corchado, J., Sanz, E.: Sistema multiagente orientado a la búsqueda, recuperación y filtrado de objetos digitales educativos. In: Actas de las VIII Jornadas de Aplicaciones y Transferencia Tecnológica de la Inteligencia Artificial, TTIA 2010 (AEPIA), pp. 65–74 (2010)
23. Schamber, L., Bateman, J.: User criteria in relevance evaluation: toward development of a measurement scale. In: Hardin, S. (ed.) Global Complexity: Information, Chaos and Control, Proceedings of the 59th Annual Meeting of the American Society for Information Science, Baltimore, MD, Learned Information, Medford, NJ, October 21-24, pp. 218–225 (1996)

Agents for Swarm Robotics: Architecture and Implementation*

F. Aznar, M. Sempere, F.J. Mora, P. Arques, J.A. Puchol, M. Pujol, and R. Rizo

Abstract. Swarm robotics is a type of robotic systems based on many simple robots interactions. Such systems enjoy many benefits such as high tolerance and the possibility of increasing the number of robots in a transparent way to the programmer; but they also have many difficulties when applied to complex problems. In this paper, we will present a hybrid architecture for swarm robotics based on a multi-agent system. The main contribution of this architecture is to make possible the use of cognitive agents to lead a robotic swarm of simple agents without losing the advantages of swarms. Moreover, the implementation of this architecture within *Real Swarm* platform and the discussion of how to apply this architecture in real systems will be presented.

Introduction

Swarm robotics can be defined as a mechanism to solve problems inspired by collective behaviours in social insects and focused in the interaction of multiple robots. Based on this metaphor of social insects, swarm robotics emphasizes aspects such as decentralized control, limited communication between individuals, local information, emergence of global behaviour and robustness. Multi-robot swarm systems differ from other multi-robot systems in [9]: 1) swarm robots are autonomous robots with a physical embodiment in the world; 2) swarm is made up of a large number of robots; 3) swarm system should consist of relatively few homogeneous groups of robots; 4) robots could be relatively incapable or inefficient; and 5) robots should only have local and limited sensing and communication abilities. These features ensure that coordination between robots will be distributed and the system will have

F. Aznar · M. Sempere · P. Arques · F.J. Mora · J.A. Puchol · M. Pujol · R. Rizo
University of Alicante
e-mail: `{fidel,mireia,mora,arques,puchol,mar,rizo}@dccia.ua.es`

* This work has been supported by the Ministerio de Ciencia e Innovacion, project TIN2009-10581.

J.B. Pérez et al. (Eds.): Highlights in PAAMS, AISC 89, pp. 117–124.
springerlink.com

a high fault-tolerance due to the redundancy of robots where each individual of the system is dispensable and can be replaced by another individual. In this sense, the system is easily scalable, allowing the aggregation or elimination of agents depending on the task. Because of these features, it is difficult to develop an architecture that is able, on the one hand, to model correctly a swarm system, and on the other hand, to coordinate the swarm to perform complex tasks.

In this paper we will revise robotic architectures for the control of both, a single robot and multiple robots, and we will assess its suitability for its use in swarm systems. Then, we will propose a hybrid multi-agent architecture to model the behaviour of a robotic swarm. This approach merges the advantages of swarm systems with multi-agent systems. We will also revise an implementation of this architecture in the platform *Real Swarm*. Finally, we will end with a discussion of the issues to be considered when implementing this architecture in real systems and we will comment our future work lines.

Robotic Architectures

In this section existing robotic architectures will be review to design an architecture with the right features to swarm robotics.

The classification of hybrid architectures is a difficult task due to the large variety of existing architectures. Although many of them are architectures for controlling a single robot, the study of these architectures can be useful to verify which division of tasks is established in them. [7] classify hybrid architectures in: managerial architectures, state-hierarchy architectures and model-oriented architectures. Managerial styles of hybrid architectures are based on the decomposition of responsibilities similar to business management (AuRA, SFX and HILARE). State-hierarchy styles of architectures organize activities by the scope of time knowledge (3T). Model-oriented architectures use a global world model (Saphira and TCA). There are many other hybrid architectures that are difficult to classify in any of the previous categories. In [8], these architectures are classified in a new category called level-organized architectures since all of them have in common that are organized in different levels that are independent of temporal features (Glair, Sharp, BERRA, SSS and Payton's Architecture).

If we try to model a swarm robotic system with the features of previous architectures we mainly run into two problems. On the one hand, all the previous architectures share high-level deliberative processes that control low-level reactive processes, either directly (as in state-hierarchy architectures) or indirectly (as in model-oriented architectures). These processes are designed for a centralized system and therefore they are difficult to implement in a fully decentralized system as required by swarm robotics. On the other hand, robustness to failures is not taken into account. In such a way that, if a deliberative module fails, the control of the robot will be lost. This is not acceptable in swarm robotics, since the failure of a system module (for example a deliberative planner) can not jeopardize the stability of whole system (that can be composed from more than one hundred of agents).

One knowledge area that has extensively studied the control of a large number of robots is multi-agent systems [4]. These systems share many features with swarm robotics [2], so it could be a good alternative to model such systems. However, multi-agent systems do not always translate easily to robotic swarms because of the particular characteristics of robotic swarms which differentiate them: their physical nature, diverse communication mechanisms and control structure [5]. In fact, if we analyse some MAS architectures, we find that some fundamental characteristics of swarm systems such as fully distributed control or fault-tolerance, are not practically taken into account.

In this paper an agent architecture to model swarm hybrid systems is presented.

Proposed Architecture

In this section an agent architecture for swarm robotics is described. This architecture has mainly two characteristics: on the one hand it considerers swarm robotics distributed nature, highlighting that the addition and elimination of robots from the system can be a transparent process, without altering the swarm functionality. On the other hand, the architecture will have a high level fault-tolerance, where the failure or lost of an agent would not compromise the system viability. There is a high probability that a robot fails in mobile robotics; hence fault-tolerance is very relevant in swarm systems. It is important to emphasize that the presented architecture follows the FIPA standard.

With all these characteristics in mind two kinds of *roles* have been defined to be developed by agents of the system, μ and Ψ *roles*.

Given the importance of fault-tolerance into the core of a swarm system, an architecture that combines an approach focused on agents (that builds fault-tolerance into agents), with other approach focused on the system (where monitorization and fault-tolerance are developed in a separate agent) has been designed.

A diagram of the architecture, with different agents and the information flow between them is presented in figure 1.

μ *role* is developed by agents responsible for interacting with physical robots (μ agent is usually placed into a physical robot). Therefore, these agents develop swarm basic tasks. So, we can consider that these agents will develop swarm tasks that are closer to reactivity.

These types of agents, once they know their address and the addresses of μ agents that interact with them, can communicate directly, without using a control agent[1]. These can be useful in situations where the agent platform cannot be accessed or it can be useful too to avoid system overloading by establishing only point to point connections. However, white pages services *AMS* and yellow pages services *DF* can be used by such agents to search other agents. These mechanisms, provided directly

[1] Many agent platforms, as FIPA based ones, require a *broker* agent, known as AMS, responsible for managing the agents of the system. The communication between μ agents, once created, do not depends on these control agents, because it can be established directly by using point to point connections.

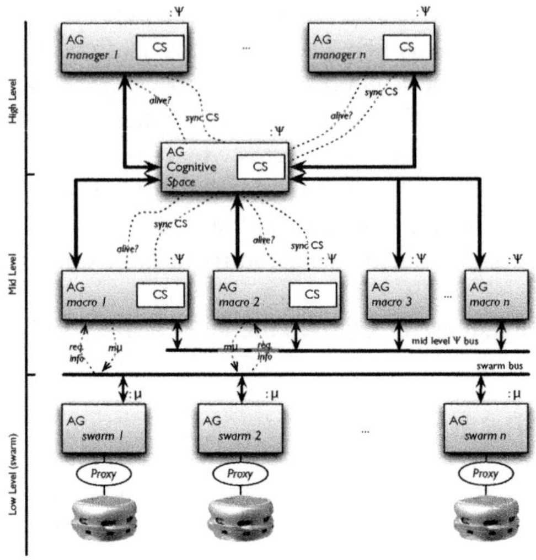

Fig. 1 Architecture diagram. Different types of agent and the most important communication channels can be seen.

by FIPA platform, are very useful, because they provide a transparent way to add and to remove agents from the platform.

μ agents can communicate only with the same type of agents. However, they provide access to basic data of the swarm task being developed and to the status of this task; by this way, Ψ agents can monitor and manage these tasks. In this way, this kind of agent are able to respond to m_μ messages using the FIPA protocol *Request Interaction Protocol Specification*.

Ψ *role* will be developed by cognitive agents. These cognitive agents are responsible for the macroscopic behaviour of the swarm and to direct the swarm towards the objective. These agents are mobile, and can be placed physically into any available robot of the swarm.

A *pure* swarm system must not have any coordinating agent, neither a centralized module acting as agent controller. However, when it is desired to develop medium or high complexity tasks, it is necessary to provide mechanisms to guide the swarm towards the required tasks directing the interaction between agents.

Ψ agents level can be medium or high. Mid-level Ψ agents are responsible to direct the swarm of μ agents to perform a given task at the macroscopic level. These agents can communicate with agents of the same role or with μ agents, to monitor and review its operation.

Ψ agents can develop high level tasks, to coordinate macroscopic agents towards their objectives. These agents can communicate only between themselves to perform their functions (they cannot communicate with μ agents).

Finally, some Ψ agents will check continuously the presence of a special agent, called *cognitive space*. Their objective will be to create this agent if it disappears from the system.

Given the importance of Ψ agents, we must assure the permanence of these agents in the system and avoid their disconnection or failure. A centralized control cannot be established, because a failure in it will cause the loss of the complete swarm. There are several mechanisms, as **replication** that provides a solution to fault-tolerance in certain parts of the system. However, due to the special characteristics of swarm robotics, even the **replication** of all agents of the system can be insufficient, because if a Ψ agent and its copy fail, the complete swarm system will stop[2].

Because of this reason the creation of a **cognitive space** agent is proposed, responsible to save and administer the state of Ψ agents of the system. The purpose of this agent is threefold: On the one hand it provides to cognitive agents a common space where they can access to shared information. This information consists of the model of the environment and the cognitive model of the system. On the other hand, it carries out task of *fault-tolerance* verification, checking the presence of all Ψ agents. If an agent of this type *dies*, the *Cognitive Space* agent is responsible for creating it. As the basic content of all *cognitive* agents is updated into *CS* the loss of information is minimal and the restoration of an agent is instantaneous. This agent has also to verify the duplicity of agents that have the same functionality to indicates its elimination. Furthermore, this agent carries out *proxy* tasks, distributing the information stored by the network of cognitive agents (it manages a database). Thus, in the most extreme case, all cognitive agents will have a copy of the content of *CS* agent, in other words, a copy of all the important information of cognitive agents, so it is possible to restore the full cognitive system without losing information, if this is necessary.

All Ψ agents synchronized with *CS*, will continuously verify the state of the cognitive agent. If this agent dies, a Ψ agent (for example, the one with a higher *Id*) will restore that agent using their internal copy of *CS*.

Real Swarm Platform: Architecture Implementation

This architecture has been implemented to validate its design and to allow us to test it in real environments. In this section *Real Swarm* platform is been presented. This platform implements the agents architecture for swarm robotics showed in this paper and also allows task executions in both, a simulated environment[3] and real robots. In figure 2, a schema of the main modules of this platform is presented.

[2] It is possible to do more than one agent replication, using a similar approach to [3]. However, we have the same problem again, if the proxy agent fails, the entire system will fail. If a proxy agent is not used we have handle constantly the synchronization between the agents and their copies. The architecture presented in this paper provides a solution to this problem.

[3] It is necessary to emphasize that one of the added difficulties of working with swarm systems is that it requires a large number of robots (between 10 and 100), so that, simulation is essential in this kind of systems.

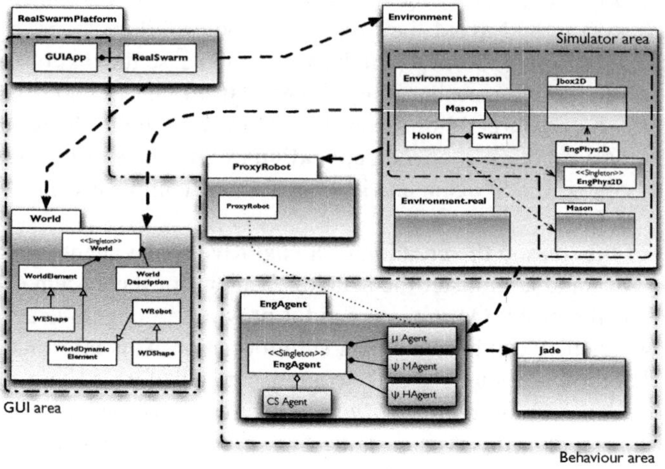

Fig. 2 Real Swarm platform design. It shows the different subsystems, main classes and packages that made it up.

RealSwarm platform is divided into several subsystems: a behaviour subsystem, a simulation subsystem and a user interface subsystem. Although in this paper we give a general introduction to the platform we will focus in the behaviour subsystem, where the previously described agent architecture is developed.

Real Swarm allows behaviour execution in both physical robots and simulated robots. Thus, MASON library is used to develop a 2D simulation subsystem. This library allow us to use large swarms and it provides a graphical test environment. See [6] for more information. Nevertheless, physical complex simulation it is not directly possible with MASON. So that, the physical library JBox2D has been used to provide to this subsystem physical capabilities to correctly simulate complex environments. The use of both, real robots or a simulated environment is transparent for the user.

Also, there is available a user interface subsystem which allows to load simulated environments and interact with robots to study its properties in real time.

The agent architecture presented in this paper is developed inside the behaviour subsystem. JADE library is used for this implementation. This library makes easier the development and the use of FIPA systems in a very efficient way. For more information see [1].

An agent engine *EngAgent* has been implemented to facilitate the creation of the different agents of the system. This engine uses *microAgent*, *macroHAgent* and *macroMAgent* classes. *microAgent* class is used to create the agents that control a robot (physical or simulated). The connexion between the agent and the robot is made with a *proxy* module, which is linked to the agent using the JADE communication function *object2agent*. *macroHAgent* agents are used to develop high level agents that are also useful in synchronization and information recovery of the *CS* agent. All *macroHAgent* need to subscribe to *CS* agent and to establish a periodic verification process of this agent.

In this way, *macroHAgent*, based on FIPA standard library *Communicative Act Library Specification* and specifically in the communicate act `subscribe`, establish a subscription with *CS* agent, so that any change in their content will be notified. However, *macroHAgent* agents are subscribed to platform events, concretely to *JADE AMS* action `dead-agent` to verify that *CS* agent is always available.

If a *CS* agent fault is detected, the highest *ID macroHAgent* agent available in the system have to request the new *CS* creation. The new *CS* agent will be waiting until it receives a message from its creator, specifying the current content of the database. Once the *CS* agent is updated, it will continue with its normal execution.

macroMAgent agents have the same assigned tasks than the previously mentioned agents but without synchronization or information recovery tasks. However, all *macroHAgent* use *CS* agent to save or recovery information in each iteration. This is very important because in a hypothetical fault of the system it is possible to recovery any type of *macro* agent.

Finally, *CSAgent* class implements the *CognitiveSpace* agent. This agent uses an internal database to facilitate the concurrent access to data to all system agents. In addition, *CS* agent verifies the existence of all macro agents using a subscription to the *JADE AMS* action `dead-agent`. If a certain agent does not exists, requests its creation using its local information.

Discussion

When we have studied the state of the art of robotic architectures we found that the design of swarms or hybrid swarms, oriented to perform complex tasks, are not considered. Moreover, robotic architectures, including multi-agent systems, often do not address agent mobility or fault-tolerance needed by a swarm system. They are mainly based on centralized control or defined with low level fault-tolerance.

The design of an agent architecture for robotic swarms is a complex task. On the one hand, to follow the basic rules of swarm robotics (such as robustness or system independence from individual robotic components) is required. On the other hand, complex tasks require high-level agents to establish system goals. The proposed architecture takes into account the needs of both, high level (cognitive) agents and low level agents (robotic swarm), providing a hybrid architecture for swarm-based robotic agents.

We consider that this architecture is an advance in modelling and designing agent-based swarm systems. In addition, the architecture implementation as a robotic simulation platform is presented. This allows anyone easily testing simulated or real robotic swarms.

Our current research focuses primarily on enhance the presented architecture in two aspects. Initially we are interested in increasing the robustness of the FIPA platform using JADE. In addition, we are testing the architecture performance for certain classic tasks of swarm robotics, in order to establish which is the optimal number of *macroHAgent* agents to balance the system in terms of robustness and functionality.

References

1. Chmiel, K., Gawinecki, M., Kaczmarek, P., Szymczak, M., Paprzycki, M.: Efficiency of jade agent platform. Sci. Program. 13(2), 159–172 (2005)
2. Cil, I., Mala, M.: A multi-agent architecture for modelling and simulation of small military unit combat in asymmetric warfare. Expert Syst. Appl. 37(2), 1331–1343 (2010)
3. Fedoruk, A., Deters, R.: Improving fault-tolerance in mas with dynamic proxy replicate groups. In: IAT 2003: Proceedings of the IEEE/WIC International Conference on Intelligent Agent Technology, p. 364. IEEE Computer Society, Washington, DC, USA (2003)
4. He, J., Lai, H., Wang, H.: A commonsense knowledge base supported multi-agent architecture. Expert Syst. Appl. 36(3), 5051–5057 (2009)
5. Higgins, F., Tomlinson, A., Martin, K.M.: Survey on security challenges for swarm robotics. In: ICAS 2009: Proceedings of the 2009 Fifth International Conference on Autonomic and Autonomous Systems, pp. 307–312. IEEE Computer Society, Washington, DC, USA (2009)
6. Luke, S., Cioffi-revilla, C., Panait, L., Sullivan, K., Balan, G.: Mason: A multi-agent simulation environment (2005)
7. Murphy, R.: An introduction to AI robotics. The MIT Press, Cambridge (2000)
8. Posadas Yagüe, J.L., Poza Luján, J.: Revisión de las arquitecturas de control distribuido. Tech. rep., Universidad Politécnica de Valencia (2009)
9. Sahin, E.: Swarm robotics: From sources of inspiration to domains of application. In: Şahin, E., Spears, W.M. (eds.) Swarm Robotics 2004. LNCS, vol. 3342, pp. 10–20. Springer, Heidelberg (2005)

An Organizational Approach to Agent-Based Virtual Power Stations via Coalitional Games

Radu-Casian Mihailescu, Matteo Vasirani, and Sascha Ossowski

Abstract. An agent-based organizational model for a smart energy system is introduced relying on dynamic coalition formation mechanisms for virtual power stations' creation and maintenance. The solution proposed introduces a distributed algorithm by formalizing an interaction scheme, deployed in a MAS environment populated by negotiation agents. On this basis, the system is able to efficiently self-organizing into coalitions representing the actual virtual power stations, capable of meeting the desired functionalities within stochastic environments.

1 Introduction

In this paper, agent technology is being deployed towards operating the Smart Grid via a bottom-up approach. The accelerated integration of renewable energy resources to the Grid entails a distributed generation solution meant at minimizing the electricity demand from large scale power plants and ensuring an equilibrated and efficient local distribution of the available energy resources(DERs) in the form of virtual power stations(VPSs). To place this into context, the US department of Energy advocates that as much as a 5% increase in efficiency of the grid would equivalate the fuel and greenhouse emission of 53 million cars [12]. The technological advances of renewable energy generators dictates for a grid capable of dealing with the variable nature of these resources.

The potential allure of the MAS paradigm to the power industry has been extensively documented thus far [6]. In this respect, several management systems have been proposed for the organization of micro-grids, which are defined as subsystems of the grid that represent an electrical and informational interconnection of generation, storage and load devices. The organization of the power system in the form of

Radu-Casian Mihailescu · Matteo Vasirani · Sascha Ossowski
e-mail: {raducasian.mihailescu,matteo.vasirani}@urjc.es,
sascha.ossowski@urjc.es

J.B. Pérez et al. (Eds.): Highlights in PAAMS, AISC 89, pp. 125–134.
springerlink.com

micro-grids has been well promoted to this date by the manifold advantages of its application [5, 8].

Setting aside from this approach that aims at imposing an architectural control, whether centralized or not, on already predefined micro-grids, our vision is intended at proposing a method for congregating the smart-grid actors to dynamically approximate optimal micro-grid configurations. The procedure is designed such that it intrinsically ensures the emergence of self-sufficient, autonomous micro-grids, thus having the ability of providing power even in the absence of a large power station. Furthermore, reducing the loads of the transmission infrastructure would result in a decrease of the network's vulnerability by minimizing the likelihood of an outage occurrence.

The foremost attributes of the resulting micro-grid organization would comprise (1) Autonomy - the systems capacity to self-manage the distributed energy resources in order of matching internal demand and providing, in case of surplus, energy towards neighbouring micro-grids; (2) Resilience - the capacity to withstand unexpected trials. In the event of an outage, the available resources of the micro-grid are prioritized towards critical loads, while only the remaining energy is being redistributed via an uncomplicated bidding procedure mediated by a specialized agent.

Hence, the concept of a virtual power station can be understood as a bundeling of DERs via an informational infrastructure. The major problem still facing the realization of VPS concerns the integration of these devices by means of a new distributed control paradigm. The management of distributed energy resources in the form of VPS poses a complex issue, mainly due to the stochastic behavior of the system and the diversity of the resources comprised. Integration at the level of VPS is essential for acquiring sufficient capacity and flexibility in order of making the systems cost effective and technically feasible. The presented approach is one of understanding the formation and coordination of virtual power stations in the sense of coalitional games. Instead of considering centralized architectures [9, 10] we point towards decentralized ones based on multiagent coordination mechanisms for systems consisting of large number of distributed power units. Moreover, our proposed scheme is intended at a dynamically and decentralized formation of the organizational VPS structure and providing mechanisms able of coping with the inherent dynamics and unpredictability of the system.

The paper is organized as follows. In Section 2 we discuss the formalization of the problem. Our agent-based organizational model is introduced in the third section with emphasis on the coalition formation algorithm proposed. In Section 4 we present experimental results and point towards future work. Section 5 concludes the paper.

2 Problem Representation

The coalition games we aim to address in our approach are the ones where the coalition formation problem is projected on an underlying network topology, given that the cost for cooperation also plays a major role. Our approach addresses essentially the interplay between agent networks and agent-based coalition formation. This class of games is primarily characterized by non-superadditivity, in the sense that gains resultant from forming coalitions are limited by the actual cost of coalition formation, thus the grand coalition is seldom the optimal structure. Furthermore, the coalitional game is subject to the dynamism of the environment. The challenge is to develop mechanisms that permit large number of autonomous agents to collectively achieve a desired functionality by permanent adaptive dynamics. Reaching an agreement for forming coalitions represents a distributed reasoning and negotiation process for the given network of agents. The cost of forming a coalition can be perceived through the negotiation process and information exchange which incur costs. In contrast to static coalition formation, dynamic coalitional games represent a more complex issue since the focus is not merely in analyzing the coalitional structure, but the main aspect under investigation is how the formation of the coalitional structure takes place through the players' interactions and its adaptability to environmental variations or externalities, which denote the evolution of the system. The two major approaches of addressing this are either to search for the coalitional structure that maximizes the total utility of the system (social welfare), or to find the structure with Pareto optimal payoff distribution for the agents involved. Generally, computing this through a centralized approach is NP-complete. The complexity of the problem addressed is due to the fact that the utility of adding or removing actors from a coalition has a dynamic valuation, as it depends on the other actors already comprising the coalition at each moment in time. Therefore, it is desirable that the coalition formation process takes place in a distributed manner, leveraging on the autonomy of the agents, which spontaneously organize into topologies and functionalities to meet the desired objectives. We believe that in order to solve these issues, the problem must be understood in the context of self-organization[1] by providing a minimum set of interaction rules that would lead to an efficient achievement of the underlined desiderata.

Returning to our initial application, the algorithm we propose consists of three phases and it is illustrated as a case study in the context of smart energy systems, the abstraction of the guiding principles being a straightforward matter. We set to investigate the integration of renewable energy resources to the grid in the form of virtual power stations by means of aggregating the power generating potential of various devices in a novel way in the context of MAS [7] . As system designers, we choose to enable the autonomous agents with the basic coordination primitives, and leave to the agents to self-organize and coordinate as the situation may demand, in a fully distributed manner.

We model the problem as a dynamic coalition formation game with the following formalization: let $M = \langle A, \beta, CS, \Phi, v \rangle$ be a multi-agent system where,

[1] The main characteristic of such processes is their ability to achieve complex collective tasks with relatively simple individual behaviors, without central control or hierarchy, working bottom up. Typically they are composed of a large number of components that interact locally. The global behavior of the system emerges from these local interactions. The system creates its complexity in the course of evolution with respect to its global constraints.

$A = \{a_1, a_2, ..., a_n\}$ represents the set of agents; the framework assumes an instance where each actor in the grid is being represented by an agent that may take one of the two profiles: provider agents($\mathcal{P}$) or load agents($\mathcal{L}$) (consumers), denoted by their profile of a day-ahead plan β, which specifies the forecasted amount of energy demanded or available: if $\beta(a_i) > 0$ then $a_i \in \mathcal{P}$ and if $\beta(a_i) < 0$ then $a_i \in \mathcal{L}$, where $\mathcal{P} \cup \mathcal{L} = A$, $\mathcal{P} \cap \mathcal{L} = \emptyset$. We will refer onwards generically, to an agent belonging to set $\mathcal{P}$ as PA and respectively, as LA for set $\mathcal{L}$. The goal of this coordination problem is obtaining a partitioning of A into a coalition structure $CS = \{S_1, S_2, ..., S_m\}$ such that the set of constraints Φ hold for every coalition, where $\bigcup_{i=1}^{m} S_i = A$, $S_i \cap S_j = \varnothing, \forall i \neq j$, while maximizing the social welfare of the system, without jeopardizing the functionality of any of the coalitions. The set of constraints considered express domain dependent conditions, which for this particular case study include a limitation of the size of coalitions. Thus, a coalition cannot exceed a particular number of component agents, which corresponds to the safety limit imposed by technological constraints. This may also be expressed as a bounded energetic potential(ψ) for each coalition: $|S_i| \leq \max$, $\sum_{a_i \in S_j} \beta(a_i) > 0$, $\sum_{a_i \in \mathcal{P}(S_j)} \beta(a_i) > \psi$. The utility function of a coalition, v, represents a multi-criterion evaluation of domain specific parameters, which will be further explained in the following section. For computing the social welfare of the system we mean the typical interpretation of averaging over the utilities of every coalition.

3 MAS-Based Algorithm for VPS

The first stage of the process concerns essentially the coalition formation mechanism, which itself consists of three phases and which would yield subsystems of the grid that represent a self-sufficient electrical and informational interconnection of generation, storage and load devices.

3.1 Coalition Initiation

Initially, during the *coalition formation phase*, neighboring PAs need to coalesce in order to attain a minimum threshold power that would qualify them as VPS. The actors can assume one of the following states: committed or uncommitted to a coalition. Function s returns the status of one agent. In the beginning, all agents are assumed to be uncommitted. Additionally, we pressume that agents' ability to communicate is bounded by topological constraints: a_i and a_j can communicate within the predefined distance Δ; $N(a_i)$ will denote the set of neighbouring agents for a_i. The decision of collaborating towards establishing a virtual power station is based on aggregating proximity rationales and other domain dependent parameters as association coefficients, as will be further detailed in the following subsections. There are numerous possible procedures for initiating a coalition (e.g. leader election [11]), though in order of placing into focus our coalition formation mechanism we have chosen a straightforward approach for accomplishing this task. Thus, those PAs whose energy availability exceeds a predefined value (ψ^*) are entitled of establishing themselves as VPS initiators and will do so with a probability p inversly proportional to the number of the agent's PA uncommitted neighbors that are also set to do so,

$$p = 1 - \frac{|\{a_i | a_i \in N(a_j), a_i \in \mathcal{P}, s(a_i) = uncommitted, \beta(a_i) \geq \psi^*\}|}{|\{a_i | a_i \in N(a_j), a_i \in \mathcal{P}, s(a_i) = uncommitted\}|}, |\bullet| \text{represents card.} \quad (1)$$

Table 1 Provider Aggregation; Initiator behavior.

```
for each a_i ∈ P randomly chosen
  if β(a_i) ≥ ψ&& s(a_i) = uncommited; with probability p
 role(a_i) = CA;
  /* coalition expansion proposal */
    for each a_j ∈ PA&&a_j ∈ N(a_i)&&s(a_j) = uncommitted
      if association_coef(i, j) > 0
      sendProposal(a_j);
      endif
    endfor
/* resolve joining coalition request */
    if incomingRequestList!=NULL
      for each a_j in incomingRequestList
        if association_coef(i, j) > 0
        List ← constructOrderedCandidateSet(a_j);
         endif
       endfor
       for (i=0; (i<max) && (energeticBalance<eMax); i++)
       a_j = popUp(List);
       s(a_j) = committed
       C_{a_i} ←— C_{a_i} ∪ {a_j}
        energeticBalance+ = β(a_j)
        endfor
    endif
  endif
endfor
```

Table 2 Provider Aggregation; PA behavior.

```
for each a_i ∈ P randomly chosen
  if β(a_i) < ψ&& s(a_i) = uncommited
    if incomingRequestList!=NULL
      for each a_j in incomingRequestList
      List ← constructOrderedCandidateSet(a_j);
      endfor
    endif
/*initiate contact*/
  for each a_j ∈ N(a_i)&&(s(a_j) = committed||(role(a_j) = CA))
    if (profile(CA(a_j)))
    List ← constructOrderedCandidateSet(CA(a_j));
     endif
  endfor
  sendProposal(popUp(List));
  endif
 endfor
```

3.2 *Provider Aggregation Phase*

The next stage to be undergone regards the *aggregation of providers*. The initiator PA assumes the role of CA (coordinator agent), which shall be performing the inter-coalition negotiations consequently to aggregating LAs for the emergent coalition. This procedure is realized in a self-organizing fashion that dynamically constructs the organizational structure via a negotiation mechanism.

With these considerations in mind the mechanism proceeds in a Contract Net-like approach with the following steps, as depicted in table 1 and table 2: CAs send requests to their neighboring PAs indicating the VPS profile that they want to realize, in terms of scale and energetic potential - the constraints imposed on the coalition formation process may vary according to the desired feature of the emerging VPS. Based upon these specifications PAs evaluate CAs' offers and select the preferred choice from their candidate set. Eventually, CAs receive offer responses and decide committing PAs based on a utility function subject to transmission costs incurred and the coalition's energetic balance. In the pseudo-code of the algorithm in table 2 the *popUp* function extracts and eliminates from the PA's list the CA that meets the criteria of being the most adequate candidate. Thus, in case the PA would not have been selected by the CA, the latter would not remain in the candidate set of the respective PA. To be noted that the decision of selecting the best candidate is carried without a complete representation of the environment, but rather based on local and possibly incomplete knowledge. This is the underlying reason for the evolutionary nature of the algorithm that iteratively approximates a solution through refinement steps.

It's worth mentioning that, while CAs are looking for suitable suppliers of energy, the PAs themselves are pursuing an active role, ongoing an exploration phase of expanding their candidate set. For instance, information regarding a coalition obtained from another PA already committed to that particular coalition could cause the initial PA to submit a joining request without being explicitly addressed by the CA. This is meant at increasing the convergence rate towards emerging coalitions.

3.3 Consumer Aggregation Phase

Once the VPS energetic potential has been ensured, the only remaining phase for the coalition formation process requires the *aggregation of consumers* (LAs), which operates in a similar manner to the previous one explained. LAs proceed by submitting their forecasted demand to the CAs in their proximity. For each such request, the CAs calculate an association coefficient that reflects the self-sufficiency of the potential coalition i assuming the joining of the particular LA j, which is denoted as a linear combination of the parameters that apply for this particular case study: security measures (which implies computing contingency analysis - C_j), transmission costs (T_j) and energetic balance (E_j):

$$\varphi_{i,j} = w_1 \cdot C_j + w_2 \cdot T_j + w_3 \cdot E_j \tag{2}$$

We note that the first two parameters (of negative values) are ought to give an indication of the network's capability of avoiding line overloads and incurred transmission costs. The former gives an indication of the impact the integration of the LA's load would produce on the system's buses in terms of the percentage of capacity drop. By transmission costs we imply the effect of the power loss, in the course of transmission, over the coalition's total power, represented by this ratio. The last parameter represents the percent increase or decrease of the coalition's energetic balance, given the desired state of offer matching demand. Formulating the association coefficient in this manner allows placing more emphasis on the contingency and transmission coefficients at the beginning of the aggregation of actors, while the energetic balance gains more significance proportionally to the number of actors involved. This is obviously a key aspect since one cannot expect attaining a reasonable energetic balance at the very beginning of the coalition formation process.

The result is normalized so that it can be interpreted as a real value between -1.0 and 1.0. The positive value indicates a favorable association while a negative one denotes the opposite. The minimal value is reserved as corresponding to unacceptable situations of congestions, where committing a particular actor would result in line overloads that violate the operating limits of the infrastructure. The utility function of the coalition, v, represents the sum of the association coefficients for the coalition members, as shown in Equation 4.18.

$$v(S_i) = \sum_{j \in S_i} \varphi_{i,j} \tag{3}$$

A resulting value of 1 signifies the ideal situation of an energetically equlibrated coalition able of supplying demand. Based on this information, a proposal response is being awarded to the most suitable LAs for joining the coalition.

LAs will conclude the procedure by selecting the coalition whose utility is mostly increased by their commitment. The decision is unequivocally accepted by the CA since it comes as a response to its precedent proposal and it is exclusively addressed to the CA. The LAs' preference for acting in this sense is justified by the fact that the utility of the coalition would have a direct effect on the price of energy being traded within the coalition as well as its reliability.

Thus, the organization of agents is fundamentally correlated with the association network amongst them that emerges from their interactions in superposition to the topological communication network. The resultant interdependence between the agents within each coalition strongly impacts the characteristics and outcome of the game.

4 Experimental Results

The experiments presented in this section aim at understanding how the underlying interaction topologies are able to foster, or else hinder, the organizational efficiency of our proposed multiagent model. We are interested in determining the effects of the communication structure on the systems behavior at large, in order of exploiting the network configurations to achieve certain goals.

For our experiments we have chosen to set aside from the less realistic scenarios of regular networks and rather consider complex topologies of non-trivial connectivity patterns. As underlying topologies for our MAS we have focused on random graphs, small-world networks and scale-free networks. The choice for these particular configurations was justified by their proliferation in real-world, as well as in agent-based scenarios [3, 2]. In order of performing a relevant comparison between these classes of networks we have considered the same average connectivity per node.

For generating the random graph structure we have used the model proposed in [4], where undirected edges are placed randomly between a fixed number of vertices, resulting in a network where each possible edge is present with equal probability. The second class under investigation addresses networks exhibiting the small-world effect. Basically, this means that the mean geodesic distance[2] scales logarithmically or slower with network size for fixed mean degree. Thus, such

[2] Shortest path through the network from one vertex to another.

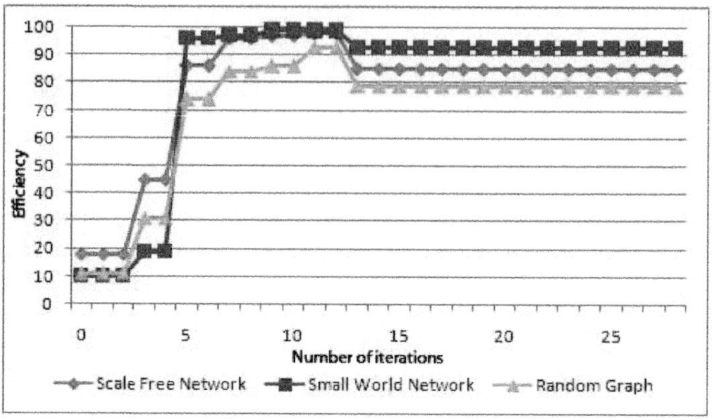

Fig. 1 Evolution of performance over time of different organizational network structures

graphs are highly clustered and have small geodesic distance. We have implemented the model introduced by Watts and Strogatz [13], which is capable of obtain small-world networks by applying the least amount of reconnecting transformations to regular graphs. The last category of networks considered are scale-free architectures, which are essentially characterized by the existence in the network of few nodes with many links. According to Barabasi, Albert and Jeong [1], constructing such topologies can be achieved by growing the network by means of preferential attachment to already existing nodes, proportional to their current degrees. It is important to say that this model generates a scale-free network that does not have the small-world property.

With the purpose of studying how will the structure affect the behavior of the system, we use as a measure of efficiency the percentage of agents that have committed to joining coalitions within a pre-set time frame. More precisely we simulate a system comprising 100 agents throughout a maximum of 30 rounds of interaction, according to the negotiation scheme proposed in Section 3. The energetic capacity for the virtual power stations is assumed at an average of 1MW (small-scale VPS). For these experiments we have considered commercially available residential DERs of 5 capacity classes varying from 50kw to 250kw. The results presented have been obtained by averaging over 20 realizations. The system proves to reach a stable organization in a short number of steps by applying the argumentation scheme proposed for all the topologies analyzed. However, in term of the efficiency metric abovementioned, there appears to be notable differences.

To begin with, one can observe from Figure 1 that the random graph topology is outperformed by both scale-free and small-world networks. In contrast, the small-world configuration is the one that dominates, exhibiting a statistically significant improvement of approximately 15% over random networks and slightly under 10% over large-scale networks. Besides, it is the only complex network that is able to reach an efficiency of above 90%. This is intuitively reasonable according to the intrinsic properties of such networks, which enable rapid spread of information due to its short geodesic distance and high clusterization. Therefore, dynamical processes are facilitated, as quick communication can be provided between distant parts of the system.

Scale-free networks appear to be an intermediate solution, as they outperforms random graphs while being surpassed by small-world networks. In spite of this, it presents a functional advantage as opposed to the other ones if we take into account how the topology affects the robustness and stability of the system. Scale-free networks exhibit resilience to random failures because of the existence of a few number of hubs. On one hand this implies that in the event of a node failure, there is a high probability that the node has a small degree and such, its removal would have a little effect on the functioning of the system at large. Furthermore, while most nodes have a low degree, they happen to lie on few paths that connect other nodes and so, their removal does not affect drastically the overall communication. On the other hand though, deliberate attacks could cause significant vulnerability of the system. Moreover, the experiments indicate that scale-free networks are able of attaining a significantly higher efficiency during the first iteration steps in comparison to the other configurations considered. This is obviously an important aspect in case a good solution needs to be reached in a bounded amount of time.

Thus, results suggest that small-world primarily and scale-free networks represent the best choices for an underlying communication topology for our MAS scenario. As the experiments have shown, the choice for a particular configuration needs though to take into account several trade-offs, according to the prioritization of the system's performance criteria.

Along these lines, as future work we plan to extend our model with an inter-coalition self-adaptation mechanism that enables for a reorganization of the coalitions in order to assure an enhanced coordination, required by the transient nature of the environment.

5 Conclusions

As a concluding remark, in this paper we have introduced a dynamic coalition formation-based model deployed in distributed environments with negotiation agents. As the coalition formation mechanism proposed occurs in networks without formal hierarchies, we have investigated the emergence of coalitions given different communication underlying complex topologies.

In particular, we have highlighted the applicability of our approach through designing a control concept capable of distributed coordination for the electricity grid in the context of multi-agent systems. This process resulted in virtual power stations that would be able to commit to a steady and robust generation profile requiring less energy from real power plants especially during high-demand periods and providing a mechanism able of reducing the complexity of the management process.

References

1. Albert, R., Barabsi, A.-L.: Topology of evolving networks: local events and universality. Phys. Rev. Lett. 85, 5234–5237 (2000)
2. Broder, A., et al.: Graph structure in the web. Comput. Netw. 33, 309–320 (2000)
3. Carley, K.M.: Smart agents and the organizations of the future. In: Lievrouw, L., Livingstone, S. (eds.) The Handbook of New Media, Sage, Thousand Oaks (2002)
4. Erds, P., Rnyi, A.: On the evolution of random graphs. Publ. Math. Inst. Hung. Acad. Sci. 5, 17–61 (1960)
5. Hatziargyriou, N.D., Dimeas, A.L.: Operation of a Multiagent System for Microgrid Control. IEEE Transactions on Power Systems (August 2005)
6. Mcarthur, S.D.J., Davidson, E.M., Catterson, V.M., Dimeas, A.L., Hatziargyriou, N.D., Ponci, F., Funabashi, T.: Multi-Agent Systems for Power Engineering Applications - Part I: Concepts, Approaches, and Technical Challenges. IEEE Transactions on Power Systems 22(4), 1743–1752 (2007)
7. Mihailescu, R., Vasirani, M., Ossowski, S.: Towards Agent-Based Virtual Power Stations via Multi-Level Coalition Formation. In: Proceedings of theInternational Workshop on Agent Technologies for Energy Systems (2010)
8. Oyarzabal, J., Jimeno, J., Ruela, J., Engler, A., Hardt, C.: Agent based micro grid management system. In: 2005 International Conference on Future Power Systems, November 18, p. 6 (2005)
9. Pielke, M., Troschel, M., Kurrat, M., Appelrath, H.-J.: Operation strategies to integrate CHP micro units in domestic appliances into the public power supply. In: Proceedings of the VDE-Kongress 2008 (2008)
10. Pudjianto, D., Ramsay, C., Strbac, G.: Virtual power plant and system integration of distributed energy resources. Renewable Power Generation IET (1), 10–16 (2007)
11. Tel, G.: Introduction To Distributed Algorithms, 2nd edn. Cambridge Univ. Press, Cambridge (2000)
12. U.S. Department of Energy, Office of Electricity Delivery and Energy Reliability, The Smart Grid: An Introduction (2009)
13. Watts, D.J., Strogatz, S.H.: Collective dynamics of small-world networks. Nature 393, 440–442 (1998)

Towards Norm Enforcer Agents

N. Criado, E. Argente, and V. Botti

Abstract. The great majority of works on norm implementation are aimed at enforcing a set of stable norms whose violation can be perfectly observed. However, this assumption is not true in open systems in which norms should evolve for dealing with unforeseen situations. As a solution to this need, we propose to develop norm enforcer agents endowed with capabilities for acquiring norms dynamically and enforcing them in uncertain environments. Specifically, a Normative BDI architecture has been extended in this paper.

1 Introduction

The development of multi-agent systems (MAS) capable of evolving and adapting to changes within their environment has been considered as a relevant open challenge. The evolution implies not only to adapt the structural components (i.e., functions, agent groups, interactive and relationship patterns between roles) and the functional behaviour (i.e., agent tasks, plans or services), but should also to adapt the behavioural norms for the agents [6].

Works on the definition of methods and theories for the dynamic evolution of norms are becoming more and more important [7]. The evolution of norms also entails the development of dynamic norm enforcing mechanisms capable of giving a computational interpretation to norms. In this sense, there is a need of developing methods for ensuring, as much as possible, norm compliance in dynamic agent societies.

The great majority of the works on norm enforcement are based on the existence of both a shared reality which is fully observed and unambiguously

N. Criado · E. Argente · V. Botti
DSIC, Universidad Politcnica de Valencia
e-mail: {ncriado,eargente,vbotti}@dsic.upv.es

J.B. Pérez et al. (Eds.): Highlights in PAAMS, AISC 89, pp. 135–142.
springerlink.com

interpreted. However, these assumptions are too much strong in dynamic domains [3]. In this sense, dynamic domains are characterized by the interaction of autonomous agents who may interact outside the system. In addition, heterogeneous agents may give different interpretations to norms causing doubts or conflicts. Accordingly, norm enforcing mechanism must evolve from static systems to more autonomous and proactive systems (or agents) capable of participating in processes for solving normative conflicts. With this aim, we propose, as a first step, the development of norm enforcer agents capable of deliberating about norm enforcement given that there is a partial and uncertain observability of both the world and the interactions among agents. The role of norm enforcing agents in complex scenarios that require conflict resolution processes will be considered by future works. Specifically, our proposal consists in extending a Normative-BDI agent architecture (or n-BDI for short) with norm enforcement capabilities. Thus, this paper is structured as follows: Section 2 contains the background of this work, Section 3 describes the norm enforcement in the n-BDI Architecture, and a conclusion is provided in Section 4.

2 Background

In this section the logic preliminaries, the normative definitions and the n-BDI architecture extended in this paper are briefly explained.

2.1 Preliminaries

Let us suppose the existence of a first-order predicate language $\mathcal{L}$ whose alphabet includes: the logical connectives $\{\wedge, \vee, \neg, \rightarrow\}$; parentheses, brackets, and other punctuation symbols; and an infinite set of variables. These variables are universally quantified implicitly. In addition, the alphabet contains non-logical predicate, constant and function symbols. Specifically, there are constant symbols that identify roles and agents. Thus, $\mathcal{R}$ and $\mathcal{A}$ are the sets containing all role and agent identifiers, respectively. The set of predicate symbols is formed by *action* predicates ($\mathbb{X}$) and *state* predicates ($\mathbb{P}$) which describe properties of the world. The enactment of roles by agents is modelled by the *play* predicate. It is a binary state predicate ($plays \in \mathbb{P}$) that describes that fact that an agent plays a role.

Let us also assume the standard definition for *wffs* (well-formed formulas). Thus, we make use of the standard notion of substitution of variables in a *wff*, where σ is a finite and possibly empty set of pairs Y/y where Y is a variable in a *wff* and y is a term. If the *wff* obtained from applying a substitution has no variable, then it is defined as grounded. Finally, a theory of $\mathcal{L}$, denoted by symbol Γ, is a set of sentences of $\mathcal{L}$.

2.1.1 Normative Definitions

This work is based on the distinction between *norm* and *norm instance* made in [5]. A *norm* is a conditional rule that defines under which conditions obligations and prohibitions should be created. The *norm instances* that are created out of the *norms* are a set of unconditional expressions that bind a particular agent to an obligation or prohibition. Moreover, a norm instance is accompanied by an expiration condition that defines the validity period or deadline of the norm instance.

Throughout this paper only prohibition norm are considered for space reasons. However, this proposal can be easily extended with obligations if we consider obligations to perform some action as prohibitions to not perform this action.

Definition 1 (Prohibition Norm). *A* ***prohibition norm*** *is defined as a tuple* $n_a = \langle D, T, A, E, C, S, R \rangle$ *where:*

- *D is the deontic modality of the norm. Since this work only consider prohibition norms, it always takes the forbidden modality* ($\mathcal{F}$)
- $T \in \mathcal{R}$ *is the target of the norm; i.e., the role to which the norm is addressed;*
- *A is a wff of* $\mathcal{L}$ *that represents the norm activation condition. It defines under which circunstancies the norm is active and must be instantiated.*
- *E is a wff of* $\mathcal{L}$ *that is the norm expiration condition, which determines when the norm no longer affects agents.*
- $C \in \mathbb{X}$ *represents the action that is forbidden.*
- $S, R \in \mathbb{X}$ *are expressions which describe the actions (sanctions S and rewards R) that will be carried out in case of norm violation or fulfilment, respectively.*

Definition 2 (Prohibition Norm Instance). *Given a theory* Γ*, an prohibition norm* $n_a = \langle D, T, A, E, C, S, R \rangle$ *is instantiated into a* ***prohibition norm instance*** $n_i = \langle D, Ta, T, E', C', S', R' \rangle$ *where:*

- $\Gamma \vdash \sigma(A)$*, where* σ *is a substitution of variables in A such that* $\sigma(A)$, $\sigma(E)$, $\sigma(C)$, $\sigma(S)$ *and* $\sigma(R)$ *are grounded;*
- $Ta \in \mathcal{A}$ *is the target agent of the norm instance such as* $\Gamma \cup \{\sigma(A)\} \vdash plays(Ta, T)$*;*
- $C' = \sigma(C)$, $E' = \sigma(E)$, $S' = \sigma(S)$, *and* $R' = \sigma(R)$.

For simplicity, it is assumed that norm instances are grounded.

2.2 Norm Compliant Agents: The n-BDI Architecture

Usually, proposals on agent architectures which support normative reasoning consider norms as static constraints that are hard-wired on agent architectures. Thus, agents are not able to learn new norms and taken into account these unforeseen norms. In order to overcome this drawback, in [2, 4], the multi-context graded BDI agent architecture [1] has been extended with recognition and normative reasoning capabilities. According to the multi-context graded BDI proposal, an agent is defined by a set of interconnected contexts $\langle \{c_i\}_{i \in I}, \Delta \rangle$. Each context $c_i \in \{c_i\}_{i \in I}$ is a tuple $\langle L_i, A_i, \Delta_i \rangle$, where L_i, A_i, and Δ_i are the language, axioms and inference rules defining the logic of each context c_i, respectively. Each context ($c_i \in \{c_i\}_{i \in I}$) is associated with a theory Γ_i. Finally, Δ is the set of bridge rules; i.e., inference rules whose premises and conclusions belong to different contexts. Thus, bridge rules take the general form:

$$\frac{c_1 : A_1, ..., c_q : A_q}{c_j : \oplus A}$$

meaning that if there exists a most general substitution σ such as for all $k \in \{1, ..., q\}$, $\Gamma_{c_k} \vdash \sigma(A_k)$ then proposition $\sigma(A)$ is inferred in context c_j. $\oplus$ and $\ominus$ are revision operators meaning that a formula is inserted or removed from a context, respectively[1].

The feature that distinguishes n-BDI agents from BDI agents in general is the availability of an explicit representation of norms and the capabilities of n-BDI agents to reason about them. Thus, a n-BDI agent [2, 4] is formed by (see Figure 1):

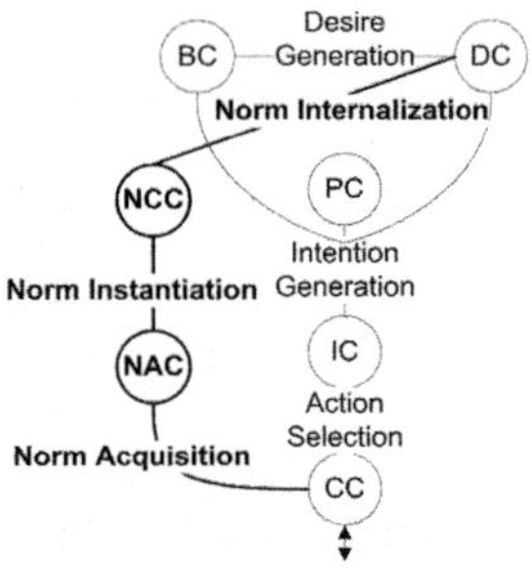

Fig. 1 The n-BDI Architecture

- *Mental* contexts [1] to characterize beliefs (BC), intentions (IC) and desires (DC). These contexts contain logic propositions such as $(\mathcal{M}\ \gamma, \delta)$. $\mathcal{M} \in \{\mathcal{B}, \mathcal{D}, \mathcal{I}\}$ is the mental modality for representing beliefs, desires and intentions, respectively; γ is a *wff* of $\mathcal{L}$; and $\delta \in [0, 1]$ represents the certainty degree associated to this mental proposition.
- *Functional* contexts [1] for planning (PC) and communication (CC).
- *Normative* contexts [2, 4] for allowing agents to have an explicit representation of norms and to consider them in their reasoning process:
  - *Norm Acquisition Context (NAC).* The process by which agents identify norms is beyond the scope of this work. Thus, the NAC is considered to

[1] Thus, $c_j : \oplus A \Leftrightarrow \Gamma_{c_j} = \Gamma'_{c_j} \cup \{\sigma(A)\}$ and $c_j : \ominus A \Leftrightarrow \Gamma'_{c_j} = \Gamma_{c_j} \setminus \{\sigma(A)\}$.

be a *black box* that contains norms that are applicable (i.e., *in force*) at a given moment. Therefore, the NAC is formed by expressions such as (n, δ), where n is a norm (see Definition 1); and $\delta \in [0, 1]$ is a real value that assigns a certainty degree to this norm. This degree is determined by the certainty to which the norm has been recognised.
  - *Norm Compliance Context (NCC).* This is the component responsible for reasoning about compliance with the set of norms that are addressed to the agent at a given moment.

- *Bridge Rules.* They connect mental, functional and normative contexts. A detailed description of these bridge rules can be found in [1]. For a more detailed description of the normative bridge rules see [2, 4].

A normative agent architecture, as the n-BDI, must allow agents to: i) determine which and how norms will be observed or violated; and ii) support agents when facing with norm violations. The original n-BDI proposal [2, 4] is focused on the former issue. However, the detection of norm violations and fulfilments and the generation of both punishing and rewarding reactions have not been considered. The present paper addresses this pending issue by extending the n-BDI proposal for performing norm enforcement tasks.

3 Norm Enforcement: Extending the n-BDI Architecture

Norm aware agents do not only require mechanisms for considering norms in their decisions, but also tools for evaluating the behaviour of others with respect to norms. Specifically, this paper proposes the extension of the n-BDI for the development of a norm enforcer agent which is in charge of controlling norms. Specifically, the extension consist on the definition of a *Norm Enforcing Context* (NEC) and additional bridge rules for reasoning and reacting to norm violations and fulfilments.

3.1 Norm Enforcing Context (NEC)

The NEC contains all norm instances regardless of the target agent. It allows the norm enforcer agent to observe the behaviour of other agents and react to it with respect to norm instances. Thus, it is formed by expressions such as: (n, δ), where n is a norm instance (see Definition 2 in Section 2.1.1) and δ is a real value in $[0, 1]$ that represents the degree with which the norm has been fulfilled. These norm instances together with the perceptions of the norm enforcer agent create motivations for punishing and rewarding target agents. Specifically, this paper proposes carrying out the norm enforcing reasoning process in four steps: norm activation, norm instance enforcement, norm instance expiration, and norm deactivation. Each one of these steps is performed by several bridge rules which are explained below.

3.2 Norm Enforcing Bridge Rules

3.2.1 Norm Activation

Each time an agent acquires the target role of a norm, if the activation condition holds, then it is active and norm instances are created. This procedure is carried out by the following bridge rule:

$$\frac{NAC : ((\langle \mathcal{F}, T, A, E, C, S, R \rangle, \delta), BC : (\mathcal{B}\ A, \delta_A), BC : (\mathcal{B}\ plays(Ta, T), \delta_{plays}), \quad f_{Activation}(\delta_A, \delta_{plays}, \delta) > \theta}{NEC : \oplus(\langle \mathcal{F}, Ta, T, E, C, S, R \rangle, 1)}$$

Thus, if there is a prohibition norm ($\langle \mathcal{F}, T, A, E, C, S, R \rangle$) whose target role is played by some agent, then norm instances will be created if the activation condition holds (i.e., $\exists \sigma$ such as $\Gamma_{BC} \vdash (\mathcal{B}\ \sigma(A), \sigma(\delta_A))$ and $\Gamma_{BC} \vdash (\mathcal{B}\ play(\sigma(Ta), T), \sigma(\delta_{plays})))$. In order to avoid the creation of norm instances if there is not enough certainty in the activation of that norm a threshold (θ) has been defined. Therefore, only those instances in which there is enough certainty ($f_{Activation}(\delta_A, \delta_{plays}, \delta) > \theta$) will be inserted in the NEC. The $f_{Activation}$ function combines the certainty ascribed to the activation condition (δ_A), the certainty assigned to the fact that the agent is playing the target role (δ_{plays}) and the norm recognition certainty (δ). Since these three events can be considered as statistically independent, we define $f_{Activation}$ as the product among the three certainty degrees ($f_{Activation} = \delta_A \times \delta_{plays} \times \delta$). Finally, the fulfilment degree of the prohibition instance is initially set to 1, since any prohibition is considered as fulfilled until the forbidden condition is detected.

3.2.2 Norm Instance Enforcement

Norm instances define an unconditional prohibition addressed to a specific agent. Therefore, when any agent carries out an action the enforcer agent must verify if this action is considered as a violation of the prohibition norm instance. If so, the norm enforcer agent will apply sanctions on the offender agent. This operation is carried out by the next bridge rule:

$$\frac{NEC : ((\langle \mathcal{F}, Ta, T, E, C, S, R \rangle, \delta), CC : (Done(Ta, C), \delta_C)}{NEC : \oplus(\langle \mathcal{F}, Ta, T, E, C, S, R \rangle, max(0, \delta - \delta_C)), DC : \oplus(\mathcal{D}\ S, \delta_C)}$$

According to this bridge rule, a prohibition instance ($\langle \mathcal{F}, Ta, T, E, C, S, R \rangle$) is violated when the target agent carries out the action which is forbidden ($Done(Ta, C), \delta_C$). Then, the fulfilment degree of the norm instance is reduced ($max(0, \delta - \delta_C)$). Finally, a desire is created for sanction the target agent ($\oplus(\mathcal{D}\ S, \delta_C)$). In this paper, enforcer agents are supposed to give more priority to the enforcement of those norm instances that have been fulfilled or violated more clearly. Thus, the desirability of the enforcing action, the sanction in this case, is defined as the certainty with which the norm condition has been perceived. How these desires are considered by the decision-making procedure is beyond the scope of the work described in this paper. Mainly, these new desires derived from norm enforcement may help the agent to select the most suitable plan. Thus, plans that have as a consequence the fulfilment of norm enforcement desires will be preferred to be intended. For a detailed description of the decision-making process in the N-BDI proposal see [4].

3.2.3 Norm Instance Expiration

Once the prohibition norm instance has expired, then if the target agent has complied with the norm it will be rewarded.

$$\frac{NEC : (\langle \mathcal{F}, Ta, T, E, C, S, R\rangle, \delta), BC : (\mathcal{B}\ E, \delta_E), \delta_E > \theta}{NEC : \ominus(\langle \mathcal{F}, Ta, T, E, C, S, R\rangle, \delta), DC : \oplus(\mathcal{D}\ R, \delta)}$$

Once the expiration condition hold ($(\mathcal{B}\ E, \delta_E)$ and $\delta_E > \theta$), then the norm instance is removed form de NEC and a desire for rewarding the agent are created ($\oplus(\mathcal{D}\ R, \delta)$). If the target agent has violated the prohibition in the past, then the fulfilment degree (δ) has been decreased. In this case, the desire of rewarding would have an insignificant degree and would never be considered.

3.2.4 Norm Deactivation

When any agent leaves a role then all norms addressed to this role do not longer affects to it and all of the norm instances must be deleted. Since these norm instances have not been expired the norm cannot be considered as fulfilled. Thus, neither reward nor sanction is applied to the agent.

$$\frac{NEC : (\langle \mathcal{F}, Ta, T, E, C, S, R\rangle, \delta), BC : (\mathcal{B}\ \neg plays(Ta, T), \delta_{\neg plays}), \delta_{\neg plays} > \theta}{NEC : \ominus(\langle \mathcal{F}, Ta, T, E, C, S, R\rangle, \delta)}$$

When the target agent of a norm instance leaves the target role ($(\mathcal{B}\ \neg plays(Ta, T), \delta_{\neg plays})$ and $\delta_{\neg plays} > \theta$), then the norm instance is removed.

4 Conclusion

The great majority of the works on norm implementation are based on two main assumptions: i) norms that control an agent society remains stable; and ii) the existence of a shared reality which is fully and unambiguously observed and interpreted. However, these assumptions are not true in open systems.

Therefore, norm enforcement mechanisms should provide solutions to systems in which the set of norms evolves along time, the interactions among agents may not be fully observed and norms can be given different interpretations. As a solution to this need we propose to develop norm enforcer agents endowed with capabilities for acquiring norms dynamically and reasoning about their enforcement in uncertain environments in which conflicts about the violation of norms may arise. In this paper, the extension of the n-BDI architecture for enforcing norms has been described. Thus, the NEC allows agents to observe and react to norm violations in uncertain environments. As future work, the norm enforcement in case of agents have given different interpretations to the same norm will be considered. Moreover, the norm-enforcing agent is being evaluated and used in the simulation of an electronic market for water-rights transfer [3] which is a well known example of a regulated environment.

Acknowledgements. This work was partially supported by the Spanish government under grants CONSOLIDER-INGENIO 2010 CSD2007-00022, TIN2009-13839-C03-01 and TIN2008-04446 and by the FPU grant AP-2007-01256 awarded to N. Criado.

References

1. Casali, A.: On Intentional and Social Agents with Graded Attitudes. PhD thesis, Universitat de Girona (2008)
2. Criado, N., Argente, E., Botti, V.: A BDI Architecture for Normative Decision Making. In: AAMAS, pp. 1383–1384 (2010)
3. Criado, N., Argente, E., Garrido, A., Gimeno, J.A., Igual, F., Botti, V., Noriega, P., Giret, A.: Norm enforceability in Electronic Institutions? In: COIN@MALLOW 2010, pp. 49–64 (2010)
4. Criado, N., Argente, E., Noriega, P., Botti, V.: Towards a Normative BDI Architecture for Norm Compliance. In: COIN@MALLOW 2010, pp. 65–81 (2010)
5. Oren, N., Panagiotidi, S., Vázquez-Salceda, J., Modgil, S., Luck, M., Miles, S.: Towards a formalisation of electronic contracting environments. In: Hübner, J.F., Matson, E., Boissier, O., Dignum, V. (eds.) COIN@AAMAS 2008. LNCS, vol. 5428, pp. 156–171. Springer, Heidelberg (2009)
6. Rodriguez, S., Perez-Lancho, B., De Paz, J., Bajo, J., Corchado, J.: Ovamah: Multiagent-based adaptive virtual organizations. In: FUSION, pp. 990–997 (2009)
7. Tinnemeier, N.A.M., Dastani, M., Meyer, J.-J.C.: Programming norm change. In: AAMAS, pp. 957–964 (2010)

Mixing Electronic Institutions with Virtual Organizations: A Solution Based on Bundles

Mario Rodrigo Solaz, Bruno Rosell i Gui, Juan A. Rodríguez-Aguilar, Vicente Julián Inglada, and Carlos Carrascosa Casamayor

Abstract. Today, software development largely consists of adapting existing functionality to perform in a new environment. In reality, integrating existing code has become a large part of the work of software developers. Looking at the different possibilities of representing human societies, the need for this integration is even higher. The societies between human beings are heterogeneus and the existing software approaches to them, do not cover all their aspects completely. If we want to translate a static organization (f.i. the Congress), from physical to electronic world, we use Electronic Institutions. In case we want to translate a dynamic organization (f.i. a fan club), we use Virtual Organizations. But, what if our organization to translate is a hybrid system where we can find features of both approaches (static and dynamic)?

This paper investigates how the Electronic Institutions (EI) and the Virtual Organizations (VO) may be integrated, in a dynamic environment, to supply a solution for these hybrid scenarios.

1 Introduction

In most of the quotidian activities we can find rules, patterns, structures, societies, etc. Some of them are rigid, immutable, others are flexible and can be variable in time. Software must faithfully represent this world variability. So, design and development of software applications at this moment, is harder than it used to be. Now,

Mario Rodrigo Solaz · Vicente Julián Inglada · Carlos Carrascosa Casamayor
Departament de Sistemes Informàtics i Computació, Universitat Politècnica de València, Camí de Vera s/n, E-46022, València, (Spain)
e-mail: {mrodrigo,vinglada,carrasco}@dsic.upv.es

Bruno Rosell i Gui · Juan A. Rodríguez-Aguilar
Unidad de Desarrollo Tecnológico en Inteligencia Artificial, IIIA-CSIC Campus de la UAB s/n, E-08193 Bellaterra, (Spain)
e-mail: {rosell,jar}@iiia.csic.es

J.B. Pérez et al. (Eds.): Highlights in PAAMS, AISC 89, pp. 143–150.
springerlink.com

software development has to manage with lots of heterogeneus data and other technical aspects as, for example, security, reliability and performance. In this way, developers' main goal is to build software applications adaptive, autonomous, flexible, robust and capable of reorganizing itself. For developers, there is a need of artifacts and tools capable of represent this variability. Fortunately, there are some technologies that cover efficiently certain aspects of these desirable feautures. The solution to this increasing complexity of the problems lies, in the mixture of technologies.

Multi-agent Systems (MAS) [2], have among its objectives, the autonomy and flexibility to meet future changes. So, it seems clear, at this new complex and highly dynamic scenario, the use of MAS can help to resolve these problems. There are two technologies, inside MAS, which can help us to represent the real world or a little piece of it. On one hand, there are the Electronic Institutions [3, 4, 8]; on the other hand there are the Virtual Organizations and its implementation: THOMAS (Me*TH*ods, Techniques and Tools for *O*pen *M*ulti-*A*gent *S*ystems) [2, 6, 9].

AMELI Electronic Institutions are the electronic equivalence to physical institutions, which guide the relations between humans. Electronic Institutions are the framework where human interactions are made, and furthermore, they define what is permitted and what is not. As is the case with real world institutions, Electronic Institutions have a fixed set of norms. This set of norms do not change during all the lifecycle of Electronic Institution. So, Electronic Institution are the best option to represent static organizations.

THOMAS Virtual Organizations are open systems formed by individuals and groups that work coordinately. The THOMAS Virtual Organizations, can be created and destroyed dynamically, and agents can enter and exit from it dynamically too. This coordinated work is restricted by institutional rules. These rules can change during the Virtual Organizations' lifecycle. So, THOMAS Virtual Organizations are the best option to represent dynamic organizations. Due to the fact that these individuals and institutions are very heterogeneus, to get them coordinated, we need organizational concepts. Modeling the organization's features as services, individuals (from this organization or from others), and other organizations have an standard interaction method.

In summary, we have two approaches to represent real world organizations and institutions: Electronic Institutions and Virtual Organizations. Each one of them is focused on different features: The Electronic institutions represent static systems and the THOMAS Virtual Organization are adapted to dynamical ones; the mixture of both technologies can enhance any case of study.

With these technologies capable to translate organizations from real world to electronic one, we only need to find the way to connect them. At this point, the Open Services Gateway Initiative (OSGi) [1] technology can contribute to the possibility of developing a hybrid system that includes the features of Electronic institutions and Virtual Organizations. The OSGi technology provides the standardized primitives that allow applications to be constructed dynamically, without requiring restarts, from small, reusable and collaborative components. So, it fits really well to our purpose. Using this technology, we can get the best of each approach, to obtain solutions richer and more adapted to real world.

In this article we combine the ideas of Electronic Institutions [3, 4, 8] and THOMAS Virtual Organizations [2, 6, 9], and we propose a new method to deal with electronic's representation of complex scenarios from real world. In our solution, we use the Electronic Institutions to represent static aspects of the problem and remains on the THOMAS Virtual Organizations the task of representing dynamic ones. In this way, we can cover all aspects of complex problems in a better way than before. Besides, with the facilities provided by OSGi technology we can solve the communication between the two approaches mentioned above to coordinate them.

The rest of the paper is structured as follows: Section 2, explains the work related with the different approaches to MAS, focusing in AMELI Electronic Institutions and THOMAS Virtual Organizations. Section 3 shows our proposal to use, jointly, AMELI Electronic Institutions and THOMAS Virtual Organizations. Section 4, presents a case study showing this integration and, finally, some conclusions of the current work are presented in Section 5.

2 Related Work

Porting organizations from real world to electronic one is really complex. A quick view over them, serves to realize about the many features they have and the quantity of aspects we have to consider in this porting task. Some of these organizations are rigid and immutable, others are flexible and can be variable in time. And software must faithfully represent this world variability. The software resource that best fits to this purpose is Multi-Agent System (MAS) [2]. The MAS, have among its objectives, the autonomy and flexibility to meet future changes. Within MAS are two ways to represent faithfully the real world, or at least part of it: AMELI Electronic Institutions and THOMAS Virtual Organizations.

With these two approaches working separately, we can model many organizations, but if we could do them to work together, the range of solved problems will be higher. At this point arrives OSGi technology to provide the nexus between both paradigms. The OSGi technology provides the standardized primitives that allow applications to be constructed dynamically, without requiring restarts, from small, reusable and collaborative components. So, using this technology, we can get the best of each approach, to obtain solutions richer and more adapted to real world.

2.1 Organizations in Multi-Agent Systems

With the production environments changing fast and continously and the markets demanding more and more exigent products, software must fit these evolving environment as fast as posible. Multi-Agent Systems are versatile and flexible, and these features convert them in an appropriate approach to solve many problems where data can be distributed and heterogeneus. MAS are based on various intelligent entities which work together to achieve a global aim. Each entity is autonomous and is capable of cooperating and interacting with other entities.

One application of Multi-Agent Systems is the definition of real organizations in the virtual world. Previous works in this field have resulted in AMELI Electronic Institutions (section 2.1.1) and THOMAS Virtual Organizations (section 2.1.2).

2.1.1 AMELI Electronic Institutions

Electronic Institutions are the electronic equivalence to physical institutions, which guide the relations between humans. Electronic Institutions are the framework where human interactions are made, and furthermore, they define what is permitted and what is not. As is the case with real world institutions, Electronic Institutions have a fixed set of norms. This set of norms do not change during all the lifecycle of Electronic Institutions. So, Electronic Institution are the best option to represent static organizations.

Eletronic Institutions' main objective is to enforce the specified rules to participating agents at runtime. To achieve this, Electronic Institutions have a middleware layer to mediate agents interactions. So, the system can filter agent's actions and decide if action is permitted or not.

2.1.2 THOMAS Virtual Organizations

THOMAS (Me*TH*ods, Techniques and Tools for *O*pen *M*ulti-*A*gent *S*ystems) [2, 6, 9], is an open architecture for the development of Virtual Organizations in the context of dynamic and open multi-agent systems. This framework is focused in organizational concepts and provides resources to manage virtual organizations in dynamic, open and large-scale environments where heterogeneus entities, like agents or services, interact. The THOMAS' functionalities, are modelled as a set of semantic web services [9] in order to provide all entities (agents or not), a standarized way to interact. These functionalities are separated into two entities: the *Organization Management System* (OMS), wich is responsible of the management of the organizations and their entities; and the *Service Facilitator* (SF), which offers simple and complex services to the active agents and organizations. Through the use of these two entities, agents in THOMAS can publish and invoke services transparently to other entities (agents or not).

Through this methodology, any agent is able to create a new Virtual Organization in THOMAS, using the set of services described above. This method makes no dependency of internal agents, so it achieves a environment fully addresed for dynamic and open systems.

2.2 OSGi Framework

OSGi [1] are the capital letters of Open Services Gateway Initiative. It aims to define open specifications of software to design compatible platforms able to provide multiple services. The OSGi specification has been defined with basic APIs to service development which allow discovering devices and services.

OSGi is developed by an alliance of different companies, providing a modular system for Java following a service-oriented pattern, where each of this modules is called *bundle*. Everyone of this bundles is a collection of classes, jars and configuration files where explicit external dependencies (if any), can be installed, loaded, updated and uninstalled dynamically within the OSGi framework.

The core component of the OSGi Specifications is the OSGi Framework and is formed by the layers (i) Bundles, that, usually, are common jar components, accompanied by a manifest file, which includes extra information; (ii) Services, that connects bundles in a dynamic way by offering a publish-find-bind model for plain old Java objects (POJO); (iii) Services Registry, also known as the API for management services (ServiceRegistration, ServiceTracker and ServiceReference); (iv) Life-Cycle, also known as the API for life cycle management for (install, start, stop, update, and uninstall) bundles; (v) Modules, that defines encapsulation and declaration of dependencies (how a bundle can import and export code); (vi) Security, that handles the security aspects by limiting bundle functionality to pre-defined capabilities and (vii) Execution Environment, that defines what methods and classes are available in a specific platform.

3 Integration Proposal

Our integration proposal ilustrate the sense of using THOMAS Virtual Organizations along with AMELI Electronic Institutions. We have designed a scenario with two groups of people. The first group represents a private readers club. The members at this club, talk about their interests on books and magazines according to their preferences. Not everyone can be a member of this club. To join the club, the new member must be endorsed by a sum of money. The second group represents all the people outside the club. These people have no access to information or conversations produced at the club. This group contains people without sufficient collateral to join the club. The main goal of second group is obtain access to information from inside the readers club. To achieve this, the club outsiders create a collaborative group, where they make a financial contributions to endorse the candidacy of one of its members. In this way, one of them can join the club and mantains informed the rest of the group.

Each group of people can be represented electronically as an organization. The club with its fixed rules and its goals, is translated to an AMELI Electronic Institution. The other association, where people can join or exit without restrictions, and make financial contributions to the group, can be translated into a THOMAS Virtual Organization. In this way, we cover the representation aspect for both groups, each one with different requisites. The other aspect to cover is comunication between groups. This can be solved using OSGi technology, which facilitates communication and integration between Java-based artifacts. The selected OSGi framework has been Apache Felix [5]. Among the reasons for this decision should be stressed its light weight, only 750 KB, and the vast amount of documentation and examples available.

In summary, we have an AMELI Electronic Institution representing the private readers club; a THOMAS Virtual Organization representing the variability of people without enough collateral to join the club; and Felix Framework to made interconnection posible.

4 Case Study

Once presented the integration proposal, this section shows how we have dealed it. Since we use an OSGi framework, we need AMELI Electronic Institutions and THOMAS Virtual Organizations to be deployed as bundles. Before this work, there was a version of Electronic Institutions that can be found in this package mode. Unfortunatelly, there was not anyone for THOMAS, so, we have built it. As in the THOMAS Framework version used both of the two internal agents are implemented as JADE agents (Java Agent DEvelopment Framework) [7], we have also needed such JADE Framework bundle. This bundle is available at their website[1]. In summary, our proposal needs three bundles' sets: the first one for EI platform; the second one for THOMAS (which contains the JADE bundled version); and one more set for the framework OSGi itself.

In our proposal, there is a hybrid agent that is capable of interfacing in a THOMAS organization and in a Electronic Institution at the same time. But also where actions on one of the worlds has implications for the other.

For this model, the *ChatBroadcasterAgent* agent connects to THOMAS and to a Chat Electronic Institution (Fig. 1). Later, using the THOMAS agents *OMS* and *SF* of THOMAS, the *ChatBroadcasterAgent* agent publishes his broadcasting service. This is a subscription service, so that subscribed THOMAS agents, are aware of the messages posted in the chat room without having to join the Electronic Institution.

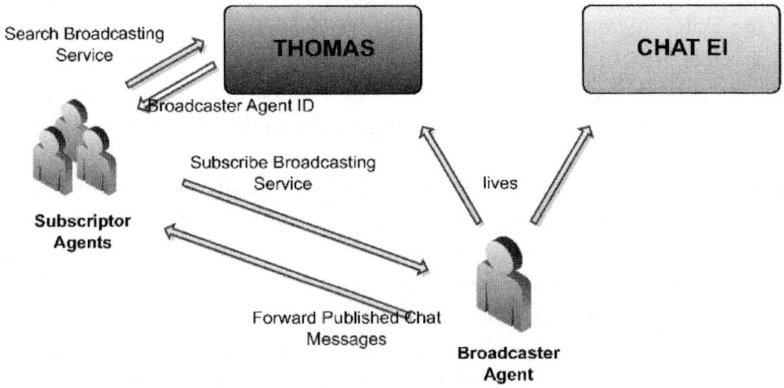

Fig. 1 Connection to THOMAS and Chat Electronic Institution.

[1] http://jade.tilab.com/

Interaction with THOMAS

The first step performed by the *ChatBroadcasterAgent* agent is the connection and interaction with THOMAS. To do this, the *ChatBroadcasterAgent* must register as a member of THOMAS (using the service *AcquireRoleProcess*). Once registered, he may request to use the service *SearchService* to find if there is a service like the one he wants to offer. After that, the *ChatBroadcasterAgent* registers the unit (*News*) and the roles (*Broadcaster* and *Subscriber*), associated with the service. They allow other THOMAS agents to use this service by calling the corresponding *RegisterUnitProcess* and *RegisterRoleProcess*. The next step is to create a profile and its corresponding THOMAS process. Then *ChatBroadcasterAgent* acquires the role *Broadcaster* in the newly created unit *News*. Once completed these actions, the broadcasting service is ready to be used by other THOMAS agents (namely *ChatSubscriptorAgentX*[2]).

The agent *ChatBroadcasterAgent* has implemented two behaviors to address interactions with other THOMAS agents. The first one is used by subscribers to alert *ChatBroadcasterAgent* about their purpose. The second one is responsible for periodically forwarding, to subscribers, all the messages that have appeared in the chat room since the last posted information.

For its part, the *SubscriptorAgentX* agent performs a series of measures similar to those of the *ChatBroadcasterAgent* agent. The first of them, is register himself as a member of THOMAS (*AcquireRoleProcess*). Once registered, he looks for a broadcast service. Once found, *SubscriptorAgentX* gets more detailed information, analyzing the profile of the service (using *getProfileProcess*) and checks what role should acquire to use this service. Knowing the unit and role, in this case *News* and *Subscriber* respectively, calls *GetProcessProcess* to obtain the data necessary to contact the service provider. As the last step of *SubscriptorAgentX*, it sends a subscription message to the provider and it waits the broadcasting of messages produced in the chat.

Interaction with Electronic Institutions

To interact with Electronic Institutions, the agent *ChatBroadcasterAgent* includes a *Behavior* that models the interaction with AMELI and launches the Chat Electronic Institution.The *ChatBroadcasterAgent* goes to *Admisson* and, if accepted by the system, then goes to *RegRoom*. When a chatroom with *java* subject is created, *ChatBroadcasterAgent* moves into it to participate. When this room closes, the *ChatBroadcasterAgent* exits from the Electronic Institution.

5 Conclusions

This paper presents a scenario where collaboration between AMELI and THOMAS has sense and shows a solution. From this solution we can extract two conclusions.

[2] Being X a number to identify the subscriptors to broadcasting service.

The first one is that our proposal is able to cover and to solve complex problems, in dynamic environments, through the mixture of paradigms (AMELI Electronic Institutions and THOMAS Virtual Organizations). This article proves it, and achieves it, in a better way than using only one of those paradigms. The second conclusion is that it is posible to translate this proposal, from teorical view to code view, and to build a software solutions based on it.

References

1. T. O. Alliance. Osgi alliance homepage, http://www.osgi.org/Main/HomePage
2. Carrascosa, C., Giret, A., Julian, V., Rebollo, M., Argente, E., Botti, V.: Service Oriented Multi-agent Systems: An open architecture. In: Autonomous Agents and Multiagent Systems (AAMAS), pp. 1291–1292 (2009)
3. Esteva, M., Rodriguez-Aguilar, J., Sierra, C., Arcos, J., Garcia, P.: On the Formal Specification of Electronic Institutions. In: Sierra, C., Dignum, F.P.M. (eds.) AgentLink 2000. LNCS (LNAI), vol. 1991, pp. 126–147. Springer, Heidelberg (2001)
4. Esteva, M., Rosell, B., Rodriguez-Aguilar, J., Arcos, J.L.: AMELI: An agent-based middleware for electronic institutions. In: Proc. AAMAS 2004, pp. 236–243 (2004)
5. The Apache Software Foundation. Apache felix homepage, http://felix.apache.org/site/index.html
6. Giret, A., Julián, V., Rebollo, M., Argente, E., Carrascosa, C., Botti, V.: An open architecture for service-oriented virtual organizations. In: Braubach, L., Briot, J.-P., Thangarajah, J. (eds.) ProMAS 2009. LNCS, vol. 5919, pp. 118–132. Springer, Heidelberg (2010)
7. Telecom Italia Lab. Java agent development framework homepage, http://jade.tilab.com/
8. Sierra, C., Rodriguez-Aguilar, J., Noriega, P., Esteva, M., Arcos, J.: Engineering multi-agent systems as electronic institutions. European Journal for the Informatics Professional 4 (2004)
9. Val, E.D., Criado, N., Carrascosa, C., Julian, V., Rebollo, M., Argente, E., Botti, V.: THOMAS: A Service-Oriented Framework For Virtual Organizations. In: 9th Int. Conf. on Autonomous Agents and Multiagent Systems (AAMAS 2010), pp. 1631–1632 (2010)

Open Issues in Multiagent System Reorganization

Juan M. Alberola, Vicente Julian, and Ana Garcia-Fornes

Abstract. In the last few years, the demand for technologies and methodologies for supporting dynamic and open Multiagent Systems has increased. Thus, several approaches for reorganization in Multiagent Systems have emerged to dynamically adapt agent organizations. Current approaches for reorganization provide support to allow agents to enter or exit the system, to change the roles played by agents, to modify the norms that regulate the system, etc. In an effort to identify the dearth of support in current approaches, this paper presents a comparison and analysis of the most relevant approaches for reorganization. Then, after an evaluation, we indicate some future research lines with regard to reorganization in Multiagent Systems.

1 Introduction

Agent technologies require support for systems that are situated in dynamic and open environments, able to adapt to these environments and capable of incorporating autonomous and self-interested components. Therefore, methodologies for dynamic systems are required to provide for all phases of the agent-based software engineering life-cycle: the identification of the problem, the design and implementation of a mechanism of adaptation, and the control and monitoring of the system once it is deployed [10].

In the last few years, many approaches for reorganization in Multiagent Systems have emerged. These approaches provide support for modifying the structure and behaviour of a Multiagent System while it is running, such as adding, removing, or substituting components. This is known as dynamic adaptation [4]. In order to assess the gaps and open issues in reorganization and self-organization in Multiagent Systems, we need to evaluate the current approaches and analyze what they provide.

Juan M. Alberola · Vicente Julian · Ana Garcia-Fornes
Departament de Sistemes Informàtics i Computació,
Universitat Politècnica de València, Camí de Vera s/n. 46022, València. Spain
e-mail: {jalberola,vinglada,agarcia}@dsic.upv.es

J.B. Pérez et al. (Eds.): Highlights in PAAMS, AISC 89, pp. 151–158.
springerlink.com

In this paper, we present a description of the most relevant current approaches for reorganization in Multiagent Systems and analyze the most relevant research issues of interest for the coming years.The rest of the paper is organized as follows. In Section 2, we describe the phases of reorganization in Multiagent Systems and define the most relevant parameters that should be analyzed for a comparison of reorganization approaches. In Section 3, we compare different reorganization approaches according to those parameters. Finally, in Section 4, we summarize the most relevant open issuess that should be addressed in the years to come.

2 Reorganization in Multiagent Systems

Reorganization in MAS defines a process that changes an organization into a new one [7]. These changes are related to the organization specification, i.e., roles, goals, services, norms, and the agent population, as well as changes in the relationships among these components. In this respect, the life-cycle of a reorganization can be defined as the process of analyzing the problems of the current organization, proposing reorganization solutions, selecting and implementing a reorganization, and evaluating this process once it is applied.

Based on the work of So and Durfee [13], we use the following phases to define the life-cycle of a reorganization process: monitoring, design, selection, and evaluation. Frameworks that support reorganization in Multiagent Systems implicitly define issues related to these phases. In the following sections, we present a description of each phase.

2.1 Monitoring

The monitoring phase defines the problems of *why* and *when* an organization needs to be reorganized. According to Dignum et al. [4], environment changes are the triggers to reorganization. Thus, we propose to classify different reorganization approaches according to whether they are *reactive* or *proactive*. A *reactive* reorganization occurs when the organization automatically responds to events that cause a reorganization such as the addition or deletion of a new role, agent, etc. These events cause the organization to make the required adjustments in order to continue to fulfill its goals. Approaches that follow this kind of reorganization focus on the need for a reorganization process to be guided by events. A *proactive* reorganization requires an implicit mechanism for reasoning about the current situation in order to decide that a reorganization is required. A *proactive* reorganization can also involve situations in which an event occurs, requiring a reorganization to be carried out after a reasoning process. Approaches that follow this kind of reorganization usually focus the reorganization process according to two different perspectives. Some approaches define a *proactive* reorganization when an event prevents the organization from fulfilling its goals. Other approaches define a *proactive* reorganization to increase the utility of the organization [1, 8].

The monitoring phase can be carried out in a *centralized* way if an agent or a specific authority is responsible for deciding that a reorganization is required. In contrast, a *distributed* way implies that a pool of agents can decide that a reorganization is required either autonomously or by means of an agreement.

2.2 Design

The design phase defines the problem of *how* a reorganization is carried out. Once a reorganization process is required, the design phase includes an analysis of the organization elements and a reorganization proposal that changes specific elements of this organization. This phase can be carried out in a *centralized* way if a single agent or a central authority is responsible for proposing the reorganization solution. A *decentralized* design involves the participation of several agents in the reorganization solution proposal. These solutions can be proposed, either *autonomously* (if each agent involved in the design phase proposes its solution according to its knowledge of the organization) or by *agreement* (if the agents involved in the design phase proposes a single reorganization solution by a joint agreement). Depending on the specific organization model, different elements can be changed in a reorganization process. Current reorganization approaches provide support for changing elements of the organization according to the requirements of the problems that they consider. This support can be divided according to the elements that are allowed to change:

- **Open System** support, allow changes in the agent population.
- **Emergence** support, allow elements that define the organization behaviour to be changed; i.e., the roles that agents play, goals, etc. Approaches that support *emergence* allow operations such as addition or deletion of roles, goals, etc.
- **Agent Dynamics** support, allow changes related to the behaviour of the agents that populate the organization. For example, these involve changes in the capabilities offered by an agent in order to be able to play a role. These changes are dependent on the agents and not on the organization.
- **Functionality Dynamics** support, allow changes in the interactions among the elements of the organization, which affect the organization functionality. For example, this support could consider changes in the services offered by a role or changes in the roles that agents play.
- **Social Dynamics** support, allow changes in the relationships between elements of the organization, which affect the organization structure. For example, the work presented in [8] allows changes the *acquaintance* relationship between a pair of agents to a *superior-subordinate* relationship.
- **Norm Dynamics** support, allow changes in the regulations of the organization. For example, approaches that support *norm dynamics* allow modifications in the specificiation of the norms that govern the organization [1, 2] as well as the addition or deletion of new norms [14].

2.3 Selection

The selection phase defines the problem of choosing *which* reorganization is finally applied. If several reorganizations have been proposed in the design phase, the selection phase chooses which one of these proposals is applied. If a single agent or a central authority is responsible for this selection, we consider the selection phase to be *centralized*. In contrast, if several agents are involved in the selection phase (for example, by a negotiation process or by social choice), we consider the selection phase to be *distributed*. Depending on whether a single reorganization is designed or several designs are proposed, several criteria can be used in the design phase to guide the design, or in the selection phase to guide the selection. We classify these criteria depending on whether or not the following issues are taken into account:

- **Goal fulfillment** criterion takes into account the fulfillment of the organization goals in order to design or to select a reorganization. In some approaches, a reorganization may be needed since the organization goals are not fulfilled. In other approaches, the organization goals are assumed to be fulfilled, and, therefore, other criteria are considered.
- **Organization utility** criterion takes into account the utility of the organization increase in order to design or to select a reorganization.
- **Global cost** criterion takes into account the reorganization costs in order to design or select the reorganization. These costs are represented by global values for all the elements of the organization. For example, a *global cost* could define the cost of adding or deleting any agent or any role.
- **Individual cost** criterion also takes into account the reorganization costs in order to design or to select a reorganization. In contrast to *global costs*, *individual costs* are represented as specific costs for each specific change. For example, an *individual cost* could define the cost of adding a specific agent or changing the specific role that a specific agent is playing.

2.4 Evaluation

According to [10], reorganization methodologies should also encompass techniques for monitoring and controlling the system once the reorganization is deployed. The evaluation phase defines the problem of analyzing *how well* a reorganization has been performed. This problem provides feedback from the reorganization in order to assess whether the utility of the reorganization is what was expected, whether the reorganization costs are what were expected, etc. This phase provides information to evaluate the quality of the reorganization that was designed and selected in order to take it into account for future reorganizations. This increases the quality of future reorganizations. This phase also provides information for considering other reorganization alternatives if the one applied does not meet expectations.

3 Analysis of Reorganization Approaches

In Table 1, we present a detailed comparison of the most relevant reorganization approaches for Multiagent Systems according to each one of the phases described in Section 2. This table shows that the monitoring phase has been considered in almost every approach. Few works consider reoranization associated to events such as the work of Deloach and Matson [3]. However, most of the approches focus the reorganization in order to maximize the organization process. Only the approaches proposed by Weyns et al. [15] and Tinnemeier et al. [14] proposes a reorganization approach that focuses on the domain, i.e., reorganization caused by changes in the environment. The table also shows that the monitoring phase is usually implemented in a distributed way, providing the specific approach with the capability for several agents to be able to detect that a reorganization process is required.

The design phase is the most frequently studied one in current reorganization approaches. Every analyzed approach provides mechanisms for proposing a reorganization solution; however, changes in the elements of the organization are not supported to the same extent. It can be observed that the elements that are mainly considered to be changed in a reorganization process are related to the *functional* and *social* dynamics. As an example, the work of Kota et al. [8] and the work of Mathieu et al. [11] consider a reorganization process that is based on changing the relationships between the agents of the system, whereas the work of Nair et al. [12] and the work of Hoogendoorn and Treur [5] consider reorganization based on changing the roles played by the agents that populate the organization. This support is related to an organization adaptation and does not take into account the addition or deletion of elements.

The selection phase is also provided by current approaches. Only a few approaches consider a distributed selection of the reorganization. The works of Hübner et al. [7] and Horling et al. [6] propose approaches in which several agents are involved in the selection of the reorganization. Thus, current approaches are mainly focused on proposing a single reorganization solution, which is finally selected. The criterion that has mainly been taken into account in the selection phase is the solution that maximizes the organization utility. However, this utility only takes into account the benefits provided by the reorganization process and not the costs for achieving it. Only Nair et al. [12] use a reward-taking policy to assess the utility of the reorganization process by taking into account the benefits of a role reallocation process and the cost associated to it. The work of Hübner et al. [7] uses a global cost associated to the number of roles and missions deleted to select a solution.

Very little effort to study the evaluation phase has been carried out so far. Therefore, it is assumed that the reorganization that is designed and selected is going to work properly. Only the work of Hübner et al. [7] measures the success level of the reorganization and incorporate it into consideration for future reorganizations by means of a Q-learning technique.

		Deloach[3]	Hübner[7]	Kota[8]	Bou[1]	Mathieu[11]	Carvalho[2]	Weyns[15]	Horling[6]	Nair[12]	Hoogendoorn[5]	Tinnemeier[14]
Monitoring	Focus	Event	Event/ Utility	Utility	Utility	Utility	-	Domain	Utility	Utility	Event	Domain
	Type	Reactive	Proactive	Proactive	Proactive	Proactive	-	Reactive	Proactive	Proactive	Reactive	Proactive
	Form	Cent.	Dist.	Dist.	Cent.	Dist.	-	Dist.	Dist.	Cent.	Dist.	Dist.
Design	Form	Cent.	Dist./ Aut.	Dist./ Aut.	Cent.	Dist./ Aut.	Cent.	Dist./ Agree.	Dist./ Aut.	Cent.	Dist.	Dist/ Aut.
	Emergence	√	√	-	-	-	-	-	-	-	-	-
	Open System	√	√	-	-	√	-	√	-	-	-	-
	Agent Dynamics	√	-	-	-	√	-	-	-	-	-	-
	Functional Dynamics	√	√	-	-	-	-	√	-	√	√	-
	Social Dynamics	-	√	√	-	√	-	√	√	-	-	-
	Norm Dynamics	-	√	-	√	-	√	-	-	-	-	√
Selection	Form	-	Dist.	-	-	-	-	-	Dist.	-	-	-
	Goal fulfillment	Design	Design	-	-	-	-	-	Design	-	Design	-
	Org. Utility	Design	Selection	Design	Design	Design	-	-	Design	Design	-	-
	Global Costs	-	Selection	-	-	-	-	-	-	-	-	-
	Individual Costs	-	-	-	-	-	-	-	-	Design	-	-
Evaluation		-	√	-	-	-	-	-	-	-	-	-

Table 1 Comparision of reorganization approaches. (*Dist=Distributed*; *Cent=Centralized*; *Aut=Autonomous*; *Agree=Agreement*)

4 Open Issues

Current reorganization approaches present some gaps in the support provided for each phase of a reorganization process. Most of the current reorganization approaches focus a reorganization on maximizing the utility. As stated in [4], a reorganization process should provide some kind of increase in utility. However, as far as we are concerned, this utility should take into account not only the gain in utility but also the cost of achieving the new organization. In a real world scenario, for example a business enterprise, the best reorganization may require firing employees who do not obtain high sales and hiring new employees that do obtain high sales. Nevertheless, this reorganization could be unachieveable since it is not feasible to fire and hire employees without taken into account the costs of this reorganization. Similarly, in Multiagent Systems, agents are active entities that are interacting and exchanging information with each other. Therefore, changes in the organization (norms, roles, interactions between agents, etc.) have costs since the agents of the organization might stop attending to the requests that they have received, might include new services to be provided, or might require new regulations to be applied to their current interactions. Otherwise, the system would has to stop and reactivate once the changes are applied. Therefore, we consider that the monitoring, design and selection phases of a reorganization model should also include a computation of the cost associated to the reorganization process. In order to make this computation accurate, the costs should be represented in the most specific way as possible. As stated in [9], human organizations may encounter problems when certain changes are required: they often take longer than expected and desired; the cost of managerial time may increase; and there may be resistance from the people involved in the change. Similarly, in Multiagent Systems, not every agent is able to change its role at the same cost (for example, the cost for an agent to change its role will not be the same if an agent is acting alone or is interacting with other agents). Nor can every new norm be added at the same cost (for example, some norms may affect every agent of the organization and other norms may only affect a few agents).

As stated in [10], social factors in the organization in Multiagent Systems will become increasingly important in an open and dynamic online world. This relates to the support for agents to be able to enter and leave societies at different times and properly assign roles, rights, and obligations. Thus, support for *open system*, *emergence*, and *agent dynamics* must be considered in reorganization models. In current reorganization approaches, there is little support to allow *agent dynamics*. This means that the skills of agents are static and are not considered to evolve over time. In other words, agents are not able to learn new capabilities or to degrade the utility of the capabilities that they offer. The phases of monitoring, design and selection cannot be improved without an evaluation once the reorganization is applied. Therefore, in current approaches it is hard to measure the utility of a future instance of an organization without considering an evaluation process which accurately assesses whether or not the final utility is what it should be, and whether or not the reorganization process has been applied in the space time that was expected. This evaluation would allow the quality of the reorganization proposed to be gradually improved. The evaluation phase would be related with the monitoring phase by providing valuable information regarding the success of a reorganization application. Efforts related to these issuess are important topics for future research and will provide reorganization models, mechanisms, and tools, that facilitate their application.

Acknowledgments. This work has been partially supported by CONSOLIDER-INGENIO 2010 under grant CSD2007-00022, the European Cooperation in the field of Scientific and Technical Research IC0801 AT, and projects TIN2008-04446, TIN2009-13839-C03-01 and PROMETEO/2008/051. Juan M. Alberola has received a grant from Ministerio de Ciencia e Innovación de España (AP2007-00289).

References

1. Bou, E., López-Sánchez, M., Rodríguez-Aguilar, J.A.: Towards self-configuration in autonomic electronic institutions. In: Noriega, P., Vázquez-Salceda, J., Boella, G., Boissier, O., Dignum, V., Fornara, N., Matson, E. (eds.) COIN 2006. LNCS (LNAI), vol. 4386, pp. 220–235. Springer, Heidelberg (2007)
2. Carvalho, G., Almeida, H., Gatti, M., Vinicius, G., Paes, R., Perkusich, A., Lucena, C.: Dynamic law evolution in governance mechanisms for open multi-agent systems. In: Second Workshop on Software Engineering for Agent-oriented Systems (2006)
3. Deloach, S.A., Matson, E.: An organizational model for designing adaptive multiagent systems. In: The AAAI 2004 Workshop AOTP, pp. 66–73 (2004)
4. Dignum, V., Dignum, F., Sonenberg, L.: Towards dynamic reorganization of agent societies. In: Proc. of Workshop on Coordination in Emergent Agent Societies, pp. 22–27 (2004)
5. Hoogendoorn, M., Treur, J.: An Adaptive Multi-agent Organization Model Based on Dynamic Role Allocation. In: Proc. of the IAT 2006, pp. 474–481 (2006)
6. Horling, B., Benyo, B., Lesser, V.: Using self-diagnosis to adapt organizational structures. In: Proc. of the 5th Int. Conf. on Autonomous Agents, pp. 529–536. ACM Press, New York (1999)
7. Hübner, J.F., Sichman, J.S., Boissier, O.: Using the MOISE+ for a Cooperative Framework of MAS reorganisation. In: Bazzan, A.L.C., Labidi, S. (eds.) SBIA 2004. LNCS (LNAI), vol. 3171, pp. 506–515. Springer, Heidelberg (2004)
8. Kota, R., Gibbins, N., Jennings, N.R.: Decentralised Structural Adaptation in Agent Organisations. In: Organized adaptation in Multi-Agent Systems, pp. 54–71 (2009)
9. Kotter, J., Schlesinger, L.: Choosing strategies for change. Harvard Business Review, 106–114 (1979)
10. Luck, M., McBurney, P., Shehory, O., Willmott, S.: Agent Technology: Computing as Interaction (A Roadmap for Agent Based Computing). University of Southampton (2005)
11. Mathieu, P., Routier, J.C., Secq, Y.: Dynamic organization of multi-agent systems. In: Proc. of the AAMAS 2002, pp. 451–452 (2002)
12. Nair, R., Tambe, M., Marsella, S.: Role allocation and reallocation in multiagent teams: towards a practical analysis. In: Proceedings of the second AAMAS 2003, pp. 552–559 (2003)
13. pa So, Y., Durfee, E.H.: An organizational self-design model for organizational change. In: AAAI 1993 Workshop on AI and Theories of Groups and Oranizations, pp. 8–15 (1993)
14. Tinnemeier, N., Dastani, M., Meyer, J.-J.: Programming norm change. In: Proc. of the 9th Int. Conference on Autonomous Agents and Multiagent Systems, pp. 957–964 (2010)
15. Weyns, D., Haesevoets, R., Helleboogh, A., Holvoet, T., Joosen, W.: The MACODO middleware for Context-Driven Dynamic Agent Organzations. ACM Transaction on Autonomous and Adaptive Systems (2010)

Pervasive Healthcare Using Self-Healing Agent Environments

Stefano Bromuri, Michael Ignaz Schumacher, and Kostas Stathis

Abstract. Pervasive healthcare systems (PHSs) are required to be constantly available to the patients accessing them. To address this issue, in this paper we present an agent-based PHS that self-heals one or more of its parts when a service disruption happens. We propose a multi-agent system (MAS) approach that utilises coordination, planning and the notion of agent environment to create a distributed system capable to heal itself even if 50% of the system is not functioning due to external causes.

Keywords: Multi-Agent Systems, Self-Healing, Planning, Pervasive Healthcare.

1 Introduction

Pervasive Healthcare [12] focuses on bringing healthcare everywhere, breaking the boundaries of hospital healthcare. In [12] Varshney defines Pervasive Healthcare Systems (PHSs) as complex systems where multiple components interact to allow large scale monitoring of physiological data of heterogeneous patients. Two main limitations affect existing PHSs: (a) failures of the distributed system are never taken into consideration and (b) the system topology is statically defined. Consequently, PHSs need fault tolerance mechanisms as system downtimes may be dangerous for patients relying on them. Space-based redundancy [5], a practice that improves the resilience of an infrastructure by replicating components, should be avoided as it is

Stefano Bromuri · Michael Ignaz Schumacher
Business Information Systems, University of Applied Sciences Western Switzerland, TechnoArk 3, CH-3950, Sierre, Switzerland
e-mail: {stefano.bromuri,michael.schumacher}@hevs.ch

Kostas Stathis
Department of Computer Science, Royal Holloway University of London, Egham Hill, EGHAM, TW20 0EX
e-mail: kostas.stathis@rhul.ac.uk

J.B. Pérez et al. (Eds.): Highlights in PAAMS, AISC 89, pp. 159–166.
springerlink.com

an expensive practice for distributed systems like PHSs. In this paper we address PHSs fault tolerance via the self-healing paradigm [7] that focuses on time-based redundancy [5], which, instead of replicating components, replicates components behaviours to ensure that if a component fails an existing component can substitute it. Mikic-Rakic et al. in [9] identified the following properties as necessary for self-healing systems: adaptability, dynamicity, awareness, observability, autonomy, robustness, distributability, mobility and traceability.

Multi agent systems (MASs) [15] represent a valid abstraction to model such systems and fulfil the self-healing paradigm requirements. In particular, the adoption of MASs facilitates the transition from a centralised computing model to a decentralised one where thousands of autonomous agents interact to achieve a common goal. Moreover, the concept of agent environment [14] has been accepted as a useful abstraction to mediate the interaction between agents and to model how the agents perceive resources and interfaces that they utilise in their interaction.

In [1] we proposed a PHS based on a distributed agent environment built on the GOLEM[1] platform [2], to support intelligent agents monitoring patients affected by diabetes. In this system we mapped the agent environment represented as a distributed rectangular grid to a real environment representing a city. In this paper we extend such a PHS by defining a novel coordination and planning algorithm that agents use to detect faults and recover the system functionalities. The contributions of this paper are: a) we introduce a practical approach based on agents to handle fault tolerance in PHSs; b) we split responsibilities between the agents and the agent environment thus simplifying the behaviour of the agents to fulfil the requirements of self-healing systems; c) we illustrate how a compact declarative specification for the agents behaviour can deal with multiple failures of the agent environment in parallel.

The rest of this paper is structured as follows: Section 2 describes the self-healing system we developed in terms of its main components; Section 3 evaluates our approach and discusses relevant related work; finally Section 4 concludes this paper and presents future directions.

2 Extending GOLEM with Self-Healing Procedures

In this paper we extend the PHS presented in [1] with self-healing procedures. The system presented in [1] is based on the GOLEM agent platform [2] whose main abstractions are agents, cognitive entities, objects, reactive entities available to the agents as resources, and containers, declaratively programmed distributed spaces where agents and objects are situated, defining a distributed agent environment. Fig. 1 shows the conceptual architecture of our extended system in relationship to its distributed topology.

We introduced an infrastructure agent in the containers of the distributed agent environment and mediation rules to mediate the messages exchanged by the agents,

[1] GOLEM stands for Generalised Onto-Logical Environments for Multi-Agent Systems.

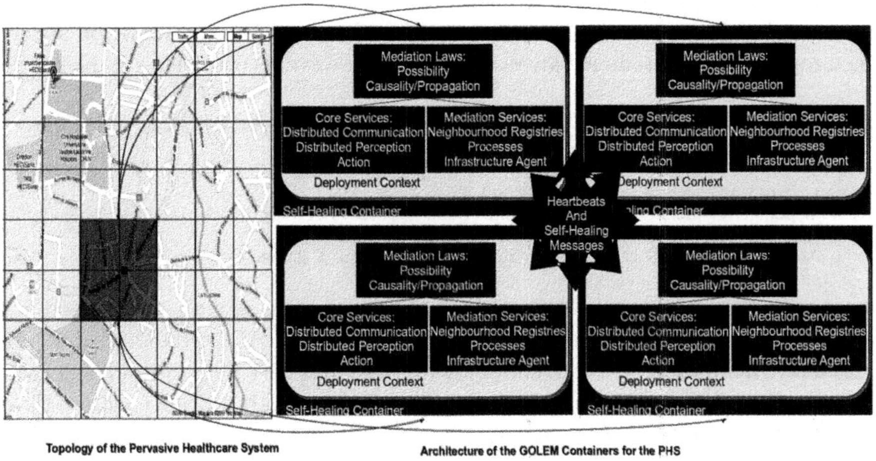

Fig. 1 Division in Containers of PHSs(on the left) and System Logic Architecture (on the right).

to propagate the events produced by the infrastructure agents and to modify the agent environment topology when a disruption takes place. The infrastructure agents use the containers interfaces to perceive topological changes and cover for dead neighbours, while the rules take care of modifying the known neighbourhood by updating the neighbourhood registries. Every GOLEM container in this setting is connected with a neighbourhood and every container shares the same architecture with the neighbour containers, following the pattern of time-based redundancy. The rationale is a system that fulfils the requirements of self-healing systems: our system adapts to changes in the topology, it is observable as we use a declarative approach to describe our entities, it is autonomous and aware as we use planning agents to deal with the failures by coordinating in a distributed environment.

2.1 *The Containers Behaviour*

We specify the observable state of the GOLEM entities and of events as C-logic structures. C-logic [3] is a convenient formalism to express complex structures as logical objects, and it has a direct translation to the Ambient Event Calculus [2] (AEC), a formalism that handles the evolution in time of logical objects by means of events in distributed settings. In our GOLEM-based PHS, we express the state of a container by means of the following C-logic structure and AEC translation:

```
container:c1[position⇒{Latitude,Longitude},side⇒50,state⇒up,neighbours⇒{container:c2,container:c3}]
                                              ⇓
happens(ev1,0). instance(c1, container,start(ev1)). object(c1,side,50,start(ev1)).
object(c1,position, Latitude,start(ev1)). object(c1,position, Longitude,start(ev1)).
object(c1,state,up,start(ev1)). object(c1, neighbour, c2,start(ev1)). object(c1, neighbour, c3,start(ev1)).
```

which means that a container represents a real world location in terms of latitude `Latitude` and longitude `Longitude`, its state is `up`, it covers a square that has a side of 50 meters and it has two neighbours c2 and c3. GOLEM containers are programmed declaratively, using the AEC formalism. Through happens/2 predicates (the 2 represents the predicate arity) we specify how events take place in the AEC:

R1) happens(ack:Event [receiver ⇒ Container], T) ←
happens(Event[actor ⇒ Agent, receivers ⇒ Containers, known neighbours ⇒ NeighbourList], T),
member(Container, Containers), holds_at(Container, neighbour, this, T).

R2) happens(inform:Event[receiver⇒ Container], T) ←
happens(Event[actor ⇒ Agent, receivers⇒SubList, cover ⇒CBroken, randomvalue ⇒ Diceroll], T),
holds_at(this, neighbourhood_list, List, T), subset(List, SubList),
member(Container, SubList), holds_at(Container, neighbour, CBroken,T).

where the happens/2 is an AEC predicate stating that an event has happened in a container of the agent environment and holds_at/4 is an AEC predicate, that provides an attribute value given an entity identifier (in this case this represents the current container), the attribute name (in this case neighbourhood_list) and the time. R1 mediates an ack event produced by an agent during the PHS normal behaviour and it states that whenever such an event happens, then this also happens in the containers within the Containers list. The happens/2 has the function of replicating in the neighbourhood of a container an event happened locally through another happens/2. Inside the ack message, there is also the known neighbour containers list at a given time, so that every agent in the distributed topology can have knowledge of the neighbours of their direct neighbours. This is similar to the successor list of the CHORD P2P algorithm [10], where given a successors list of length r, and a disruption probability p for a single node, then the CHORD ring disruption probability is p^r, meaning that the ring resilience can be improved by increasing the successors list length. In our case, the probability that the PHSs cannot restore the area covered by a node is p^8, when keeping a list of neighbours of neighbours in a grid like topology, where if needed the resiliency of our PHS can be improved by increasing the neighbourhood knowledge.

R2 mediates disruptions happening in the distributed settings. Once an agent fails to send an ack events to a neighbour container that is down due to external causes, the agent sends an inform event to all the containers that have a neighbouring relationship with the unresponsive container and it starts the healing procedure that we will discuss later. Additional predicates have been defined to update the neighbours list when a communication fails and when a neighbour is substituted by another container. For the moment, the tasks of joining a network and redeploying a failed container are handled by a human actor. We will address these issues in future work.

2.2 *The Infrastructure Agents*

A GOLEM agent consists of a declarative module embedded in an agent body, which is situated in a container to perceive the events happening in it. The infrastructure agent cognitive model is based on two cycles, one to process the events sensed by the body and one to plan and act in the environment. The pseudo code for the two agent mind cycles is reported here (CSP stands for Conditional-STRIPS-Planner, an extension of the STRIP planner [4] to handle conditional plans):

procedure ACTING-Cycle(time)
  static:KB, a knowledge base; ACTION-QUEUE, a queue of actions accessible by the agent body;
    $p_1, p_2 \ldots p_k, where \forall p_i, p_i \in Plans \subset KB$, a set of plans;
  $currentstate \leftarrow$ STATE-DESCRIPTION(KB,time); $goal \leftarrow$ NEXT-GOAL($currentstate,time$);
  if $\nexists p_i | p_i.goal = goal$ **then** $p_k \leftarrow$ CSP($currentstate,goal$); ADD(Plans, p_k);

  $p_{exec} \leftarrow$ NEXT-EXECUTABLE-PLAN(Plans,$time$),

  if($p_{exec} = nil$) **then** NOW(timenew); ACTING-Cycle(timenew);
  else $currentact = p_{exec}.nextact$,
    if(CONDITIONAL?($currentact$)) **then**
      if(CHECK-KB(KB,IF-PART[$currentact$]))
        then $p_{exec} \leftarrow$ THEN-PART[$currentact$]; ADD(Plans, p_{exec}); NOW($tnew$); ACTING-Cycle($tnew$);
      else $p_{exec} \leftarrow$ ELSE-PART[$currentact$]; ADD(Plans, p_{exec}); NOW($tnew$); ACTING-Cycle($tnew$);
    else ADD-ACTION(ACTION-QUEUE, $currentaction$); NOW($tnew$); ACTING-Cycle($tnew$);

procedure PERCEPTION-Cycle($time$)
  static:KB; PERCEPTION-QUEUE; $p_1, p_2 \ldots p_k where \forall p_i, p_i \in Plans \subset KB$;
  $percept \leftarrow$ NEXT-PERCEPT(PERCEPTION-QUEUE, $time$);
  UPDATE-KB(KB, $percept, time$); NOW($timenew$), PERCEPTION-Cycle($timenew$)

The PERCEPTION-Cycle/1 reads the PERCEPTION-QUEUE for percepts coming from the environment. Such percepts are used to update the knowledge base KB about the state of the containers that are known in the agent environment. A plan is represented in the agent mind as a C-logic object. For example, the following plan expressed in AEC:

```
plan:p1[goal ⇒ cover:g1[container⇒ c2], diceroll ⇒ 3000,next_action ⇒ ac1, delay_action ⇒ 0,
  sequence ⇒ {inform:ev4[cover ⇒ c2, diceroll ⇒ 3000],wait:ev5[delay⇒ 6], if_then_else:if1 }].

if_then_else:if1[if ⇒ check_winner(ag1,p1,3000),
                 then ⇒ { modify_topology:ev6[cover⇒ c2], end_plan:ev8 }, else ⇒ { end_plan:ev9 } ]

check_winner(A,P, Diceroll)← now(Time),
     not (holds_at(P,competitor, Comp,Time),holds_at(Comp, diceroll, Diceroll*,Time), Diceroll* >Diceroll).
```

specifies a plan p1, with a goal g1 to cover for a container c2. In this plan the actions performed are: an inform event ev4, then a wait ev5 action with a 6 seconds delay, to attend for messages incoming from other agents. These actions are pushed by the ACTING-Cycle/1 in the ACTION-QUEUE/2 of the agent body, that produces them in the container, which then mediates the actions according to the rules previously defined. The inform event is sent to the neighbourhood of the dead container. Inside this event there is a random value diceroll that is used by the agents to compete for the coverage of a container (in this case c2). After waiting for the delay, the planner finds an if1 object as the next action to perform. This is a C-logic object representing an if-then-else structure, handled by the CONDITIONAL?/1 predicate in the planner, that checks for the if condition check_winner/3 in the if-then-else structure. The planner then executes either the then sequence of actions or the else sequence of actions according to the result of the check_winner/3 condition. If agent ag1 is the competition winner, it covers for the dead container and ends the plan, otherwise the plan is destroyed. Furthermore the NEXT-EXECUTABLE-PLAN/2 predicate distinguishes between plans that have been frozen due to the introduction of a delay, and plans whose execution can continue, allowing for plans with different goals to be executed in parallel. Despite the communication taking place between the agents, two or more containers may end up covering the same area. When the inconsistency is perceived, the interested agents start a resolution protocol similar to the one to cover for a dead neighbour. High level planning agents have the advantage of a compact definition of the agent mind as plans are structures that can be instantiated and destroyed according to the interaction state, while reactive agents would require a low level verbose agent mind, that is difficult to debug in distributed settings.

3 Evaluation and Related Work

To evaluate our system we deployed a 10x10 grid of self-healing GOLEM containers in a dual core Intel Centrino 2, 2.66 Ghz with 3Gb of RAM.

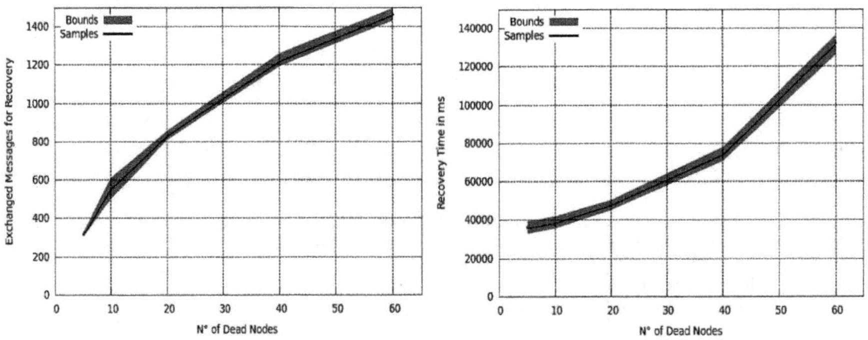

Fig. 2 Performance Evaluation of a 10x10 Pervasive Agent Environment

A PHS should be distributed on a grid and deploying it on a single host gives only a partial view of the real performances of the system. In real settings we foresee that some of the containers will require more resources when representing hot spots (i.e. super markets or hospitals), while unpopulated areas will require less resources. This is a matter of future work, but we recognise the need to study the system in real settings. Still, this evaluation offers important insights about the system performances. As evaluated by Urovi et al. in [11], a GOLEM container can support up to 50 users, meaning that a 100 containers grid, as in this evaluation, supports up to 4000-5000 users. This is a meaningful scale for a real system deployed in medium sized cities where the number of patients needing constant monitoring is on the order of hundreds. We took an average of 10 samples for each of the two curves in Fig. 2, that evaluate respectively the number of messages to recover from failures with an increasing number of containers and the total downtime with respect to the number of dead containers. These are two critical parameters: if too many messages are exchanged this can impact the PHS performances, and if there are long downtimes, emergencies may happen when the system is unavailable. When producing the curves in Fig. 2 we made these assumptions: a) since the system is deployed in a single host, the delays of a real network are not taken into consideration b) we assumed that the containers die all at the same time, that is a pessimistic assumption, as the probability of this happening is very low; c) the system presents failures after every container had time to learn about its neighbours.

The first part of Fig. 2 has a logarithmic behaviour because the more nodes in a neighbourhood die, the less nodes take part to the competition for covering a dead node, producing less messages. Consequently, the second part of Fig. 2 behaves like a quadratic curve as the alive infrastructure agents have to execute more plans to cover a bigger area. Finally, the introduction of parallel plans helps agents to minimise the downtime as whenever a dead neighbour is detected, a new plan is instantiated to cover it. Also, introducing agent communication, mediated by the agent environment, allows to minimise a) the number of messages exchanged in the environment b) the uncontrolled growth of the area controlled by a container and c) the conflicts arising in the healing protocol.

Self-healing is a important topic within the distributed system community. Mikic-Rakic et al. in [9] propose the PRISM model for self-healing and fault tolerance. Such a model has a set of components connected by means of communication ports that exchange synchronous and asynchronous events and that rely on metal-level components for the self-healing procedures. With respect to PRISM, the infrastructure agents are meta-level components communicating and sharing knowledge about the topology of the distributed system. From the stand point of self-healing systems, in [6] Selvin et al. and Kondacs in [8], propose to utilise a bio-inspired approach to rebuild geometric shapes. These works demonstrate that using nature inspired models allows to have a very resilient service capable of self-healing 99% of its components. Our approach is similar to the one proposed in [8] and [6] except that we have further constraints on the number of messages exchanged and on the recovery time. Another contribution that uses agents is the one of Haesevoets et al. in [13], where the MACODO system defines laws in the agent environment to handle the consistency of the agent roles. As in [13], we separate the concerns of handling self-adaptation between the agent environment and the agents by means of cognitive agents dealing with the changes of the environment.

4 Conclusion and Future Works

In this paper we presented a pervasive healthcare system where agents reorganise the agent environment to self-heal from a fault of one or more of the containers composing it. We utilise planning agents that can reason in parallel about multiple faults and that produce plans and interact to cover for missing containers in the environment. The novelty of the approach resides in using planning agents combined with a complex declarative agent environment that simplifies the interaction between the agents controlling the distributed system. We evaluated the system downtime and the number of message exchanged to recover from a growing number of dead containers, discovering that the system scales up and it can recover in useful time even when more than 50% of the system is down. Future works include deploying the system in real setting and testing it with real users as well as extending the algorithm to define the topology of the environment dynamically at deployment time. Another interesting issue for future work is how to deal with patients roaming in a distributed network that is self-healing from a disruption.

References

1. Bromuri, S., Schumacher, M.I., Stathis, K.: Towards distributed agent environments for pervasive healthcare. In: Proceedings of the Eighth German Conference on Multi Agents System Technologies, MATES 2010 (2010)
2. Bromuri, S., Stathis, K.: Distributed Agent Environments in the Ambient Event Calculus. In: DEBS 2009: Proceedings of the Third International Conference on Distributed Event-Based Systems. ACM, New York (2009)
3. Chen, W., Warren, D.S.: C-logic of Complex Objects. In: PODS 1989: Proceedings of the Eighth ACM SIGACT-SIGMOD-SIGART Symposium on Principles of Database Systems, pp. 369–378. ACM Press, New York (1989)
4. Fikes, R., Nilsson, N.J.: STRIPS: A New Approach to the Application of Theorem Proving to Problem Solving. In: IJCAI, pp. 608–620 (1971)
5. Gärtner, F.C.: Fundamentals of fault-tolerant distributed computing in asynchronous environments. ACM Comput. Surv. 31, 1–26 (1999)
6. George, S., Evans, D., Davidson, L.: A biologically inspired programming model for self-healing systems. In: WOSS 2002: Proceedings of the First Workshop on Self-Healing Systems, pp. 102–104. ACM, New York (2002)
7. Ghosh, D., Sharman, R., Rao, H.R., Upadhyaya, S.: Self-healing systems – survey and synthesis. Decision Support Systems 42(4), 2164–2185 (2007); Decision Support Systems in Emerging Economies
8. Kondacs, A.: Biologically-inspired self-assembly of two-dimensional shapes using global-to-local compilation. In: IJCAI 2003: Proceedings of the 18th International Joint Conference on Artificial Intelligence, pp. 633–638. Morgan Kaufmann Publishers Inc., San Francisco (2003)
9. Mikic-Rakic, M., Mehta, N., Medvidovic, N.: Architectural style requirements for self-healing systems. In: Proceedings of the First Workshop on Self-Healing Systems, WOSS 2002, pp. 49–54 (2002)
10. Stoica, I., Morris, R., Liben-Nowell, D., Karger, D.R., Kaashoek, M.F., Dabek, F., Balakrishnan, H.: Chord: a scalable peer-to-peer lookup protocol for internet applications. IEEE/ACM Trans. Netw. 11, 17–32 (2003)
11. Urovi, V., Bromuri, S., Stathis, K., Artikis, A.: Towards runtime support for norm-governed multi-agent systems. In: Lin, F., Sattler, U., Truszczynski, M. (eds.) KR. AAAI Press, Menlo Park (2010)
12. Varshney, U.: Pervasive Healthcare Computing: EMR/EHR, Wireless and Health Monitoring. Springer Publishing Company, Incorporated, Heidelberg (2009)
13. Weyns, D., Haesevoets, R., Helleboogh, A., Holvoet, T., Joosen, W.: The macodo middleware for context-driven dynamic agent organzations. ACM Transactions on Autonomous and Adaptive Systems 5(1), 3.1–3.29 (2010)
14. Weyns, D., Omicini, A., Odell, J.: Environment as a first class abstraction in multiagent systems. Autonomous Agents and Multi-Agent Systems 14(1), 5–30 (2007)
15. Wooldridge, M.: MultiAgent Systems. John Wiley and Sons, Chichester (2002)

Supporting Dynamics Multiagent Systems on THOMAS

Alfonso Machado and Vicente Julian

Abstract. Nowadays, one of the goals of multi-agent systems (MAS) is to construct systems capable of autonomous, flexible decision-making and cooperate with other systems within a society. This society should take certain characteristics into account, such as distribution, constant evolution, or a flexibility that allows its members (agents) to enter and exit. Given the characteristics of these open environments, particularly their dynamism, it is essential to find a new approach to support the evolution of these systems and to facilitate their growth and update at execution time. In general, it is necessary to define the standards and platforms required for the interoperability and adaptability of the agents that meet these requirements. This article attempts to present THOMAS as a MAS architecture specifically addressed to support dynamic agent societies.

Keywords: Multi-Agent Systems, Virtual Organizations, Adaptive System, Dynamic System.

1 Introduction

Nowadays there is a clear trend towards using methods and tools that can develop multi-agent systems (MAS) capable of performing dynamic self-organization when they detect changes in the environment. Virtual Organizations (VO) [4] are a means of understanding system models from a sociological perspective. It is based on the principles of cooperation among businesses within a shared network. It takes advantage of the distinguishing elements that provide flexibility and a quick response capability and these form a strategy aimed at customer satisfaction. Virtual

Alfonso Machado · Vicente Julian
Departamento de Sistemas Informáticos y Computación,
Universidad Politécnica de Valencia, Camino de Vera S/N 46022, Valencia (Spain)
e-mail: {amachado, vinglada}@dsic.upv.es
http://www.dsic.upv.es

J.B. Pérez et al. (Eds.): Highlights in PAAMS, AISC 89, pp. 167–174.
springerlink.com

Organizations have been usefully employed as a paradigm for developing agent systems [2] [4] One of the advantages of organizational development is that systems are modeled with a high level of abstraction, so the conceptual gap between real world and models is reduced. In addition this kind of system offers facilities to implement open systems and heterogeneous member participation [10]. Organizations should not only be able to describe structural composition (i.e. functions, agent groups, interaction and relationship patterns between roles) and functional behavior (i.e. agent tasks, plans or services), but should also be able to describe behavioral norms of agents, the entry and exit of dynamic components, and formation, which is also dynamic, for agent groups. Given the characteristics of these open environments, particularly their dynamism, it is essential to find a new approach to support the evolution of these systems and to facilitate their growth and update at execution time. According to this need, this paper presents how THOMAS (MeTHods, Techniques and Tools for Open Multi- Agent Systems) [3] [6], a MAS architecture for Virtual Organizations, is specifically addressed to support dynamic agent organizations.

2 Dynamic on Multiagent Systems

The Organizational Dynamic is related to the input/output of agents, with the acquirement of roles by part of the agents, the creation of groups and with the management of behavior. In the definition of the dynamic of an organization we must specify: (i) with respect to the entry of agents: when agents are allowed to enter the organization; what their position will be in the organization; the process of expulsion of agents with anomalous behaviors; (ii) with respect to the acquirement of roles: how the agents will adopt a specific role; the association of agents with one or more roles; (iii) with respect to the dynamic creation of groups: the definition of federations, coalitions, congregations, etc and (iv) finally, with respect to the management of behavior: how control the compliance of agents's behavior with the rules of society. For a society, the idea of adaptation is its capacity to get involved with the environment and to be part of this symbiosis which allows to work together. In order to support the adaptation of an agent society, we need mechanisms to allow the dynamism of the entities of the MAS that represents this society. Particularly, we need:

- *Organization Dynamics Formation*. It means that the system allows to create and delete organizations. Theses operations do not have to be a drawback in the evolution of the system, or it was needed to relaunch the system. The origin of these changes could be owed to an agent wants to create a new organization, or that the system believes that organization is not useful any more, or is empty.
- *Services Requirements Discovery*. it is allowed to modify and make changes to the offered services. In the same way new services are advertised or are composed by a set of services, it must be possible to eliminate them dynamically without changing the running system.

- *Norm Management.* A Multiagent System was completely dynamic if it supports dynamism in the management of norms. Management of norms means that it can be added, deleted or modified norms in the system and if any norm change or modify, the system has to be able to adapt to changes.

None of existing approaches [2][5][7][8][9] is capable of coordinating tasks by member agents of the organization for solving a common problem. Moreover, they do not consider that task planning should adapt to change in the environment either. The architecture that we show in this study is OVAMAH (Multiagent-based Adaptive Virtual Organizations). OVAMAH is based on THOMAS, which focuses on defining structure and norms.

3 Dynamic on THOMAS

In order to model open and dynamic virtual organizations, it becomes necessary to have a different infrastructure. This infrastructure should use agent technology in the development process and to apply decomposition, abstraction. In addition, while keeping in mind all of the requirements cited in the previous section. THOMAS is the name given to an open multi-agent architecture. It is based on a services oriented approach and primarily focuses on the design of virtual organizations. The architecture is essentially formed by a set of services that are modularly structured. THOMAS consists of three main components:

- *Service Facilitator* (SF), it provides a place where autonomous entities can register service descriptions as directory entries.
- *Organization Manager Service* (OMS), it is responsible for specifying and administrating its structural components (role, units and norms) and its execution components (participating agents and the roles they play, units that are active at each moment).
- *Platform Kernel* (PK), it provides the basic services on a multi-agent platform system and incorporates mechanisms for transporting messages that facilitate interaction among agents.

The advantages of using these components for a dynamic and decentralized open system are: (i) Allowing dynamic: new organizations can be created at execution time, allowing the development of MAS which emerge or change dynamically. Moreover, the organizations may also be destroyed when its purpose is reached; (ii) Improving the way emergent behaviors such as composed services may arise: new, relevant and complex services can be composed at runtime, composing the new registered services with the existing ones; TTT (iii) Improving the localization techniques and composition of services: entities may publish the services they demand (not only the ones they offer), due to the dynamic of an open system, when an entity arrives at the system and discovers that it is able to provide this demanded service, it registers as a provider. (iv) Allowing to express a normative control:

the OMS is in charge of controlling the role enactment process. It also stores all norms defined in the system and provides some services for adding or deleting norms. Below, we explain the methods that affect the creation of organizations, roles management, service registration and control of norms. All these methods allow dynamic in THOMAS.

3.1 Organizations

Organization in THOMAS are structured by means of organizational units. The organizational units represent groups of entities (agents or other units), which are related in order to pursue a common goal. These organizational units have an internal topology (i.e. hierarchical, team, plain), which imposes restrictions on agent relationships and control (ex. member, supervisor, subordinate). In THOMAS, a virtual unit has been defined in order to represent the "world" system in which agents participate by default. The OMS creates organizations within this virtual unit, by means of registering organizational units, which can also be composed of more units.

- **Add.** An agent execute this method for registering a unit in the OMS. The method adds a new unit to the list of active units which they can enter and exit new agents and new units may register that inherit from this one. If an agent wants to invoke this functionality, it may invoke the method Register Unit, as follows:

```
result = OMSservices.registerUnit("TourismMarket",
      "hierarchy", "Offer Tourism Services", "virtual");
```

- **Delete.** To delete a unit of the list of active units, only if the unit's role is `creator` and is performed by a single agent. Deleting a unit of the system implies that there has been a change in the structure of the system as an entity has been deleted due to some event, it is a clear example of the possibilities of dynamic systems. Deregister unit method would be invoke by an agent as follows:

```
result = OMSservices.DeregisterUnit("TourismMarket");
```

3.2 Roles

A role in THOMAS represents a position within the unit in which it is defined. It is related to some interaction norms, imposed by the unit structure and its specific position within the unit, as well as some behavior norms, which specify its functionality (services that it needs and offers), restrict its actions (prohibition, obligations and permissions) and establish the consequences of these norms (sanctions and rewards). A role has a number of features such as visibility (internal or external), position (member or supervisor) in a hierarchical or team unit and privacy (private or public). In THOMAS, a role can be dynamically added, deleted, acquired or leaved by an agent, how THOMAS allow do that are explained below.

- **Add.** Register a role within an unit adds a new role to the list of roles that an unit. This action allows agents to acquire a role within the organization which will be associated with a number of features as mentioned above. Its use is as follows:

```
result = OMSservices.registerRole("MarketManager",
"TourismMarket","external","supervisor","private","member");
```

- **Delete.** The method removes the role from the list of roles that is associated an unit. Since that time, other agents can not play that role. To perform this action, you do not have any agent who playes that role, or any rule referres to the role, because the system would be inconsistent. How to use the method, it is as shown as follows:

```
result = OMSservices.DeregisterRole("MarketManager");
```

- **Acquire.** To acquire a role within an organization means that the agent will perform this action may perform certain tasks associated with that role, and/or who will occupy a position within the unit. An agent may have multiple roles and can perform various tasks within an organization or even can have different levels in the system. The method is used as follows:

```
result = OMSservices.acquireRole("MarketManager",
                                  "TourismMarket");
```

- **Leave.** To leave a role, or not to play a role in an organization implies that an agent has decided to stop playing a role within a unit, and all involve it, such as to carry out some norms, or to offer some services. The use of this method is as follows:

```
result = OMSservices.leaveRole("member", "virtual");
```

3.3 Services

THOMAS services are implemented as semantic web services. Each service has a WSDL. This document describes network services as a set of endpoints operating on messages that contain either document-oriented or procedure-oriented information. The operations and messages are described abstractly and then bound to a concrete network protocol and message format to define an endpoint. Furthermore, for each service registered in the SF, there should be two OWL-S documents, one with the profile description and the other one with the process and grounding description. This division avoids redundant information in the SF whenever a service is provided by two providers. In this case, both providers offer the service with the same profile but different implementation. These OWL-S description facilitates the automation of web service tasks, including automated service discovery, execution, composition and interoperability. We store this information how we can see as follow:

```
ProfileDescription NewClientProfile = new ProfileDescription(
    "http://localhost:8080/.../NewClientProfile.owl",
    "NewClient");

ProcessDescription NewClientProcess = new ProcessDescription(
    "http://localhost:8080/.../NewClientProcess.owl",
    "NewClient");
```

- **Add Profile.** When an agent adds a profile description in the SF, this involves to add a new entry in the SF from the list of services needed, used or required by the system. To add a description of the service implies that you can find the service and finds out if someone offers it or not. When an agent finds the service can run it or add as a provider. To add a profile to the SF method is executed as follows:

```
result = SFservices.registerProfile(NewClientProfile);
```

- **Delete Profile.** This method is used when an agent decides to delete it because its no longer need, or because someone in the system no longer deemed necessary. This method remove a Profile Description of the list of services offered in the system. When someone deletes a Profile Description, anyone is not able to find any possible description of that service. Only you can delete the service if there are not providers. A possible example of its use is as follows:.

```
result = SFservices.deregisterProfile(NewClientProfile);
```

- **Add Process.** This method adds a service implementation. The service implementation is described in the associated profile which is provided by an agent that offers the service. The execution of this method involves that when you want to know who implements a particular service, you see the agent that has registered the Process. One possible use of this method would be as follows:

```
result = SFservices.registerProcess(NewClientProcess);
```

- **Delete Process.** This method removes an implementation (*Process Description*) of a service of SF, which was offered by an agent. When the method executes, it removes the agent that offers it and the implementation of the list of service providers associated with the Process Profile. How to use the method, it would be as follows:

```
result = SFservices.removeProvider(NewClientProcess);
```

- **Search Service.** This method is executed by an agent when it wants to find a service in the system. The search engine used is a search syntax. Search syntax engine finds all those services that are related with description of the service that the agent wants to request. Then, these services give back the agent who executed the search. The method returns a *Profile Description* list where each element of the list correspond to a type of service and description. An example of usage is as follows:

```
outputs = SFservices.searchService("NewClient");
if (outputs.size() == 0) { // Not find the service
} else { // The service is found }
```

3.4 Norms

The THOMAS architecture allows defining norms that prescribe agent rights and duties in terms of who can provide a service, when and under which circumstances. In addition, norms can be viewed as a coordination skill for organizing MAS, since they specify the desired behavior of the society members. Therefore, norms have been divided into: (i) organizational norms, related to services offered by the OMS to the members of the organization, so then establishing organizational dynamic such as role management; and (ii) functional norms that define role functionality in terms of services that can be requested/provided, service requesting order, service conditions and interaction protocols that should be followed.

- **Add Norm.** To add a new rule in the system. The method adds it to the list of active rules. When the rule is added, it must be fulfilled. The method is used as follows:

```
result = OMSservices.registerNorm("norm1"
  "FORBBIDEN ... acquireRole_MESSAGE(CONTENT(ROLE_'Payee'))");
```

- **Delete Norm.** This method removedes a rule from the active list of rules. The method is used as follows:

```
result = OMSservices.deregisterNorm("norm1");
```

4 Conclusions

This paper has explained how THOMAS is an architecture specifically addressed to support dynamic agent societies. This feature allows the evolution of MAS systems and facilitates their growth and update at runtime. This architecture represents a solid step towards obtaining truly dynamic virtual organizations[1]. The framework is available for download on the project's web-page[2]. It implements the whole set of services described in the architecture, with basic support for norm management. This version has been used to validate the feasibility of this approach and is being applied to different scenarios such as tourism, leisure activity management in a mall, and health emergencies.

[1] A detailed document of this specification can be downloaded from http://www.fipa.org/docs/THOMASarchitecture.pdf

[2] http://users.dsic.upv.es/grupos/ia/sma/tools/Thomas/

References

1. Artikis, A., Kaponis, D., Pitt, J.: Multi-Agent Systems: Semantics and Dynamics of Organisational Models. In: Dynamic Specications of Norm-Governed Systems, IGI Globa (2009)
2. Boissier, O., Gateau, B.: Normative multi-agent organizations: Modeling, support and control. In: Normative Multiagent Systems (2007)
3. Carrascosa, C., Giret, A., Julian, V., Rebollo, M., Argente, E., Botti, V.: Service Oriented MAS: An open architecture (Short Paper). In: Decker, Sichman, Sierra, Caste franchise (eds.) Proc. of 8th Int. Conf. on Autonomous Agents and Multiagent Systems (AAMAS 2009), Budapest, Hungary, pp. 1291–1292 (2009)
4. Ferber, J., Gutknecht, O., Michel, F.: From agents to organizations: An organizational view of multi-agent systems. In: Giorgini, P., Müller, J.P., Odell, J.J. (eds.) AOSE 2003. LNCS, vol. 2935, pp. 214–230. Springer, Heidelberg (2004)
5. Gasser, L., Ishida, T.: A dynamic organizational architecture for adaptive problem solving. In: Proc. of AAAI 1991, pp. 185–190 (1991)
6. Giret, A., Julián, V., Rebollo, M., Argente, E., Carrascosa, C., Botti, V.: An open architecture for service-oriented virtual organizations. In: Braubach, L., Briot, J.-P., Thangarajah, J. (eds.) ProMAS 2009. LNCS, vol. 5919, pp. 118–132. Springer, Heidelberg (2010)
7. Hales, D.: Group Reputation Supports Beneficent Norms. The Journal of Artificial Societies and Social Simulation (JASSS) 5(4) (2002)
8. Hubner, J., Sichman, J., Boissier, O.: S-Moise+:A middleware for developing organized multi-agent systems. In: Boissier, O., Padget, J., Dignum, V., Lindemann, G., Matson, E., Ossowski, S., Sichman, J.S., Vázquez-Salceda, J. (eds.) ANIREM 2005 and OOOP 2005. LNCS (LNAI), vol. 3913, pp. 64–78. Springer, Heidelberg (2006)
9. Lakkaraju, K., Gasser, L.: Norm Emergence in Complex Ambiguous Situations. In: Proceedings of the AAAI Workshop on Coordination, Organizations, Institutions and Norms AAAI, Chicago (2008)
10. Mao, X., Yu, E.: Organizational and social concepts in agent oriented software engineering. In: Odell, J.J., Giorgini, P., Müller, J.P. (eds.) AOSE 2004. LNCS (LNAI), vol. 3382, pp. 1–15. Springer, Heidelberg (2005)

Adaptive Multi Agent System for Guiding Groups of People in Urban Areas

Anais Garrell, Oscar Sandoval-Torres, and Alberto Sanfeliu

Abstract. This article presents a new approach for guiding a group of people using an adaptive multi agent system. For the simulations of the group of people we use social forces, with theses forces human motion is controlled depending on the dynamic environment. To get the group of people being guide we use a set of agents that work cooperatively and they adapt their behavior according to the situation where they are working and how people react. For that reason, we present a model that overcomes the limitations of existing approaches, which are either tailored to tightly bounded environments, or based on unrealistic human behaviors. In particular we define a Discrete-Time- Motion model, which from one side represents the environment by means of a potential field, and on the other hand the motion models for people and robots respond to realistic situations, and for instance human behaviors such as leaving the group are considered. Furthermore, we present an analysis of forces actuating among agents and humans throughout simulations of different situations of robot and human configurations and behaviors. Finally, a new model of multi-robot task allocation applied to people guidance in urban settings is presented. The developed architecture overcomes some of the limitations of existing approaches, such as emergent cooperation or resource sharing.

1 Introduction

The interest on developing social and cooperative agents has significantly increased throughout the recent years. In this work we present a new approach for guiding people in open areas of urban settings using multiple agents acting in a cooperative way. One of the agents is the *leader*, as a human tour-guide. It is placed at the front of the group and its role is to estimate the trajectory of both the people and the rest of agents. The other agents, called *shepherds*, are responsible for guiding the people,

Anais Garrell · Oscar Sandoval-Torres · Alberto Sanfeliu
Institut de Robòtica i Informàtica Industrial - Universitat Politècnica de Catalunya
e-mail: {agarrell,osandoval,sanfeliu}@iri.upc.edu

J.B. Pérez et al. (Eds.): Highlights in PAAMS, AISC 89, pp. 175–184.
springerlink.com

preventing any person leaving the group, and following the path given by the leader, considering in every instant people reactions using social forces [1].

Furthermore, in this research, we go one step ahead, presenting a method to optimize locally the tasks assignment to agents for doing their missions. Agents assignation are done by analyzing the minimum work required to do such task, where the function to minimize is based on one hand, by agents motion (which will be applied with robot on the future), and, on the other hand, by the impact of such motions on peoples displacement.

Moreover, an orientation where the main question is not about the division of tasks between agents is presented. In the developed approach the participation to solve a task is not limited to a single agent. Agents will try to participate in the tasks that give them more benefits, even when the task is already being done by someone else. In many cases the tasks can be performed by more than one agent. This feature has not been explored so far by other existing architectures.

This paper contents has been distributed as follows. We start presenting the representation model of the environment and people behavior. In section III the cost function for the rescuing people task is presented. In section IV the MRTA model we are presenting is applied to the task of people guidance. And last but not least, the results and conclusions are presented in sections V and VI respectively.

2 Modeling the Environment and People Motion

For modeling the environment where the agents (future robots) work, we have developed a model called Discrete Time Motion model (DTM) which has two components: The Discrete Time component and the Discrete Motion component. The former estimates position, orientation and velocity of the robots and persons, and the position of the obstacles at a time instance k. It will be used to estimate the intersection of the people with the obstacles and detect if someone is leaving the group. The Discrete Motion component estimates the change of position, orientation and velocity of people and robots between two time instances k and $k+p$. It will be used to compute the robots' trajectory to reach the goal while preventing people leaving the group.

2.1 The Discrete Time Motion Model

The first task of the Discrete Time component is to estimate position, orientation and velocity of the robots and persons. This is done with a standard particle filter formulation [2].

Then, the Discrete Time component aims to represent the areas where the robots will be allowed to move, by means of potential fields. To this end, we define a set of functions that describe the tension produced by the obstacles, people and robots over the working area. These tensions are computed based on the area defined by a security region surrounding each one of the persons, robots and obstacles.

In order to decide the trajectories the robots will follow we will define a potential field over the working area, and perform path planning in it. In particular, the goal the robots try to reach will generate an attractive force pulling the robots towards it. On the other hand, the obstacles will generate a repulsive potential pushing a given robot away . We parameterized all these attractive and repulsive forces by Gaussian functions. For more detail of this model see [6].

$$T_p(\mu_p, \Sigma_p)(x) = \frac{1}{|\Sigma_p|^{1/2}(2\pi)^{n/2}} e^{-\frac{1}{2}(x-\mu_p)^T \Sigma_p^{-1}(x-\mu_p)} \quad (1)$$

2.2 *Modeling People Motion*

In order to model people's motion we will use the concepts introduced by the works of Helbing et al. [1], this research studies the dynamics of pedestrian crowds from the "social" point of view. More specifically, they describe the motion of pedestrians based on social forces which are the result of the internal motivations of the individuals to perform certain motions. For more information see [1].

Let us now explain mathematically. People usually take the shortest path, which may be formally represented as the shape of a open polygon with edges $\mathbf{r}_\alpha^1 \ldots \mathbf{r}_\alpha^n :=$ $\mathbf{r}_\alpha^0$, where α refers to a given person and $\mathbf{r}_\alpha^0$ the destination he/she wants to reach.

The desired motion direction $\mathbf{e}_\alpha(t)$ of a pedestrian α will then be: $\mathbf{e}_\alpha(t) :=$ $\frac{\mathbf{r}_\alpha^k - \mathbf{r}_\alpha(t)}{\|\mathbf{r}_\alpha^k - \mathbf{r}_\alpha(t)\|}$ where $\mathbf{r}_\alpha(t)$ is the *current position* and $\mathbf{r}_\alpha^k$ is the subsequent edge of the polygon that will be reached. A deviation of the desired speed, v_α^0, from the current velocity, $\mathbf{v}_\alpha^0(t) := v_\alpha^0 \mathbf{e}_\alpha(t)$, may also exist due to deceleration or obstacle avoidance processes:

$$\mathbf{F}_\alpha^0(\mathbf{v}_\alpha, v_\alpha^0 \mathbf{e}_\alpha) := \frac{1}{\tau_\alpha}(v_\alpha^0 \mathbf{e}_\alpha - \mathbf{v}_\alpha) \quad (2)$$

where τ_α is a relaxation term. In practice we set the term τ to 0.5 for all the pedestrians [1].

3 Adaptation Model for Rescuing People

One of the biggest issues when guiding a group of people using multi agents, it is the possibility that a person or some people escape from the formation, in this case the agents have to adapt to the new situation to solve the new task. To this end, we will speak in term of robot instead of agent because all this theory will be applied to robots. The cost function, described below, speaks in Work terms, and it can be divided into two blocks: *(i)* Robot work motion, and *(ii)* Human work motion. In order to know what robots' tasks are, we have considered the following situations: *(i)* One robot has to look for the person (or people) that can potentially escape from the crowd formation and push him (or them) to regroup him (or them) into group, *(ii)* one robot has to go behind the people in order to push them in case that the crowd formation is broken down while the Leader guides the formation.

Firstly, the leader robot computes a path planning and moves to the next point. We also assume that there exists a *drag force* that will attract people behind the robot. Here, the robot has only to move from the present position to the next one of the guiding path. The *Pushing task* occurs when the robot pushes a person that has gone away in order to reach the crowd formation. This task can be also applied when a robot pushes a person (or people) who is (are) going behind the crowd formation in order to regroup people when the formation is broken down. Finally, *Crowd traversing task*, where the robot has to move through the formation to achieve the estimated position of the person that goes away from the crowd formation. In order to compute the dragging, pushing and crowd traversing forces, we use the equations defined in previous works on human behavior with other individuals [1]. Working with autonomous mobile robots, the robot i work motion is expressed by:

$$f_i^{mot} = m_i a_i; \quad W_i^{mot} = f_i^{mot} \Delta s_i \tag{3}$$

where m_i is the mass of the i-th robot, a_i its acceleration and Δx_i the space traversed by the robot to achieve its goal.

In this problem it is necessary to consider the *dragging, pushing* and *crowd intrusion forces* that robot's motion produces and that can affect to people. This component is called *Human Work Motion*, and it is the expense of people's movements as a result of robot's motions. As it has been mentioned several times in this paper, the group follows the robot guide/leader, and there is a set of robots that help to achieve their goal.

The dragging force is necessary when the leader robot guides the group of people from one place to another. It acts as an attractive force, hence the force applied by robot leader i to each person j is:

$$f_{ij}^{drag}(t) = -C_{ij}\mathbf{n_{ij}}(t) = -C_{ij}\frac{x_i(t) - x_j(t)}{d_{ij}(t)}; \quad d_{ij}(t) = \|x_i(t) - x_j(t)\| \tag{4}$$

where $d_{ij}(t)$ is the normalizated vector pointing from person j to robot i at instant t. C_{ij} reflects the attraction coefficient over the individual j, and it depends on the distance between the robot leader and person j. Thus, the dragging work that robot leader applied to each individual is defined by:

$$W_{drag} = \sum_{\forall \text{ person j}} f_{ij}^{drag} \Delta s_j \tag{5}$$

Where Δs_j is the distance traveled by the person j.

The *Pushing force* is given by the repulsive effect developed by shepherding robot on the group of people, for regrouping a person (or the broken crowd) in the main crowd formation. This repulsive force is due by the intrusion of the robot in the people's living space, which is five feet around humans. The territorial effect may be described as a repulsive social force:

$$f_{ij}^{push} = A_i exp^{(r_{ij}-d_{ij})/B_i} \mathbf{n_{ij}} \left(\lambda_i + (1+\lambda_i) \frac{1+cos(\varphi_{ij})}{2} \right) \tag{6}$$

Where A_i is the interaction strength, $r_{ij} = r_i + r_j$ the sum of the radiis of robot i and person j, usually people has radii of one meter, and robots 1.5 m, B_i parameter of repulsive interaction, $d_{ij}(t) = \left\| x_i(t) - x_j(t) \right\|$ is the distance of the mass center of robot i and person j. Finally, with the choice $\lambda < 1$, the parameter reflects the situation in front of a pedestrian has a larger impact on his behavior than things happening behind. The angle $\varphi_{ij}(t)$ denotes the angle between the direction $\mathbf{e}_i(t)$ of motion and the direction $-\mathbf{n_{ij}}(t)$ of the object exerting the repulsive force. We can write pushing work by:

$$W_{push} = \sum_{\forall \text{ person in } \Omega_i} f_{ij}^{push}(t) \Delta s_j \tag{7}$$

Where Ω_i is the set of people in which one of the helper robots have reached the living space.

And last but not least, the *Traversing force* is determined by the forces applied by the robot when is traversing the crowd. For security reasons, we have considered in this research that the value of this force is infinity, so we will ensure that a robot will not cross the crowd in order to avoid any damage.

The cost function for agent (robot) i, given a specific task, is the following one:

$$W_i = \delta_{mot} W_i^{mot} + \delta_{drag} W_i^{drag} + + \delta_{push} W_i^{push} + \delta_{trav} W_i^{trav} \tag{8}$$

$$\text{where } \delta_k = \begin{cases} 1 \text{ if this task is assigned} \\ 0 \text{ if this task is not assigned} \end{cases}$$

Where k could be *pushing, dragging, traversing* or *motion*. For each period of time, the leader and shepherded agents (robots) will be given a task in the guiding mission, which will imply one or several robot motion works and human robot works.

Finally, the task assignment for the agents (robots) will be the one which minimizes the minimum assigned work cost required to do the global task. It is computed by the following way:

$$C = argmin\{W_{total}(c)\}, \; \forall \text{ configuration c} \tag{9}$$

where the *Configurations* mean how the tasks are distributed among the agents, for each configuration c agents compute W_{total} which is the addition of all W_i for all agents i that are working cooperatively.

4 Multi Agent Coordination

In the previous section, we have described a const function that it would be used when one person or more people escape from the formation. In the present section we will define a new approach of Multi Robot Task Allocation (MRTA). Our proposal addresses the challenge of people guidance in a way no previously explored, using MRTA. In agents and robotics systems there are many approaches to task allocation, but almost all solve tasks that does not involve human interaction. Other shortcoming in current architectures is that most of them assign one task to one robot, limiting the capacities of robot teams to work cooperatively. A good review and analysis of current multi-robot task allocation architectures is [3] of Gerkey and Matarić.

In general, existing proposals attempt to allocate one task to one robot. Only a few consider the case for cooperating to solve a task, and with the exception of ASyMTRe [5] in which cooperation is somehow preset on the system definition, cooperation only occurs when there are special situations (i.e. errors) during execution.

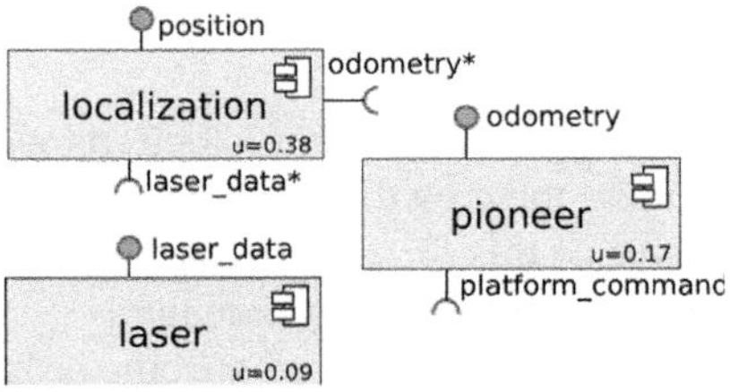

Fig. 1 A component includes uncertainty u and required ports (marked with an asterisk).

Our architecture, Selfish Task Allocation (STA), follows the Component Based Development, meaning that all the software to be used inside STA must be constructed as a component (see figure 1, therefore it must have a name and include input and/or output ports. Furthermore we add uncertainty to each component and the property *required* to each port.

Each component $C_i = (n^{C_i}, P^{C_i}, u^{C_i})$, has a name n^{C_i}, a list of ports $P^{C_i} = \{P_1^{C_i}, P_2^{C_i}, P_3^{C_i}, ..., P_m^{C_i}\}$, and an uncertainty value u^{C_i}. Each port $P_j^{C_i} = (t^{P_j^{C_i}}, w^{P_j^{C_i}}, r^{P_j^{C_i}})$ is described by an *information type* $t^{P_j^{C_i}}$ that represents the kind of information transported in the port, $w^{P_j^{C_i}}$ is a binary value that stores the *direction* of the port (input or output), and $r^{P_j^{C_i}}$ that indicates if a port is *required* or not to accomplish that component.

As each piece of software inside a robot should be designed as a component, a robot can be conceived as a collection of components $R = \{C_1, C_2, C_3, ..., C_n\}$.

Since our approach addresses the problem as a task allocation system, we need to formalize the list of tasks $T = \{T_1, T_2, T_3, ..., T_o\}$ and the description of each of that tasks $T_k = (n^{T_k}, g^{T_k}, s^{T_k})$.

Where n^{T_k} represents the name of the task, g^{T_k} the geographic information, if any (for example in people guidance the position of the group to be guided and the goal of the group), and s^{T_k} the status of the task (unsolved, attempted or solved).

We decided to tackle the problem of multi-robot task allocation with a different point of view, instead of use a complex group algorithm to distribute tasks, our approach uses the single robot task selection algorithm, allowing each robot to select the task that better fits its capabilities. When all the robots selects their best task, a task management algorithm coordinates the actions of the robots on the same task and task allocation just emerge. This section explores the Single Robot Task Selection (SRTS) algorithm we develop for our architecture. We understand SRTS as the algorithm used to define which is the best task for each robot.

Our approach for single-robot task selection is inspired by the proposal of Tang and Matatić, ASyMTRe [5], which in turn is inspired in information invariants. Like previous proposals (information invariants and ASyMTRe), our proposal called Selfish Task Allocation (STA) *"allows robots to reason about how to solve a task based upon the fundamental information needed to accomplish the task"* [5].

The Single-Robot Task Selection (SRTS) algorithm, part of STA, begins when it receives the list of tasks T, then $\forall T_k \in T$, the SRTS algorithm searches if $\exists C_i$ s.t. $n^{C_i} = n^{T_k}$, once C_i is found the algorithm tries to *activate* that component satisfying the following rules:

1. C_i can be activated iff $\forall P_j^{C_i} \in P^{C_i}$ where $r^{P_j^{C_i}} = true$ can be connected to a *compatible port* (see rule 2). At the same time components providing those ports must be activated.
2. Two ports $P_a^{C_i}$ and $P_b^{C_i}$ are compatibles iff they have the same information type $t^{P_a^{C_i}} = t^{P_b^{C_i}}$ and opposite direction $w^{P_a^{C_i}} \neq w^{P_b^{C_i}}$.

As result of the search some solutions could be obtained, we represent each solution as the list of the components involved in the solution of the task $S_l = \{C_x, C_y, ...\} \subset R$. And as can be inferred, if there exists more that one algorithm to solve the same problem (particle filter- and odometry-based localization) or if two or more components provide the same information type, could exist many ways to satisfy the required port of a component, and therefore many alternative solutions for each task $S = \{S_1, S_2, ..., S_l\}$ between all these alternatives, to select the task to be performed we use the solution with less uncertainty in connections u_{T_x} and displacement $u_D(pos, g_{T_k})$.

$$T_s = T_x, \text{ s.t. } u_{T_x} = \min_{\forall T_k \in T} u_{T_k} \tag{10}$$

with,$u_{T_k} = u_D(pos, g^{T_k}) + u_{S_m}$, s.t. $u_{S_m} = \min_{\forall S_l \in S} u_{S_l}$, where, u_{S_m} represents the solution S_l with less uncertainty, $u_D(pos, g^{T_k})$ the displacement uncertainty, uncertainty of reach the position of the task g_{T_k} from the robot position *pos* (only for mobile robots), u_{T_k} the uncertainty of T_k when the solution with less uncertainty is selected, and T_s the finally selected task.

Once the robot selects the task that it can perform with less uncertainty (T_s), it tries to tackle that task, and probably other robot is already working on it. We propose an algorithm to face up the possible issues generated for the multi-robot task tackling.

We distinguish two main challenges when more than one robot acts over one single task, the first related to task-specific behavior (i.e. the position of each robot in cooperative box pushing) and the second related to detect when robots become hindering each other.

Our approximation to the task-specific coordination challenge lies on the assignment of an *id* to each robot that joints to solve the same task, the id assignment and the number of robots participating in the task is managed in a distributed manner and is only valid inside that task.

The id and the number of robots into each task must be used by the component designer to define the robot specific behavior for each task.

To face up the challenge of detect when there are too many robots tackling the same task, we propose the use of performance functions. While acting, each robot continuously calculates the performance of the team in the specific task

$$p_t^{T_k} = P^{T_k}(V_{t-1}^{T_k}, V_t^{T_k}) \tag{11}$$

Where, $V_t^{T_k}$ represents the value of environment variables involved in T_k at moment t. $P^{T_k}()$ The specific performance function for T_k. And $p_t^{T_k}$ the performance of T_k at time t.

As mentioned before, in our proposal of task selection each robot chooses the task that better fits their capabilities. But this selection mechanism means that many times the robots select the same task. Here it is when performance metrics make sense. When a robot tries to participate in a task where other robots are already working, it is accepted in test mode, all the robots adapt their behavior to the new number of robots and they continue tackling the task and getting performance information, but after a predefined time, the performance of the team is compared with the performance stored before the new member inclusion, if it was increased then the robot is considered now as part of the team, else the robot is asked to leave the task.

5 Implementation and Results

The results we will present correspond to different synthetic experiments, some of them within the previous map. In these experiments, the dynamical models of the persons –we considered a group of 5 persons– will follow the models described in the Section described before. In figure 3 some instants of time of a group of agent working cooperatively solving the task of guiding a group of people is shown.

We perform some experiments where one group was guided, we include in the simulated persons the ability to randomly leave the formation, to test our cost function and to study which is the recovery component and prove the behavior of the proposed architecture in group splitting. In fig. 2 it can be seen that when people leave the group a recover task is added to the list and some robot reacts to solve this task.

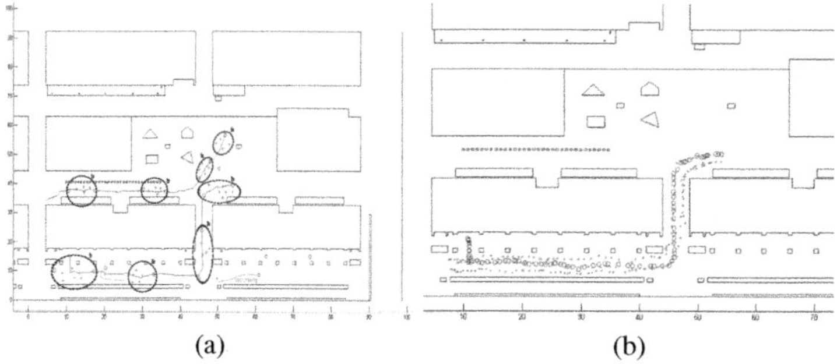

Fig. 2 (a) Two different group of people are being guided by groups of cooperative agents. Several instants of time and the entire trajectory are shown. (b) Entire trajectory of people and robots in a guiding people mission.

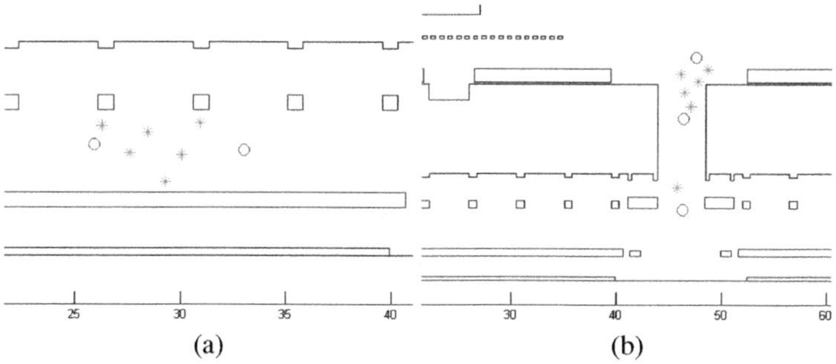

Fig. 3 (a) A group of people is being guided by two agents (red circles), in this occasion no human escapes from the formation, the task is being done correctly. (b) Two agents are guiding the group while a third agent is rescuing somebody who tries to escape.

6 Conclusions

We have presented a new model to guide people in urban areas with a set of multi agents that work cooperatively and are able to adapt their behavior depending on people motion. In contrast to existing approaches, our method can tackle more realistic situations, such as dealing with large environments with obstacles, or regrouping people who left the group. For that reason, this work can be applied in some real robots applications, for instance, guiding people in emergency areas, or acting as a robot companion.

References

1. Helbing, D., Molnar, P.: Social force model for pedestrian dynamics. Physical Review E 51 (1995)
2. Arulampalam, M.S., Maskell, S., Gordon, N., Clapp, T., Sci, D., Organ, T., Adelaide, S.A.: A tutorial on particle filters for online nonlinear/non-GaussianBayesian tracking. IEEE Transactions on Signal Processing (2002)

3. Gerkey, B.P., Matarić, M.J.: A formal analysis and taxonomy of task allocation in multi-robot systems. The International Journal of Robotics Research (2004)
4. Lien, J.M., Rodriguez, S., Malric, J.P., Amato, N.M.: Shepherding Behaviors with Multiple Shepherds. In: Proc. IEEE Int. Conf. Robot. Autom, ICRA (2005)
5. Tang, Y., Parker, L.E.: Towards Schema-Based, Constructivist Robot Learning: Validating an Evolutionary Search Algorithm for Schema Chunking. In: ICRA 2008 (2008)
6. Garrell, A., Sanfeliu, A., Moreno-Noguer, F.: Discrete Time Motion Model for Guiding People in Urban Areas using Multiple Robots. In: IEEE/RSJ IROS 2009 (2009)

Agent Adaptation through Social Opinion

Juan A. Garcia-Pardo and C. Carrascosa

Abstract. A dynamic environment whose behavior may change in time presents a challenge that agents located there will have to solve. Changes in an environment e.g. a market, can be quite drastic: from changing the dependencies of some products to add new actions to build new products. The agents working in this environment would have to be ready to embrace this changes to improve their performance which otherwise would be diminished. Also, they should try to cooperate or compete against others, when appropriated, to reach their goals faster than in an individual fashion, showing an always desirable emergent behavior. In this paper a reinforcement learning method proposal, guided by social interaction between agents, is presented. The proposal aims to show that adaptation is performed independently by the society, without explicitly reporting that changes have occurred by a central authority, and even without trying to recognize those changes.

1 Introduction

Learning is in general desirable since it will allow an agent to fit better in the environment, either because by design they were not 100% accurate in their parameters and they had to adjust them, or just because some of the parameters of the environment were unknown or vary through time.

When talking about learning in multi-agent systems there are some situations in which there is no other way to know how the environment is going to react to agents' actions but to actually perform them. In these cases multi-agent learning (*MAL*) reinforcement learning (RL) techniques [1] are useful to discover how the universe in which the MAS is located works. This phase aims to learn the parameters which steer the *mental model of the universe*, and usually ends when some error measure between the *real* behavior of the universe and the *expected* or *modeled* one is small enough to neglect it. The exploitation phase —profiting from what it was learned— can mainly be used from this point on.

Juan A. Garcia-Pardo · C. Carrascosa
Universidad Politecnica de Valencia, Camino de Vera s/n. 46022 – Valencia (Spain)
e-mail: {jgarciapardo,carrasco}@dsic.upv.es

J.B. Pérez et al. (Eds.): Highlights in PAAMS, AISC 89, pp. 185–194.
springerlink.com

One approach to RL in the MAL community is taken from the Game Theory framework [2, 3], in which the agents in the system act rationally to achieve their goals. The algorithms found in the research papers from this area usually focus on the equilibrium (Nash-equilibrium) of the solution, i.e. the solution is *good enough* for all the parties involved (cooperating or competing). One of the problems of these kind of approaches is the consideration of rationality of the agents, which may not be true [4]. In fact, humans do not usually behave rationally [5], in the sense described by Nash [6] equilibrium.

On the other side, there are cooperative learning algorithms which allow to obtain an optimal policy common to the whole set of learning agents (e.g. see [7]), with no need from them to behave rationally in the Nash sense, but the algorithm must know the whole set of agents which are learning the policy under discussion, and has some other constraints. The environment covers the *mechanical* part (the one which is independent to the MAS) and also the *social* part (actually, the MAS itself, including every agent capable of performing changes in the environment, both social or mechanical).

When the environment is dynamic learning is going to be essential. The agents will require to learn a set of parameters which were not present at design time. Changes in the environment usually require them to learn again. But some changes require something beyond a parameter estimation. They require adaptation by the agents. An example of this adaptation would appear in some *self-healing* systems [8].

If the agents could self-adapt to the environment, depending only on the interactions with the agents that inhabit in the same universe, and re-adapt every time significant changes would appear, the system would auto-control itself, and would not need a centralized logic trying to figure out whether there was a significant change or not, and which actions would be the most appropriate to take to improve the outcome.

Our goal in this paper is to study the behavior of self-adaptive agents immerse in an unknown dynamic environment, which is open and heterogeneous. Enabling self-adaptation through social interaction between certain agents will show that it is possible to have adaptive agents with no central control at all.

Section 2 presents an introduction to the problematic of adapting in dynamic systems and a proposal based on social reinforcement to tackle the problem of detecting significant changes. Section 3 proves the convergence of Q-learning using that approach. Sections 4.1 and 4.2 introduces the testbed used to check the feasibility of such approach and the simulation details; some results made in simulation can be seen in section 4.3. Lastly, Section 5 presents the evaluation of the results and indications of which is going to be the next steps in this research.

2 Adaptation in Highly Dynamic Systems: Social Reinforcement

In any open multi-agent system that allows heterogeneous agents, the adaptation of part or the whole society is a difficult task [9]. Adaptation should occur when changes that arise are significant enough to yield losses higher than the cost of the adaptation itself to the new conditions. Detection of the importance of changes in the environment is not naive, since the optimal policy is unknown a priori, without which we can not obtain reference values to compare both the individual behavior of each agent and the one of the whole society.

For all non-static environments, we can say that there are certain changes that require dynamism and adaptability of agents. The changes that could lead to an adaptation of the agents can be divided into the following taxonomy:

- *Mechanical Changes:* they happen in the environment, such as changes in rules or characteristics of the environment where agents are located.
- *Social Changes:* they happen in the organization (in the agents):
  - A new social agent appears
  - An existing social agent disappears
  - A new non–social agent appears
  - An existing non-social agent disappears

We will give the name *Social Agents* to those which have been prepared to automatically adapt themselves to the changes through the communication of their opinion, and give the name *Non–Social Agents* to the rest, whether they are human controlled or [partially] autonomous, but not ready for self-adaptation through the social opinion mechanism.

Changes in the transition function and in the rewards can happen, despite our efforts to find a suitable representation, particularly when the environment changes. For example, in a market there may constantly appear new ways to build new products, such as by introduction of new objects (functions) which allow actions unknown until that moment, or by changing prices or dependencies. All this *mechanical* changes will have an effect.

Trying to adapt a multi-agent system in a *bottom-up* way, so, trying to exhibit emergent behavior through the design of the individuals has been always desirable, as long as the the system as a whole gets adapted. We will try to address the problem of detecting significant changes in the system, in an autonomous way, through the *social opinion reinforcement* (R_Σ).

Each agent would care only of maximizing its own benefit until asked to review its behavior. There would be no "central authority" managing the adaptation: the *social reinforcement* would take care of this. By having an agent that is capable of observing the outcome of their actions in the universe, and providing this agent with *social reinforcement* —the latter significant of the opinion that the rest of the organization has about its actions— we can take advantage of the received information, which could show us the social approval with respect to the actions taken by the agent, as an indicative value of the utility of this agent to the society.

These socially sensible agents would use social approval (or disapproval) to determine whether to change its behavior or to continue with the policy which they consider the optimal one, π^*. The parameter of exploration k can be modified from the social reinforcement, being now function of a set of values, which in turn are a function of the variables of the environment: $K : \{R_\Sigma\} \to \mathbb{R}$; $R_\Sigma : X \to \mathbb{R}$

We need to define the social reinforcement in order to compute the exploration probability. The social reinforcement is computed as shown in (1).

$$r_{\Sigma,t}(s_t,a_t) = \sum_a \rho_a \ \tau_a \ r_{\Sigma,t,a}(s_t,a_t) \tag{1}$$

According to this definition the social reinforcement value is not bounded. The parameter τ_a represents the trust that this agent has in agent a being proper and correct in its reinforcement. Usually $0 \leq \tau_a \leq 1$. The parameter ρ_a represents the reputation of agent a according to the society. Depending on the value of ρ_a and τ_a the opinion of agent a will be more important or not, for each agent a in the society.

A proposal for the k evaluation function, which takes into account the social reinforcements both positive and negative, is shown in equation (2). The calculation of the exploration parameter k is now done using a series similar to the sigmoid function:

$$k_{t+1} = \frac{k_t}{k_t + 2^{k_t(r_{\Sigma,t}(s_t,a_t)-2)}} \tag{2}$$

Which is bounded between 0 and 1.

The series expressed in equation 2 converges for monotonic values of $r_{\Sigma,t}$ as shown in Sec. 3.

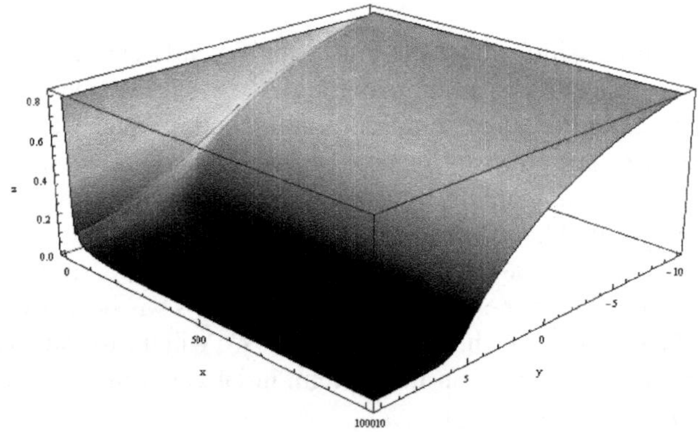

Fig. 1 Evolution of k_t (z-axis) as time (x-axis) and Σ (y-axis) change. $k_0 = 0.8$

The effect of this k function coupled with the optimal policy π^* searching algorithm is shown by simulation, in Sec. 4.1. A picture the behavior of the k_t series is shown in Fig. 1.

3 Convergence of Learning

Convergence of the learning process is proven under some assumptions: (1) The agent can represent the universe without taking into account other agents. Al least the agent has function of representation which can represent the universe with no ambiguity with a high probability, probability that increases with time. (2) The agent knows the set of actions that are available at any time. (3) The agent can observe the response of the system to its actions.

To prove that the Q-learning algorithm converges now, the k_t series has to converge as well. Being the k_t series convergent, the Q-learning algorithm will behave as usual, thus converging to the optimal policies (under the former assumptions) if all the states are visited enough. Lets write (2) using $\sigma \equiv r_{\Sigma,t}(s_t, a_t)$ for the sake of clarity:

$$k_{t+1} = \frac{k_t}{k_t + 2^{k_t(\sigma-2)}} \tag{3}$$

In the limit when $t \rightarrow \infty$ we expect $k_{t+1} = k_t$, thus

$$\begin{aligned}
k_t &= \frac{k_t}{k_t + 2^{k_t(\sigma-2)}} \\
1 &= (1-k_t)2^{-k_t(\sigma-2)} \\
2^{\sigma-2}(\sigma-2) &= 2^{\sigma-2}(\sigma-2)2^{-k_t(\sigma-2)}(1-k_t) \\
2^{\sigma-2}(\sigma-2) &= 2^{(1-k_t)(\sigma-2)}(1-k_t)(\sigma-2) \\
ln(2)2^{\sigma-2}(\sigma-2) &= e^{(1-k_t)(\sigma-2)ln(2)}(1-k_t)(\sigma-2)ln(2) \\
(1-k_t)(\sigma-2)ln(2) &= W(ln(2)2^{\sigma-2}(\sigma-2)) \\
k_t &= 1 - \frac{W(ln(2)2^{\sigma-2}(\sigma-2))}{(\sigma-2)ln(2)} \qquad (4)
\end{aligned}$$

In (4) the $W(x)$ stands for the Lambert-W function such that for every number $x \in \mathbb{R}$, $x = W(x)e^{W(x)}$.

Since the convergence of the exploration probability is proven, the convergence of the Q-learning algorithm follows easily:

An agent will cast a *ballot* in the range $[-1, +1]$ depending on the rewards obtained by the *mechanical* part of the universe. Using for example the following rule while in a given state s and after executing action a: When $Q_{n+1} > Q_n$ cast $+1$; when $Q_{n+1} < Q_n$ cast -1; in case $Q_{n+1} = Q_n$: cast $+1$ if $\delta_{n+1} = \delta_n$ and cast -1 otherwise. This all in the case the environment is deterministic. When it is not, and has an stochastic transition function, the expressions $\delta_{n+1} = \delta n$ have to be replaced by the corresponding ones $p(\delta_{n+1}|\delta n)$. The higher the probability of the transition, the closer the value would be to $+1$.

4 Tests and Validation

4.1 Study Case

To study the effects of social learning on a practical level we have proposed a simple collaborative game: a market economy in which the production line has different types of units, and the agents should build some or consume certain products for their survival. The agent enters the system with certain amount of money and life points; these values hold for every agent entering the system. Agents with life below 0 disappear: it is considered by the system that the agent is unable to produce more than what it consumes. This is called a *failure* of an agent —it failed to be productive—. In the event of a player disappearance, the market automatically removes all products which were unsold and belong to that player. If the agent reenters the market, it spawns with the initial values of *life* and *money*.

Any product may have n dependencies of n_i, $i \in [0, p]$ units of m different products, and need some specific time to be built. Finally, the price of the products is fixed by each one of the agents. To get a product from the market the player must pay in advance. Negative balances are not allowed, but agents are able to modify the price of the products they built even when they are already on the market.

In this environment the actions of other players are not observable, but its results may be: they cannot tell if an agent is making a product, but they notice —if they are willing to— that the agent built something when it puts the product on sale on the market.

The environment may change the production rules of the system at any time. The variations allowed are the creation and deletion of product types, changes on the dependencies –both the amount and the type– and the modification of the base time needed for production. Those will be what we called *mechanical changes* before. The *social changes* encloses the possibility of agents entering and leaving the market at any time.

The environment can simulate supply chain processes, where demand can vary and rules can be altered. This way, a good agent would innovate and adapt to the variations automatically, and —with some time— come back with a good behavior competing and collaborating with other agents inside the environment.

4.2 Simulation

Reinforcement learning in a continuous domain environment cannot be applied unless some conditions are given. For the simulation the agents use a well-known technique for grouping states (Soft State Aggregation, or SSA) which allow them to represent an infinite space of continuous values, with D dimensions, in a D-dimensional finite space (clusters) [13]. For the study case the environment state space is seen as a D-dimensional matrix, grouping the states ($\in \mathbb{R}^D$) in an exponential way. The discrete position (or index of the cluster) i for the continuous variable x is computed as:

$$i = \min(log_2(x+1), \omega) \qquad (5)$$

Bounding i on some number (ω) which is to be considered near to the maximum that is to be seen for that variable, for D dimensions that are taken into account for the environment representation. Therefore the total set of states which every agent must represent is $\omega^D \times |A|$. The number of actions $|A|$ is known by every agent (although it may change with time).

Two different types of agents will be differentiated; those who can receive and understand social opinions –executing Social-Welfare RL– and those which not –using standard Q-learning–.

Learning of the Q-table was done individually for both the social-aware agents and the non-social. Parameter ω was fixed at 5, which means that the last state for each dimension will represent values of $x \in [15, +\infty]$. The dimensions (or variables) taken into account to represent the environment are the number of products of each kind in the market plus agent's *life* and *balance*.

Three types of products where loaded: *Wheat*, with no dependencies and needing 1 cycle. *Flour*, requiring 2 units of *flour* and 2 cycles. And *Bread*, eatable, providing 10 units of life, requiring 2 units of *flour* and 2 cycles.

The actions available to the agents are the creation of any kind of the products, plus another one called *eating*. The total space of representation needed by each agent is only of $(5)^{(3+2)} \times 4 = 12500$ states. Learning stage has taken $1,000,000$ cycles, with a probability of exploration $k = 1$, hopping to explore as many states as possible, as many times they could. Every agent starts with exactly the same Q-matrix at the beginning of the simulation. The experiment includes three different scenarios, thus there are two significant changes. The scenarios are:

1. No changes in the environment.
2. Relaxing the production rules: *Flour* needs 1 unit of time and 1 unit of *Wheat*; *Bread* needs 1 unit of time and 1 unit of *Flour*.
3. Hardening the production rules: *Wheat* needs 2 units of time; *Flour* needs 2 units of time and 3 units of *Wheat*; *Bread* stays as originally, needing 2 units of time and 2 units of *Flour*.

As a remark, it is important to control the value of κ –to avoid machine precision problems– such that never reaches 0. For that matter, the computation is modified as:

$$\kappa = \begin{cases} \varepsilon & \text{if } \kappa < \varepsilon \\ 1 - \varepsilon & \text{if } \kappa > (1 - \varepsilon) \\ \kappa & \text{otherwise} \end{cases}$$

for a given $\varepsilon \to 0$ which is representable by a machine.

It is expected to observe a better adaptation of the social-aware agents compared to the non-social ones, which use traditional reinforcement learning (SSA Q-learning).

4.3 Results

Two different experiments were carried out: the first one with 4 agents in the system, and the second one with 32. Two rounds per experiment were done: one with classic Q-learning agents (using SSA to represent the environment), and another round with social aware RL agents (also using SSA). In both experiments we measured the number of failures and the welfare of the society, for the two types of agents. Both values are normalized per number of agents. In each figure the two vertical lines represent the moment in time when a change in the environment was made.

The *failures* represent agents which were in the system for a time yielding loses instead of producing welfare. It is the equivalent of business having loses in the economy: they start with an initial amount of resources but they have a limited time until they start yielding benefits. Otherwise they would confront bankruptcy. Notice that the lower the number of failures, the better the system as a whole –it means the system has coped with many different agents and few of them had problems with their policies–.

The *welfare* represent the wealth of the society. The higher, the better. The final aim of the system is, actually, to produce as much wealth as possible. The agents, which do not cooperate explicitly, face the problem of concurrency in the environment, hence the fact that the double the number of agents does not mean the double of wealth.

In figure 2 it is shown that social agents have fewer failures than non-social ones. As expected in the theory, the more agents the system has, the easier is to notice changes; but also when there are many agents, it is more probable that they interfere with each other, which would make harder to find a policy in a short period of time – policies get more complicated as agents are added in the society–. That explains also the data shown in figure 3, where a big difference between the two types of agents appear. When having more agents in the system it is easier to have them spread out of the different states, so when the majority agree that they should behave adaptively, it is more probable that the different states of the system have been already explored.

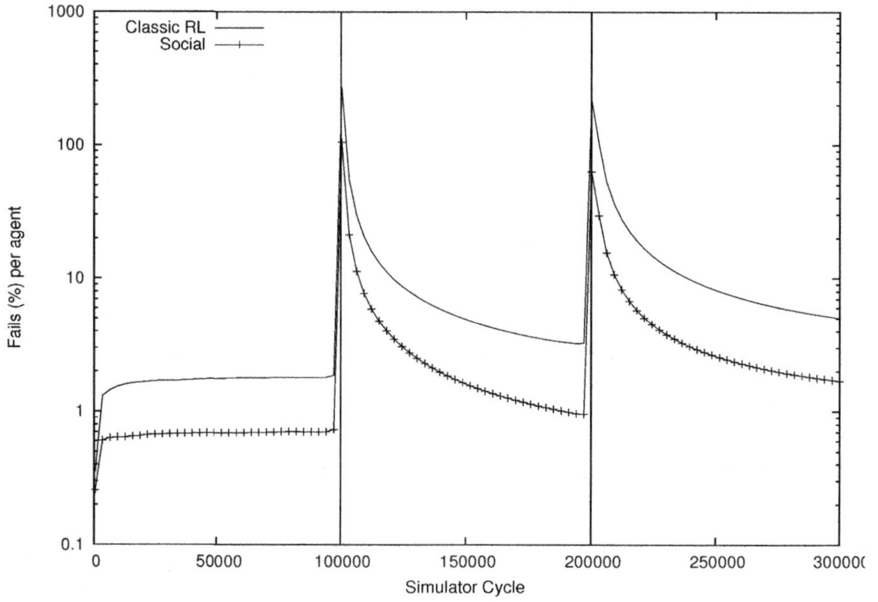

Fig. 2 System with two significant changes, failures per agent comparison. 32 agents in the system.

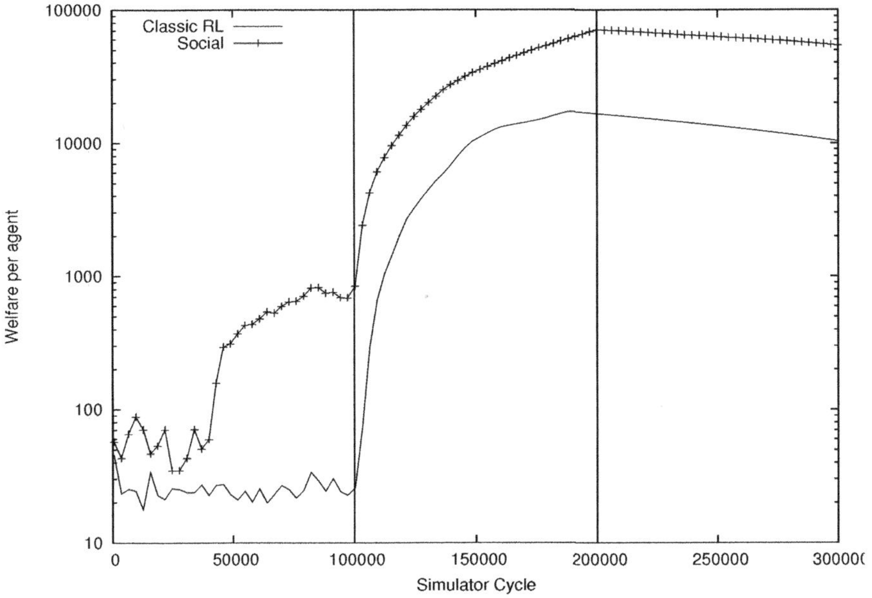

Fig. 3 System with two significant changes, welfare per agent comparison. 32 agents in the system.

5 Conclusions and Future Work

This paper shows that the communication of the opinion inside the society is a good mechanism to trigger innovation when a change requires it. The mechanism works better when the number of agents increases, which is very promising in terms of scalability of MAS using this method as the learning algorithm.

The convergence of the algorithm has been proven, but the agents can converge towards a non-global optimum. Where this may be a problem for some environments, it can be alleviated with agents which are even more pessimistic than the one showed in this paper —being more pessimistic would make the agent more innovative, thus exploring the environment longer—.

Some further work should be carried out to integrate the social welfare mechanism in other RL algorithms. Also tests in zero-sum games would clarify whether this mechanism works well in pure competitive games. Some scenarios —rock, paper, scissors; matching pennies; etc.— are a good benchmark in zero-sum games. Related to this line of research is the possibility of computing the similarity of goals of two agents given their social opinions. Intuitively we can say that when two agents have similar opinions, their goals are similar, with increasing probability as time increases. This similarity would be semantically similar to a composition between trust and reputation between both agents: a measure of similarity.

Acknowledgements. This work has been partially funded by TIN2009-13839-C03-01, TIN2008-04446, PROMETEO/2008/051, GVPRE/2008/070 projects, CONSOLIDER-INGENIO 2010 under grant CSD2007-00022.

References

1. Sutton, R.S., Barto, A.G.: Reinforcement learning i: Introduction (1998)
2. Vidal, J.: Learning in multiagent systems: An introduction from a game-theoretic perspective. In: Adaptive Agents and Multi-Agent Systems, pp. 562–562
3. Akchurina, N.: Multiagent reinforcement learning: algorithm converging to nash equilibrium in general-sum discounted stochastic games. In: AAMAS 2009: Proceedings of The 8th International Conference on Autonomous Agents and Multiagent Systems, pp. 725–732 (2009)
4. Shoham, Y., Powers, R., Grenager, T.: Multi-agent reinforcement learning: a critical survey. In: AAAI Fall Symposium on Artificial Multi-Agent Learning (2004)
5. López-Paredes, A., Hernández-Iglesias, C., Gutiérrez, J.P.: Towards a new experimental socio-economics: Complex behaviour in bargaining. Journal of Socio-Economics 31(4), 423–429 (2002)
6. Hu, J., Wellman, M.P.: Nash Q-learning for general-sum stochastic games. The Journal of Machine Learning Research 4, 1039–1069 (2003)
7. Melo, F.S., Ribeiro, M.I.: Coordinated learning in multiagent MDPs with infinite state-space. In: Autonomous Agents and Multi-Agent Systems, pp. 1–47
8. Ghosh, D., Sharman, R., Raghav Rao, H., Upadhyaya, S.: Self-healing systems–survey and synthesis. Decision Support Systems 42(4), 2164–2185 (2007)
9. Helleboogh, A., Vizzari, G., Uhrmacher, A., Michel, F.: Modeling dynamic environments in multi-agent simulation. Auton. Agents Multi-Agent Syst. 14(1), 87–116 (2007)
10. Dignum, V., Dignum, F., Sonenberg, L.: Towards dynamic reorganization of agent societies. In: Proceedings of Workshop on Coordination in Emergent Agent Societies, pp. 22–27 (2004)
11. Hu, J., Wellman, M.P.: Multiagent reinforcement learning in stochastic games
12. Claus, C., Boutilier, C.: The dynamics of reinforcement learning in cooperative multiagent systems. In: Proceedings of the Fifteenth National Conference on Artificial Intelligence, pp. 746–752. AAAI Press, Menlo Park (1998)
13. Singh, S.P., Jaakkola, T., Jordan, M.I.: Reinforcement learning with soft state aggregation. In: Advances in Neural Information Processing Systems 7, pp. 361–368. MIT Press, Cambridge (1995)

Virtual Organizations in Information Fusion

S. Rodríguez, D. Tapia, F. De Paz, and F. De la Prieta

Abstract. Advantages of intelligent approaches such as Multi-Agent Systems (MAS) and the use of Wireless Sensor Networks (WSN) within the information fusion process are emerging, especially on context-aware scenarios. However, it is necessary to go further and propose better ways to deal with the avalanche of data that these scenarios provide. In this sense, virtual organizations of agents are an interesting alternative because they provide the necessary capacity to handle open and heterogeneous systems. This paper presents a framework that defines a method for creating virtual organization of software and hardware agents.

Keywords: Multi-agent systems, fusion information, virtual organizations, wireless sensor networks.

1 Introduction

Scientific community within the information fusion area remains heirs to the traditions and techniques of sensor fusion, which is primarily concerned with the use of sensors to provide information to decisions systems. This has led to most models of information fusion processes are directed by data fusion in which the central core are the sensors and data. An alternative way is to take a intelligent approach as MAS within the fusion process are emerging. Agents are suitable for fusion because they can represent autonomous fusion entities by modeling their capabilities, expertise and intentions [14].

Nowadays, there are small, portable and non-intrusive devices that allow agents to gather context-information in a dynamic and distributed way [2]. However, the integration of such devices is not an easy task. Therefore, it is necessary to develop innovative solutions that integrate different approaches in order to create open, flexible and adaptable systems.

MAS allow the participation of agents within different architectures and even different languages. The development of open MAS is still a recent field of the MAS paradigm and its development will allow applying the agent technology in new and more complex application domains. However, this makes it impossible to trust agent behavior unless certain controls based on social rules are imposed. To

S. Rodríguez · D. Tapia · F. De Paz · F. De la Prieta
Departamento Informática y Automática, Universidad de Salamanca, Salamanca, Spain
e-mail: {srg,dantetapia,fcofds,fer}@usal.es

J.B. Pérez et al. (Eds.): Highlights in PAAMS, AISC 89, pp. 195–202.
springerlink.com

this end, developers have focused on the organizational aspects of agent societies to guide the development process of the system.

This article describes an agent approach to fusion applied to dynamic contexts. To this end, it is used the HERA (*Hardware-Embedded Reactive Agents*) platform and OVAMAH (*Adaptive Virtual Organizations: Mechanisms, Architectures and Tools*) platform. In HERA agents are directly embedded on the WSN nodes and their services can be invoked from other nodes(including embedded agents) in the same WSN or other WSN connected to the former one. And thanks to OVAMAH, the framework can incorporate the self-adaptive organizational capabilities of multi-agent systems and create open and heterogeneous systems. The article is structured as follows: next section the related approaches. Section 3 shows the system proposal, including the description of the HERA platform, the heart of the system. Sections 4 present some results and conclusions obtained.

2 Related Approaches

Information fusion can be understood as the study of techniques oriented at the combination and integration of highly heterogeneous data. The heterogeneity of the data may be due to various reasons such as the representation format, conceptual structure, or the source of the data. One of the relatively recent approaches is the use of MAS for information fusion.

Recent years have given way to a number of MAS architectures that utilize data merging to improve their output and efficiency. Such is the case of Castanedo et al.[4], that propose the CS-MAS architecture to incorporate dynamic data fusion through the use of an autonomous agent, locally fused within the architecture. Other models, such as HiLIFE [9], which cover all of the phases related to information fusion by specifying how the different computational components can work together in one coherent system.

These kinds of systems, despite having all the advantages of MAS, aremonolithic. In an environment in which the data heterogeneity is a key feature, it is necessary to use systems with advances capacities for learning and adaptation. In this sense within MAS, an approach that is gaining more weight in recent times is to consider organizational aspects [3], and more concretely those based on the virtual organizations (VO).

Currently, there are not virtual organization-based applications oriented to fusion information, however we can find some approaches that try to propose advances in this way. For example, the e-Cat System [8] focuses on the distribution and the integration of information. This system is based on enhancing the skills or abilities of members of the organization, through the definition of the different types of skills and relationships between them. This organization is aimed at ensuring the maximum independence between the different partnerships created and the information privacy. Another example, perhaps more centralized in the fusion of information, is the KRAFT (*Knowledge Reuse and Fusion / Transform*) architecture[11], which proposes an implementation of agents where organizational aspects are considered to support the processes of heterogeneous knowledge management.

The approach proposed in this article presents an innovative model where MAS and VO are combined to obtain a new architecture especially oriented to construct information fusion environments.

3 Proposed Framework

This section discusses some of the most important problems of existing approaches that integrate agents into WSNs, including their suitability for constructing intelligent environments. It also presents the proposed integration of information fusion systems that use the capabilities of MAS. With HERA, agents can invoke services in WSN, even in low resource devices. And OVAMAH provides the capability of incorporate the organizational capabilities of MAS.

One of the most prevalent alternatives in distributed architectures are MAS. An agent, in this context, is anything with the ability to perceive its environment through sensors, and to respond through actuators. A MAS is defined as any system composed of multiple autonomous agents incapable of solving a global problem, where there is not global control system, the data is decentralized and the computing is asynchronous [14]. There are several agent frameworks and platforms [13] that provide a wide range of tools for developing distributed MAS. The development of agents is an essential component in the analysis of data from distributed sensors, and gives those sensors the ability to work together. Furthermore, agents can use reasoning mechanisms and methods in order to learn from past experiences and to adapt their behavior according the context [14]. All these capacities make the agents appropriated to be applied in information fusion.

The most well-known agent platforms (like Jade) offer basic functionalities to agents, but designers must implement nearly all organizational features, like communication constraints imposed by the organization topology. In order to model open and adaptive VO, it becomes necessary to have an infrastructure than can use agent technology in the development process and apply decomposition, abstraction and organization techniques. OVAMAH [3] is the name given to an abstract architecture for large-scale, open multi-agent systems. It is based on a services oriented approach and primarily focuses on the design of OV.

Any intelligent fusion system has to take into account the information about the context (people and their environment), which can be gathered by sensor networks. Each element of sensor network is called a node. Each sensor node is habitually formed by a microcontroller, a transceiver for transmission and a sensor or actuator mechanism. There are wireless technologies such as IEEE 802.15.4/ZigBee and Bluetooth that enable easier deployments than wired sensor networks. Whilst traditional networks aim at providing high QoS transmissions, WSNs protocols concentrate their efforts on energy saving.

The combination of agents and WSNs is not easy due to the difficulty in developing, debugging and testing distributed applications for devices with limited resources. The interfaces developed for these distributed applications are either too simple or do not even exist. Therefore, there are researches [10] that develop methodologies for the systematic development of MAS for WSNs but they are too specific and usually based on Mobile Agents. Other studies [12], try to reduce the redundancy of the data gathered by sensors from different types of networks by using mobile agents that pack the data. Other research packs the data gathered by the sensor network, reduces the reception and delivery times and saves energy in devices with a low autonomy capacity [1]. Along the lines of fusing or packing there are models that try to state the quality of the fused data and the cost of communication between the transmitter and the receiver.

The HERA platform tackles some of these issues by enabling an extensive integration of WSNs and optimizing the distribution, management and reutilization of the available resources and functionalities in its networks. HERA is an evolution of the SYLPH platform [5]. SYLPH follows a SOA model for integrating heterogeneous WSNs in intelligent systems. HERA contemplates the possibility of connecting wireless sensor networks based on different radio and link technologies. HERA allows the agents embedded into nodes to work in a distributed way and does not depend on the lower stack layers related to the WSN formation or the radio transmission amongst the nodes that form part of the network. HERA can be executed over multiple wireless devices independently of their microcontroller or programming language. HERA allows the interconnection of several networks from different wireless technologies (e.g. ZigBee/Bluetooth). This facilitates the inclusion of context-aware capabilities into intelligent fusion systems because developers can dynamically integrate and remove nodes on demand.

The HERA Agents Layer allows running HERA agents. To communicate with each other, HERA agents use a specific communication language called HERACLES. Each HERA agent is an intelligent piece of code running over a node. There must be at least one facilitator agent in every agent platform. This agent is the first created in the platform and acts as a directory for searching agents. In HERA, this agent is the HERA-SDN (HERA Spanned Directory Node).

3.1 Information Fusion Background

What is important in a system like this is decide in what manner to combine or fuse information. In general, data can be fused at different levels [7]: (i) *sensor level fusion,* where multiple sensors measuring correlated parameters can be combined; (ii) *feature level fusion*, where analysis information resulting from independent analysis methods can be combined; and (iii) *decision level fusion*, where diagnostic actions can be combined. These levels generally depend on many factors, in order to provide the most generic and expandable system that can be applied to a wide variety of engine applications with varied instrumentation and data sources, we have chosen to perform the information fusion at the three levels as shows the Figure 1.

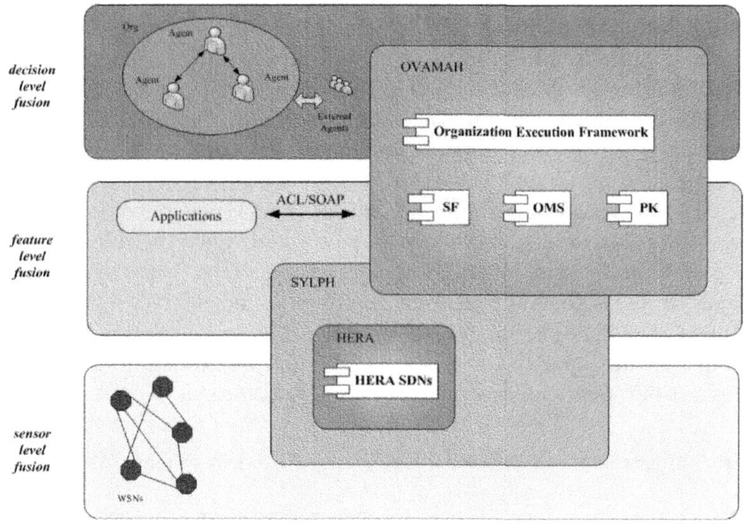

Fig. 1 Proposed framework

On the one hand, at level sensors, with HERA it is possible working with different WSN in a transparent way to the user. A node in a specific type of WSN (ZigBee) can communicate with a node in another type of WSN (Bluetooth).

On the other hand, at a higher level of features and decision, detecting changes in the environment and its consequent action in the system can be managed on the platform thanks to the services and functions that make the agents of the OV thanks to OVAMAH platform. The principal objective of the high level fusion 3 is to transform multiple sources of several kinds of sensors and performance information into a monitoring knowledge base. Embedded in this transformation process is a fundamental understanding of node of WSN functions, as well as a systematic methodology for inserting services to support a specific action according to information received by the node.

The architecture proposes a new and easier method to develop distributed MAS, where applications and services can communicate in a distributed way, independent of a specific programming language or operating system. The core of the architecture is a group of deliberative agents acting as controllers and administrators for all applications and services. The functionalities of the agents are not inside their structure, but modeled as services. This approach provides the systems with a higher ability to recover from errors and a better flexibility to change their behavior at execution time.

4 Results and Conclusions

The case study delineates the scope and potential VO in the design and development of information fusion processors for deployment in multi-agents environments. An organization is implemented by using the model proposed in [6]. The simulation within the virtual world represents ae-health environment, the roles that have been identified within the case study are: Communicator, SuperUser, Scheduler, Admin, Device Manager, Incident Manager.

In order to evaluate the impact of the develop of the MAS using an organizational paradigm, it is necessary to revise the behavior of the MAS in terms of performance. A prototype was constructed based on OVAMAH that could be compared to the previous existing system [6]. The MAS shown in this study is not open and the re-organizational abilities are limited, since the roles and norms cannot be dynamically adapted. As can be seen in the Table 1, the system proposed in this paper provides several functional, taxonomic, normative, dynamics and adaptation properties. The organizational properties are a key factor in architecture of this kind, but the capacity for dynamic adaption in execution time can be considered as a differential characteristic of the architecture.

Table 1 Comparison of organizational and no organizational systems

Features		No organization system	Organizational system
Functional	BDI Model	yes	yes
Taxonomic	Group		yes
	Topology		yes
	Roles		yes
	Interactions	yes	yes
Normative	Norms		yes
Dynamics	Agent Joining	yes	yes
	Role Enactment		yes
	Behaviour control	yes	yes
	Org. Joining		yes
Adaptation	Taxonomic		yes
	Normative		yes
	Functional	yes	yes

In order to test the HERA platform,a distributed WSN infrastructure with HERA running over it has been development. This experiment consisted of trying to start a platform with HERA over a ZigBee SYLPH network infrastructure. The infrastructure consisted of a ZigBee network with 31 nodes (sensors of actuators).The nodes were distributed in a short-range simple mesh, with less than 10 meters between any router and the coordinator. Each time the ZigBee network was formed, nodes were powered on different times, so that the mesh topology was different each time. However, they were some constraints: the maximum depth was 5 and the maximum number of neighbors or children for each node was 8.

After the entire network was correctly created the coordinator and SDN tried to instance a HERA-SDN. HERA-SDN instanced itself and started the HERA platform registering a special SYLPH service called "HERA" on the SDN. Then, 10 of the 30 SYLPH nodes tried to instance one HERA agent, each of them in the HERA platform. Once the HERA-SDN and the 10 HERA agents were successfully instantiated, HERA-SDN started to "ping" every of the ten HERA Agents with a request HERACLES frame including an inform-if command and waiting for a inform frame as a "pong" response. This experiment was run 50 times to measure the success ratio of the platform start and the agent instantiation. However, if the

SYLPH network could not be correctly created, or the HERA platform could not be completely started and created, these runs were also discarded and not taken into account as forming part of the 50 runs. If any HERA agent crashed it was immediately restarted. HERACLES messages were registered to measure when a ping-pong failed and if a HERA agent had to be restarted.

The results (Table 2) indicate that it is necessary to improve SYLPH creation and the instantiation of HERA Agents. In the first case, a better ARQ (Automatic Repeat Request) mechanism could increase SSP-over-WSN transmissions. In the second case, it is necessary to debug the implementation of the agents and fix errors. In addition, the robustness of the HERA agents should be improved by introducing a mechanism to ping and keep running.

In summary, this paper proposes a new perspective for information fusion where intelligent agents can manage the workflow. These intelligent agents collaborate inside a model based on VO. The agents take advantage of their capacities of learning and adaptation to provide information fusion models. Moreover, HERA facilitates and speeds up the integration between agents and sensors. A totally distributed approach and the use of heterogeneous WSNs provides platform that is better capable of recovering from errors, and more flexible to adjust its behavior in execution time.

Concluding, within the proposed framework, a new infrastructure supporting seamless interactions among hardware and software agents, with capabilities able to recognize and self-adapt to a diversity of environments is being designed and developed.

Table 2. Results of the HERA performance experiments.

Total runs	55
SYLPH created correctly	53
HERA started correctly	50
All 10 HERA agents correctly instantiated	50
Total pings tried	7200
Ping-pongs not completed	15
Total restarted HERA agents in an hour	8

Acknowledgments. This work has been partially supported by the project MICINN TIN 2009-13839- C03-03.

References

[1] Alonso, R.S., García, Ó., Zato, C., Gil, Ó., de la Prieta, F.: Intelligent Agents and Wireless Sensor Networks: A Healthcare Telemonitoring System. In: Proceedings of PAAMS 2010, Salamanca, Spain (2010)

[2] Borrajo, M.L., Corchado, J.M., Corchado, E.S., Pellicer, M.A., Bajo, J.: Multi-Agent Neural Business Control System. Information Sciences (Informatics and Computer Science Intelligent Systems Applications An International Journal) 180(6), 911–927 (2010)

[3] Carrascosa, C., Giret, A., Julian, V., Rebollo, M., Argente, E., Botti, V.: Service Oriented MAS: An open architecture (Short Paper). In: Decker, Sichman, Sierra, Castelfranchi (eds.) Proc. of 8th Int. Conf. on Autonomous Agents and Multiagent Systems (AAMAS 2009), Budapest, Hungary, May 10-15, pp. 1291–1292 (2009)
[4] Castanedo, F., Patricio, M.A., García, J.M., Molina, J.: Data Fusion to Improve Trajectory Tracking in Cooperative Surveillance Multi-Agent Architecture?, Information Fusion. An International Journal. Special issue on Agent-based Information Fusion 11, 243–255 (2010), doi:10.1016/j.inffus.2009.09.002
[5] Corchado, J.M., Bajo, J., Tapia, D.I., Abraham, A.: Using Heterogeneous Wireless Sensor Networks in a Telemonitoring System for Healthcare. IEEE Transactions on Information Technology in Biomedicine. Special Issue: Affective and Pervasive Computing for Healthcare 5518, 663–670 (2009)
[6] De Paz, J.F., Rodríguez, S., Bajo, J., Corchado, J.M., Corchado, E.: OVACARE: A Multi-Agent System for Assistance and Health Care. In: Setchi, R., Jordanov, I., Howlett, R.J., Jain, L.C. (eds.) KES 2010. LNCS (LNAI), vol. 6279, pp. 318–327. Springer, Heidelberg (2010)
[7] Hall, D.L., Llinas, J.: An Introduction to Multisensor Data Fusion. Proceedings of the IEEE 85(1) (January 1997)
[8] Hübner, J.F., Sichman, J.S., Boissier, O.: Using the Moise+ for a cooperative framework of mas reorganisation. In: Bazzan, A.L.C., Labidi, S. (eds.) SBIA 2004. LNCS (LNAI), vol. 3171, pp. 506–515. Springer, Heidelberg (2004)
[9] Sycara, K., Glinton, R., Yu, B., Giampapa, J., Owens, S., Lewis, M., LTC Charles Grindle: An integrated approach to high-level information fusion. Information Fusion 10(1), 25–50 (2009)
[10] Kwon, Y., Sundresh, S., Mechitov, K., Agha, G.: ActorNet: An Actor Platform for Wire- less Sensor Networks. In: Proceedings of the Fifth International Joint Conference on Autonomous Agents and Multiagent Systems, pp. 1297–1300. ACM, Hakodate (2006)
[11] Preece, A., Hui, H., Gray, P.: KRAFT: Supporting virtual organizations through knowledge fusion. In: Artificial Intelligence for Electronic Commerce: Papers from the AAAI 1999 Workshop, pp. 33–38. AAAI Press, Menlo Park (1999)
[12] Tapia, D.I., Alonso, R.S., De Paz, J.F., Corchado, J.M.: Introducing a Distributed Architecture for Heterogeneous Wireless Sensor Networks. In: Omatu, S., Rocha, M.P., Bravo, J., Fernández, F., Corchado, E., Bustillo, A., Corchado, J.M. (eds.) IWANN 2009. LNCS, vol. 5518, pp. 116–123. Springer, Heidelberg (2009)
[13] Tynan, R., O'Hare, G., Ruzzelli, A.: Multi-Agent System Methodology for Wireless Sensor Networks. Multiagent and Grid Systems 2(4), 491–503 (2006)
[14] Wooldridge, M.: An Introduction to MultiAgent Systems. Wiley, Chichester (2009)

Greenhouse Indoor Temperature Prediction Based on Extreme Learning Machines for Resource-Constrained Control Devices Implementation

A. Paniagua-Tineo, S. Salcedo-Sanz*, E.G. Ortiz-García,
A. Portilla-Figueras, B. Saavedra-Moreno, and G. López-Díaz

Abstract. In this paper we present an Extreme Learning Machine approach for a real problem of indoor temperature prediction in greenhouses. In this specific problem, the computational cost of the forecasting algorithm is capital, since it should be implemented in resource-constrained devices, typically an embedded controller. We show that the ELM algorithm is extremely fast, and obtains a reasonable performance in this problem, so it is a very good option for a real implementation of the temperature forecasting system in greenhouses.

1 Introduction

Greenhouses are structures thought to recreate an indoor microclimate which allows the crop development by protecting plants from adverse outdoor conditions. In greenhouses provided with the appropriate equipment, this microclimate can be maintained by means of climate control. Environmental factors like inside air temperature, humidity or CO_2 concentration can be controlled by including systems such as heating ventilators or roof and side automatic openings [1, 4, 5, 7].

In modern greenhouses, these automatic systems are usually controlled by embedded controllers, which are usually resource-constrained systems. As shown in Figure 1, a given prediction-controller based on temperature

A. Paniagua-Tineo · S. Salcedo-Sanz · E.G. Ortiz-García · A. Portilla-Figueras
B. Saavedra-Moreno
Departamento de Teoría de la Señal y Comunicaciones, Universidad de Alcalá
e-mail: sancho.salcedo@uah.es

G. López-Díaz
Departamento de Ingeniería, Fundación Tecnova, Parque Científico Tecnológico de Almería.

* Corresponding author.

J.B. Pérez et al. (Eds.): Highlights in PAAMS, AISC 89, pp. 203–211.
springerlink.com

prediction needs of a forecasting system to obtain a prediction of the temperature and then decide actions to be executed, such as automatic roof opening/closing, heating starting etc. In general, the use of embedded controllers implies that any forecasting algorithm included in these controllers must have a very low computational cost.

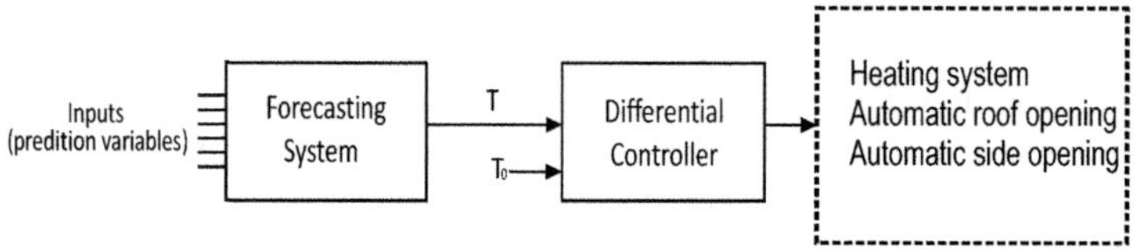

Fig. 1 Prediction-Controller diagram.

The problem of the majority of forecasting algorithms based on soft-computing techniques is that they provide very good approaches in terms of performance, but with the drawback of a significant computation time. Recently, a new kind of neural networks called Extreme Learning Machines (ELM) has been proposed [6]. One of the main characteristics of the ELM approach is that it has an extremely fast computation time, with a very reasonable performance. Thus, the ELM could be perfect as forecasting algorithms for embedded systems.

In this paper we show the performance of these networks in a problem of indoor temperature prediction in a greenhouse. We show a comparison with a Support Vector Machine (SVM), in terms of performance and computation time to show that the computation time of the ELM is assumable to be implemented in embedded controllers, and the difference in performance with the SVM is small.

The rest of the paper is structured in the following way: next section summarizes the most important characteristics of the ELM algorithm. Section 3 is the experimental part of the paper, where we briefly describe the SVM algorithm used for comparison purposes and the main results obtained with the ELM in real data for three days in a greenhouse of Almería, Spain. Section 4 closes the paper giving some final remarks.

2 Extreme Learning Machines

An artificial neural network (ANN) is a massively parallel and distributed information processing system, successfully applied to model a large amount of nonlinear problems [2]. ANNs learn from a given set of examples, by constructing an input-output mapping to perform predictions of future samples. ANN are known to be universal approximators of a large class of functions, with a high degree of accuracy. Single hidden layer neural networks are widely used in forecasting problems, including problems of prediction of temperature in greenhouses [5]. One of the most used ANN is the so-called Multi-Layer Perceptron (MLP), which consists of an input layer, a hidden layer and an output layer, as shown in Figure 2.

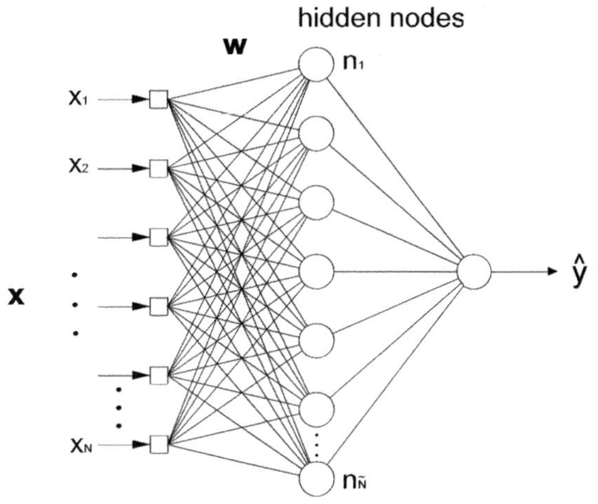

Fig. 2 MLP example diagram.

Each neuron in the hidden layer is an autonomous processing node, which processes the arriving input information in the following way:

$$y = g\left(\sum_{j=1}^{n} w_i x_j - \theta\right), \tag{1}$$

where y is the output signal, x_j, for $j = 1, 2, \ldots, n$ are the input signals, w_j is the weight associated with the j-th input and θ is a bias variable. The transfer function g is usually considered as the logistic function:

$$g(x) = \frac{1}{1 + e^{-x}}. \tag{2}$$

The training process of a MLP consists of, given a training inputs set $\mathbf{X} = [\mathbf{x}_1, \ldots, \mathbf{t}_N]$, and a training out vector $\mathbf{T} = [\mathbf{t}_1, \ldots, \mathbf{t}_N]$, determining the optimal set of weights, and in some cases the network's structure, which minimizes a measure of error. In the network's training process the first step consists of obtaining the so-called *hidden layer output matrix* $\mathbf{H}$ (output of the different neurons for all the input vectors):

$$\mathbf{H} = \begin{bmatrix} g(\mathbf{w}_1 \cdot \mathbf{x_1} + b_1) & \cdots & g(\mathbf{w}_m \cdot \mathbf{x_1} + b_m) \\ \vdots & \cdots & \vdots \\ g(\mathbf{w}_1 \cdot \mathbf{x_n} + b_1) & \cdots & g(\mathbf{w}_m \cdot \mathbf{x_n} + b_m) \end{bmatrix}_{n \times m} \quad (3)$$

where m stands for the number of neurons in the hidden layer. Of course, the training process implies then to tune the network's weights and biases.

Several methods such as gradient-based learning approaches, Levenberg-Marquardt algorithm, evolutionary computing techniques etc., have been previously used to train MLPs. The problem of these methods is that in the majority of cases they require lots of iterations in order to obtain good training performance.

The Extreme Learning Machine (ELM) is a novel and fast learning method recently proposed in [6]. It has been applied to a large number of classification and regression problems. The beauty of this technique is its simplicity together with surprising results comparable to the best known actual techniques. The structure of an ELM is similar to a MLP, however the ideas behind ELM are completely novel. In the next subsection we describe the main concepts needed to describe the ELM approach.

2.1 The ELM Algorithm

In [6] it was proven that if the activation function is infinitely differentiable in any interval, then the hidden layer output matrix $\mathbf{H}$ is invertible and $||\mathbf{H}\beta - \mathbf{T}|| = 0$. They also proved that given any small positive value $\epsilon > 0$, there exists an integer $\tilde{N} \leq N$ that satisfies that $||\mathbf{H}_{N \times \tilde{N}} \beta_{\tilde{N} \times m} - T_{N \times m}|| < \epsilon$. Using these results, the ELM algorithm has been proposed as a very fast approach to train ANNs. It can be described as follows: given a training set $\aleph = \{(\mathbf{x}_i, \mathbf{t}_i | \mathbf{x}_i \in \mathbb{R}^n, \mathbf{t}_i \in \mathbb{R}^m, i = 1, \cdots, N\}$, an activation function $g(x)$ and number of hidden nodes ($\tilde{N}$), (note that we have followed here the notation given in [6], which is different than the one used to introduce matrix $\mathbf{H}$ in Equation (3)).

1. Randomly assign inputs weights $\mathbf{w}_i$ and bias b_i, $i = 1, \cdots, \tilde{N}$.

2. Calculate the hidden layer output matrix $\mathbf{H}$, defined as

$$\mathbf{H} = \begin{bmatrix} g(\mathbf{w}_1 \cdot \mathbf{x_1} + b_1) & \cdots & g(\mathbf{w}_{\tilde{N}} \cdot \mathbf{x_1} + b_{\tilde{N}}) \\ \vdots & \cdots & \vdots \\ g(\mathbf{w}_1 \cdot \mathbf{x_N} + b_1) & \cdots & g(\mathbf{w}_{\tilde{N}} \cdot \mathbf{x_N} + b_{\tilde{N}}) \end{bmatrix}_{N \times \tilde{N}} \quad (4)$$

   Usually the Sigmoid function (Equation (2)) is considered as the activation function $g(x)$.
3. Calculate the output weight vector β as

$$\hat{\beta} = \mathbf{H}^{\dagger}\mathbf{T}, \tag{5}$$

where $\mathbf{H}^{\dagger}$ stands for the Moore-Penrose inverse of matrix $\mathbf{H}$ [6], and $\mathbf{T}$ is the training output vector, $\mathbf{T} = [\mathbf{t}_1, \ldots, \mathbf{t}_N]^T$. By calculating the output weight with the Moore-Penrose inverse matrix, we ensure that the norm of these weights is the smallest from the group of possible solutions.

Note that the number of hidden nodes ($\tilde{N}$) is a free parameter of the ELM training, and must be estimated for obtaining good results. Usually, scanning a range of $\tilde{N}$ values is the solution for this problem.

3 Experiments

3.1 Dataset Description

Our study is focused on the prediction of indoor temperature for three different days in a greenhouse situated in Almería (Spain). The dataset is formed by samples obtained from the greenhouse with different conditions, for example, with different aperture of the windows which allow to control and maintain the indoor temperature. The prediction of indoor temperature can be carried out by means of different variables. In this paper, we consider the following:

- Outdoor temperature
- Outdoor relative humidity
- Wind speed
- Wind direction

All these variables are sampled with a time-interval of $\Delta T = 35s$, obtaining a dataset for each day of approximately 2470 samples. This dataset is permuted and divided by using an 80% and 20% for each day, given several train and test subsets which allow to carry out the experiments described below.

3.2 Alternative Algorithm for Comparison: Support Vector Machines for Regression

One of the most important forecasting statistic models are the support vector machines for regression (SVMr) [9]. The SVMr are appealing algorithms for a large variety of regression problems due to they do not only take into account the error approximation to the data, but also the generalization of the model, i.e., their capability to improve the prediction of the model when new data are evaluated by it. In the classic model presented in [9] the following optimization problem is solved in order to obtain the final model:

$$\min\left(\frac{1}{2}\|\mathbf{w}\|^2 + C\sum_{i=1}^{l}(\xi_i + \xi_i^*)\right) \tag{6}$$

subject to

$$y_i - \mathbf{w}^T\phi(\mathbf{x_i}) - b \leq \epsilon + \xi_i, \quad i = 1, \ldots, l \tag{7}$$

$$-y_i + \mathbf{w}^T\phi(\mathbf{x_i}) + b \leq \epsilon + \xi_i^*, \quad i = 1, \ldots, l \tag{8}$$

$$\xi_i, \xi_i^* \geq 0, \quad i = 1, \ldots, l \tag{9}$$

The Sequential Minimal Optimization (SMO) is commonly used to solve this optimization problem (through its representation in its dual formulation) because it allows to obtain small training times. On the other hand, the optimization problem needs defining the hyper-parameters C, ϵ and the Gaussian Kernel Function parameter γ in order to improve the accuracy of the obtained models. The definition of these hyper-parameters can be carried out by means of the methodolgy studied in[8], which is based in the bounding of the hyper-parameters to reduce the search space for the grid search algorithm which finds the optimal set of parameters. This methodology also allows to obtain good training times with a very good performance for several typical applications.

3.3 Results

In order to evaluate the performance of the ELM model in indoor temperature forecasting, we compare it with the use of the SVM model described in the previous section, knowing the good performance that this model presented in several dataset in the literature. The prediction of indoor temperature is carried out by using a different interval of time between the variable to predict (indoor temperature) and the input variables described in Section 3.1. Note that this prediction interval is a multiple of sampling interval ΔT. In this way it is possible to evaluate the period of time in which the obtained models are good enough to be used.

To carry out this comparison we evaluate two characteristics of the models: their accuracy and their computational cost. The accuracy is measured by means of the root mean square error (RMSE) in several test dataset obtained with several permutations of the original data. Note that in the case of ELM the training algorithm has an important random component. Thus, we have launched 30 times each training for the ELM in each dataset. Since we have 20 different permutations of the data, we obtain 600 different experiments for each day under study. Note that in the case of the SVMr, the training algorithm is deterministic, so to obtain 600 values we repeat 30 times the result obtained by the SVMr in each one of the 20 permutations.

Table 1 shows the accuracy comparison of the ELM and SVM models in the 600 experiments evaluated for the three days and nine intervals of time in each one (multiples of ΔT). In this table we see that the accuracy obtained by the SVM model is better than ELM in all experiments, however, note that the difference is less than 0.5 degrees, which can be an acceptable loss considering that it is only the 2% of the indoor temperature real value.

Table 1 Accuracy comparison based on RMSE mean and standard deviation for the SVM and ELM models

		SVM		ELM	
Day	$\#\Delta T$	Mean	Std	Mean	Std
1	1	0.3927	0.1112	0.4591	0.0356
	2	0.3859	0.1107	0.4681	0.0518
	3	0.4028	0.1067	0.5155	0.1319
	4	0.3888	0.1090	0.4851	0.0914
	5	0.4072	0.1116	0.5179	0.1005
	6	0.3979	0.1085	0.4937	0.0875
	7	0.4259	0.1116	0.5415	0.1358
	8	0.4113	0.1103	0.5377	0.1072
	9	0.4182	0.0943	0.5388	0.1040
2	1	0.5118	0.4211	0.7093	0.2816
	2	0.5138	0.4382	0.7424	0.3421
	3	0.5086	0.4474	0.6964	0.2994
	4	0.4751	0.4617	0.6885	0.3130
	5	0.5209	0.4684	0.7468	0.4074
	6	0.5102	0.4732	0.7250	0.3389
	7	0.5226	0.4652	0.7766	0.4189
	8	0.5814	0.4883	0.7163	0.3228
	9	0.5129	0.5115	0.7263	0.3546
3	1	0.5860	0.1279	0.7477	0.1961
	2	0.6196	0.1281	0.7928	0.1793
	3	0.6141	0.1278	0.7650	0.1932
	4	0.5744	0.1298	0.7433	0.2232
	5	0.5876	0.1275	0.8248	0.3706
	6	0.5848	0.1352	0.8134	0.2882
	7	0.5895	0.1296	0.8984	0.4316
	8	0.6285	0.1380	0.8195	0.2339
	9	0.6475	0.1300	0.8500	0.2873

On the other hand, we evaluate the computational cost needed to obtain the ELM and SVM models. In an approximate way, the complexity of the ELM training algorithm is obtained by means of the complexity of the Single Value Decomposition algorithm used to calculate the Moore-Penrose inverse of matrix ($\mathbf{H}$), i.e., $O(\tilde{N}N^2)$, where N is the number of samples in the train set and $\tilde{N}$ is the number of neurons in the hidden layer. On the contrary, the complexity of the SVM is estimated to be approximately quadratic with the number of samples $O(N^2)$ for the LIBSVM algorithm [3]. However, it is necessary to consider the search of the hyper-parameters described in the previous section, which converts the complexity in $O(k^3N^2)$ where k is the number of samples evaluated for each hyper-parameter. Although both complexities depends on the term N^2, note that the number of samples in the hyper-parameter search space is usually very high in order to obtain a good accuracy of the model, and thus it is higher than the number of neurons, making the complexity of the ELM smaller than the SVM. A similar result can be seen by means of the training times obtained in the experiments, which are shown in Table 2. It is easy to see that the training time for the ELM model is very small comparing to the SVM model. Considering than

Table 2 Computation time comparison between SVM and ELM algorithms.

Day	#ΔT	SVM	ELM
	1	1692.8960	0.0549
	2	1716.4780	0.0569
	3	1727.6735	0.0570
	4	1708.0055	0.0564
1	5	1735.1140	0.0567
	6	1704.5385	0.0568
	7	1706.8455	0.0571
	8	1711.0835	0.0569
	9	1727.5260	0.0569
	1	1834.3850	0.0559
	2	1826.1055	0.0572
	3	1821.2745	0.0575
	4	1823.4880	0.0573
2	5	1827.4065	0.0572
	6	1817.6915	0.0566
	7	1803.0515	0.0568
	8	1819.6505	0.0569
	9	1794.6880	0.0569
	1	1772.1300	0.0565
	2	1777.8630	0.0569
	3	1726.2805	0.0565
	4	1768.8230	0.0566
3	5	1773.0845	0.0564
	6	1751.7630	0.0563
	7	1747.5840	0.0541
	8	1657.6415	0.0568
	9	1623.1290	0.0565

the RMSE difference between the SVM and ELM is smaller than a half degree and the training time is very small, we conclude that the ELM is a very good soft-computing technique to solve the indoor temperature prediction in a system with limited computational resources.

4 Conclusions

In this paper we have presented a very fast approach to predict the indoor temperature prediction in greenhouses. This forecasting system can be implemented in embedded systems to control automatic actions in the greenhouse, such as the roof opening, the heating starting, etc. The presented approach is based on a extremely fast neural network recently proposed, known as Extreme Learning Machine (ELM). We have shown that the ELM is able to obtain reasonable results in terms of temperature prediction, only slightly worse than a Support Vector Regression algorithm, but with a much better performance in terms of computational cost, which is a key point to implement this system in a real resource-constrained controller.

References

1. Bennisa, N., Duplaix, J., Enéa, G., Halouac, M., Youlal, H.: Greenhouse climate modelling and robust control. Computers and Electronics in Agriculture 61, 96–107 (2008)
2. Bishop, C.M.: Neural networks for pattern recognition. Oxford University Press, Oxford (1995)
3. Bottou, L., Chapelle, O., DeCoste, D., Weston, J. (eds.): Large Scale Kernel Machines. MIT Press, Cambridge (2007)
4. Coelho, J.P., Moura Oliveira, P.B., Boaventura Cunha, J.: Greenhouse air temperature predictive control using the particle swarm optimisation algorithm. Computers and Electronics in Agriculture 49, 330–344 (2005)
5. Ferreira, P.M., Fariab, E.A., Ruano, A.E.: Neural network models in greenhouse air temperature prediction. Neurocomputing 43, 51–75 (2002)
6. Huang, G.B., Zhu, Q.Y., Siew, C.K.: Extreme learning machine: Theory and applications. Neurocomputing 70, 489–501 (2006)
7. Leal Iga, J., Leal Iga, J., Leal Iga, C., Ayala Flores, R.: Effect of air density variations on greenhouse temperature model. Mathematical and Computer Model 47, 855–867 (2008)
8. Ortiz-García, E., Salcedo-Sanz, S., Pérez-Bellido, A., Portilla-Figueras, J.A.: Improving the training time of support vector regression algorithms through novel hyper-parameters search space reductions. Neurocomputing 72, 3683–3691 (2009)
9. Smola, A.J., Schölkopf, B.: A tutorial on support vector regression. Statistics and Computing (1998)
10. Teitel, M., Tanny, J.: Natural ventilation of greenhouses: experiments and model. Agricultural and Forest Meteorology 96, 59–70 (1999)

Propagation of Agent Performance Parameters in Wireless Sensor Networks

J.C. Cuevas-Martinez, J. Canada-Bago,
J.A. Fernández-Prieto, and M.A. Gadeo-Martos

Abstract. Wireless sensor networks are composed of resource-constrained sensor nodes, powered by batteries, with limited CPU and memory, and wireless communication. In spite of the fact that sensor nodes are resource-constrained devices, several soft computing technologies have been adapted to them. In order to save battery, sensor nodes work in cycles based on awake and sleep modes. In this work we propose a method, based on a differential decision system, to calculate dynamic parameters that control the awake-sleep cycle in a multi-agent sensor structure and their propagation to other sensor nodes in a network. As an application of the proposed system, a sound pressure monitoring application is presented. Results have shown that the proposed method utilizes less work cycles than continuous measuring systems, saving battery and improving the lifetime of sensor nodes, with a reasonable lost of precision.

Keywords: Resource-constrained devices, multi-agent systems, wireless sensor networks.

1 Introduction

Wireless Sensor Networks (WSNs) [1] are composed of a large number of sensor nodes, powered by batteries, where each node consists of a processing unit with limited computational capability and memory, wireless communication, probes and actuarors. Their range of application is very wide: intelligent agriculture, industrial control and monitoring, environmental monitoring systems, surveillance, health monitoring, traffic monitoring, etc.

In order to prolong the sensor lifetime, sensor nodes usually operate in a work cycle in which they, first, execute the application (measure, calculate outputs, actuate, etc.), decide if it is necessary to connect with a base station or other sensors,

J.C. Cuevas-Martinez · J. Canada-Bago · J.A. Fernández-Prieto · M.A. Gadeo-Martos
Telecommunication Engineering Department
University of Jaen, Alfonso X El Sabio 28, 23700 Linares, Jaen (Spain)
e-mail: {jccuevas,jcbago,jan,gadeo}@ujaen.es

J.B. Pérez et al. (Eds.): Highlights in PAAMS, AISC 89, pp. 213–221.
springerlink.com

and then are configured in a sleep mode during an interval, which is calculated for each application and traditionally remains constant.

Although considerable research has been devoted to different WSN applications [2], rather less attention has been paid to adapting the **Sleep Mode Intervals (SMIs)** in each work cycle from the evaluation of the surrounding environment. In most cases, the estimated SMI is calculated with energy, coverage or routing restrictions [3]. The objective of this work is to obtain dynamic parameters related to awake-sleep cycle using a multi-agent sensor structure and their propagation among sensor nodes in a WSN in order to improve their lifetime. These dynamic parameters will allow sensor nodes to adapt their work cycles to the current conditions, reducing the number of work cycles and battery consumption, and therefore, increasing their lifetime. On the other hand, sharing the parameters with neighbour sensor nodes will improve the complete network behaviour.

The remainder of the paper is organized as follows. The following section deals with related work. Section 3 shows the sensor multi-agent structure and the system proposed to obtain dynamic parameters that will be propagated among sensors in a WSN. Section 4 presents the system adaptation to an application devoted to sound pressure monitoring. Section 5 shows the experimental results. Finally, conclusions are drawn in Section 6.

2 Related Work

WSNs [4] have become a new important area under study due to their possibilities, like mobility, distributed processing, monitoring and control of dangerous processes. However, WSNs have powerful constraints, mainly when sensor nodes are isolated (limited power source, computational capacity, wireless interference, routing, etc) that the applications for WSNs have to take into account. Therefore, that is one of the reasons of using intelligent systems to manage sensor nodes. Moreover, WSNs represent an ideal scenario to integrate intelligent agents that can accomplish complex applications despite the constraints of the WSNs [5].

Multi-agent theory and its applications have been under study for several years [6] and have become a real solution for a wide variety of complex problems [7]. In this work, the multi-agent system is embedded inside sensor nodes to manage certain decisions of the sensor. That is the case of WISMAP [8], a WSN application management protocol that defines a special multi-agent based framework over WSNs. WISMAP encloses communications, application process, data format, resource hierarchy and agent interaction inside and outside the sensor node. That framework shows that multi-agent systems can perfectly suit into a sensor node to manage WSNs resources efficiently. And one of those important resources is the battery, so sensor nodes usually have to stay most of the time in power saving modes and only in active mode for very short periods of time. Therefore, SMI is one of the parameters under study nowadays in WSNs.

3 Propagation of Agent Performance Parameters

The SMI represents an important aspect in WSNs due to the fact that it permits to prolong the lifetime of sensor nodes. Traditionally, the SMI is calculated for each application remaining constant during its lifetime or, in other cases, is calculated by sensor nodes individually. The main objective of this work is the propagation of dynamic parameters related to awake-sleep cycle among the sensor nodes in a WSN in order to improve their lifetime. Basically, each sensor node calculates its own next cycle SMI, changes to sleep mode for this time, evaluates it when returning to awake mode, and shares its value and evaluation with the rest of sensor nodes by mean of transmitting them via a base station (computer with an access point to the WSN). The shared values of SMIs and their evaluation allow sensor nodes to improve and adjust their values.

The next section presents the multi-agent structure used by sensor nodes that it is an enhancement of WISMAP structure. This structure includes a new subsystem with two new methods, one to obtain the SMI, and the other to evaluate the SMI. Section 4 presents an application of sound pressure monitoring.

3.1 Multi-Agent Structure for WSNs

The sensor node software is based on the multi-agent structure defined by WISMAP. This structure is composed of three agents: management, application control, and communication. The main objectives of the **management agent** are the execution of other agents and the control of the sensor sleep-awake cycle, in which it calculates a new SMI, programs the sensor node in sleep mode, and allows it to return to awake mode. Executed by the management agent, the **application control agent** allows the sensor node to control the execution of different applications (measurement, actuators, FRBS, etc). The **communication control agent** incorporates the application protocol, which allows sensor nodes to communicate with other sensor nodes, neighbouring sensors, and with a base station.

3.2 Differential Decision System

The proposed method to calculate the next SMI, the **Differential System (DS),** is based on the difference between the variable value measured in the present cycle and its values obtained in previous cycles, the region where the values fit and the battery level. The method consists of three steps:

1º) The sensor node measures the value of the object variable and calculates the difference between the new value and the measured in the previous cycle.

2º) After that, the sensor node verifies if the present value belongs to a different region than previous values. Besides, the sensor node may take into account if the present value belongs to critical regions.

3º) Taking into account the difference of values, change of zone, critical regions, battery level, and previous SMIs, the sensor calculates the next SMI.

The parameters to calculate the SMI depend on the application and have to be adapted to its objective. Section 4 presents one application of sound pressure monitoring where a particular SMI is calculated using the DS.

3.3 Proposed Success Conditions

We propose that the evaluation of the SMI can be based on a set of success conditions that are evaluated by each node when it returns to awake mode. Sensor nodes can take into account internal conditions, such as previous SMIs, scheduled tasks and the battery level, and external information, such as variable values, variable tendencies, alarms, and the neighbor or network SMI. On the one hand, critical regions can be defined in the range of input variables. For example, success conditions can include scenarios such as long increments in the value of the variable with small SMIs, short increments with large SMIs, large SMIs with low battery levels, and small SMIs in critical regions. On the other hand, failures conditions can include scenarios such as long increments with large SMIs, short increments with small SMIs, small SMIs with low battery levels, and large SMIs in critical regions. As we have indicated previously, the way on which the next SMI is evaluated depends on the application and has to be adapted to its objective.

4 Sound Pressure Monitoring Application

We have to remark that the main purpose of the solution explained above is to propagate the best SMI obtained among all the sensor nodes to avoid unnecessary awake cycles in every sensor node that is monitoring a particular magnitude in applications where power supply for sensors are not suitable or even impossible. Thus, the applications that are covered by this system have to be studied and modelled separately and must show an inertial behaviour like, for example, the temperature of a room, or the revolutions per minute of an engine.

The magnitude used to test the DS is the sound pressure. This magnitude is used to measure the loudness of sound in one area for a certain period of time. Sound pressure is a typical magnitude in acoustical pollution [9] but with the right tuning can be used to control other systems like working machinery [10]. Engines or moving parts produce similar sound patterns; therefore several ranges of sound pressure can be assigned to different loads of work or failures.

4.1 Differential System Parameters and Calculus

The parameters of the DS can be divided among three categories, related to the difference in measured values, sound pressure level and battery level:

- Maximum SMI (SMI_{max}): if the new SMI> SMI_{max} then SMI= SMI_{max}.
- Minimum SMI (SMI_{min} if the new SMI< SMI_{min} then SMI= SMI_{min}.
- Region hop (*rh*): if the new measure goes into a different region shown in Table 1 then *rh*= abs(current region value - previous region value).

Table 1 Differential system thresholds

Sound pressure	Lower critical	Lower normal	Upper normal	Upper critical	
	45 dB	55 dB	65 dB	75 dB	
Region value	-2	-1	0	1	2

- Steady (*st*): *st* grows one unit if the new measure does not change of region.
- Difference (d): the difference between the previous measure and the new one.
- Previous SMI (ps): this is the previous estimated SMI.
- New SMI (ns): the new SMI that is modified by the three phases of the DS.

Then, with *rh* and *st,* the base increment *i* is calculated as follows:

$$i = i - (rh * 0.1) + (st * 0.05) \qquad (1)$$

- Delta (Δ) is the adjustment to *ps* due to the difference between the new measure and the previous one, and it can take three values:
  - $\Delta = +10\%$ if $d < 10$ dB
  - $\Delta = -10\%$ if $10 \text{ dB} \le d < 20$ dB
  - $\Delta = -20\%$ if $d \ge 20$ dB
- Increment of duty time (*B*) due to battery level (β) is the last modifier applied to *ps* to obtain *ns*. *B* can take the following values:

Table 2 Increment of duty time (*B*) due to battery level (β)

β	$\beta > 75\%$	$50\% > \beta \le 75\%$	$25\% > \beta \le 50\%$	$10\% > \beta \le 25\%$	$\beta \le 10\%$
B	+ 0%	+ 10%	+ 30%	+ 60%	+ 100%

With all the modifiers the final *ns* value is:

$$ns = (ps * i)\,(1 + \Delta)\,\,(1 + B) \qquad (2)$$

4.2 Quality Evaluation

To achieve the main goal of the system described in this paper, the propagation of performance parameters, every sensor calculates its own quality evaluation based on the previous SMI and the new region value of the taken measure. The evaluation gives an integer number *Q*, from $-\infty$ to $+\infty$. The algorithm to calculate Q uses two SMI values as bounds to define what is short (SMI_{short}) or long (SMI_{long}) compared to the possible SMI range. The algorithm of Q evaluation is the following:

- IF rh = 0 AND ps is NOT short THEN Q = Q + 1 ELSE Q = Q - 1
- IF rh=1 AND ps is NOT long THEN Q = Q + 1 ELSE Q = Q - 1
- IF rh=2 AND ps is short THEN Q = Q + 1 ELSE Q = Q - 2

Where:

short: SMI<SMIshort and SMIshort= SMImin + (SMImax - SMImin)/4
long: SMI>SMIlong and SMIlong= SMImin + (SMImax - SMImin)/2

When the level of quality reaches the stability threshold, established empirically in Q = 5, the sensor transmits to the base station the information of the SMI calculated and its evaluation. This message is stored in the base station and compared with other received from sensor nodes in the same area that the sending one. If the evaluation value is higher than the other values, the next time the base station would receive a sensor message, the response message will also carry that evaluation value and the SMI calculated by the previous sensor. Then, the SMI received will be used as the new SMI for the next estimation. In this approach, sensor nodes controlling the same area can collaborate to reach the most appropriate SMI and save processing time when surrounding sensors calculates a better SMI.

5 Results

The systems explained in previous sections have been tested in several scenarios to analyze the error accumulated in the measures when the SMI is not constant and short compared to the inertia of the system. The main purpose of these tests is to demonstrate that the DS can take enough measures to detect the evolution of inertial magnitudes, with less work cycles than Continuous Measurement Systems (CMSs). It is expected that the error introduced by DS will be insignificant and the sensor node lifetime will be prolonged.

5.1 Description of the Experiment

The system has been tested with pseudo-random sound pressure signals and real sound pressure measures obtained from working computers. Those kinds of systems are highly inertial and could model common systems like engines or conveyor belts, where sound should keep a constant pattern along time, only changing when the working duty increases or when there is a failure.

The range of the sound pressure values goes from 40 dBA to 90 dBA. Normal working values are around 60 dBA, so 40 dBA is close to silence, and 90 dBA is the louder sound pressure expected. In order to test the system and to get the sound pressure values, we have designed and implemented a simple analog circuit, equipped with an electret microphone, which has been incorporated to a Sun SPOT sensor.

5.1.1 Simulations

The pseudo-randomly generated signals, that represent inertial systems, have values of sound pressure from 40 dBA to 90 dBA.Every test shows one day of continuous working, with three different measures taken: every second (reference signal), every 20 seconds (CMS), and the measures taken from the DS. Each simulation is calculated with three battery charges, 99.9%, 50% and 10%.

5.1.2 Real System

In order to measure the quality of DS and avoid the effects of different battery consumption among sensor nodes, only one sensor was used to take the real sound pressure samples. The real system used is an opened computer case with one additional fan that can be manipulated to simulate failures. The total processing time of a sensor node with the DS running is around 1800 ms in a Sun SPOT. The normal processing time of a sensor node testing only the instant sound pressure takes around 1450 ms, so the penalty of the DS system can be assumed due to the reduction in the number times to wake up.

5.2 Experimental Results

The results of the experiments show that the DS can follow the evolution of the system with a reasonable loss of precision and with less awake cycles than the CMS. Figure 1 illustrates the evolution of both systems.

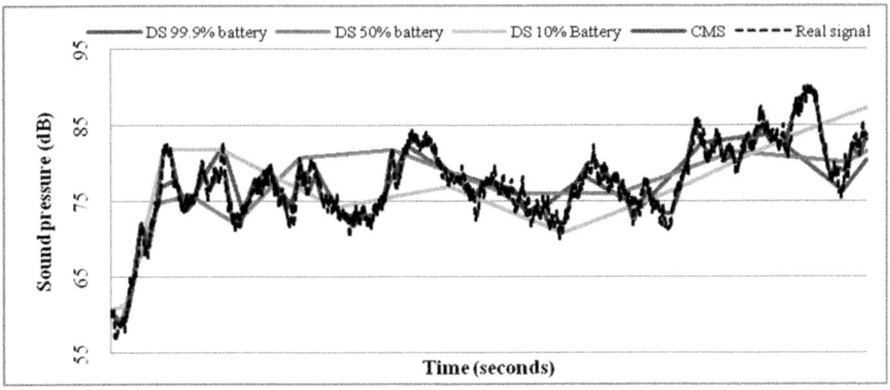

Fig. 1 Interpolated signals with pseudo-random measures for 2000 seconds

To obtain the interpolated signals the CMS woke up the sensor 100 times and the DS only 25 times (with 99.9% of battery charge), that is around a 65 % of power consumption reduction taking into account the total processing time.

5.3 Error Measures

It is clear that with fewer measures the possibility of losing significant values increases. Nevertheless, in WSNs the cost of losing measures can be more acceptable than excessive power consumption but only when the global error introduced by the DS was similar to the CMS error. The first error indicator is the absolute error of the interpolated signals generated through the measurement points of both systems compared to the real signal. The evolution of the absolute error for the

Table 3 Absolute error

System	DS 99.9% battery	DS 50 % battery	DS 10% battery	CMS
Absolute error	2,316%	3,693%	3,996%	1,127%

same period than Figure 1 is shown in Table 3. These experiments have been repeated for more than a hundred of different signals giving always similar values.

5.3.1 Quality Evaluation

As can be observed in Table 4, the best quality is obtained when the battery charge is almost full, and battery regulation does not apply yet (see section 4.2). Nevertheless, the result for 50% battery is good enough to reach 5 at the end of the interval.

Table 4 Evaluation of Q for the same period than Figure 1

	Instant (s)	20	208	407	790	1004	1259	1570	1749	2000
Battery	99.9 %	0	-1	1	4	6	7	9	8	10
	50.0 %	0	-1	1	1	1	2	3	4	5
	10.0 %	0	-3	-2	-3	-4	-5	-6	-6	-5

6 Conclusions and Future Research

This work has presented a method to obtain dynamic parameters related to the awake-sleep cycle in sensor nodes in order to prolong the lifetime of sensor nodes. Besides, a sound pressure monitoring has been developed to test the method. The results demonstrate that simple differential methods, with low computation cost, can save battery in continuous sensing applications with a reasonable loss of precision. On the other hand, further research is necessary in DS to model other real magnitudes and to tune the increment of the delta evaluation system and reduce the adaptation time of sleep mode interval to appropriate values. Future work will focus on the design and development of a FRBS devoted to infer the SMI.

Acknowledgments. This work has been partially supported by the University of Jaen and Caja Rural de Jaen (Spain) (project UJA 08_16_18) and the Ministerio de Ciencia e Innovación (project TEC2009-13619).

References

[1] Akyildiz, I.F., Su, W., Sankarasubramaniam, Y., Cyirci, E.: Wireless sensor networks: A survey. Computer Networks 38(4), 393–422 (2002)

[2] Yick, Y., Mukherjee, B., Ghosal, D.: Wireless sensor network survey. Computer Networks 52(12, 22), 2292–2330 (2008)

[3] Cohen, R., Kapchits, B.: An optimal wake-up scheduling algorithm for minimizing energy consumption while limiting maximum delay in a mesh sensor network. IEEE/ACM Trans. Netw. 17(2), 570–581 (2009)
[4] Karlsson, B.: Intelligent Sensor Networks - an Agent-Oriented Approach. In: Workshop on Real-World Wireless Sensor Networks (2005)
[5] Cuevas-Martinez, J.C., Gadeo-Martos, M.A., Fernandez-Prieto, J.A., Canada-Bago, J., Yuste-Delgado, A.J.: Wireless Intelligent Sensors Management Application Protocol-WISMAP. Sensors 10(6), 8827–8849 (2010)
[6] Wooldridge, M.: An Introduction to Multiagent Systems. Wiley, New York (2002)
[7] Karl, H., Willig, A.: Protocols and Architectures for Wireless Sensor Networks. John Wiley & Sons, Chichester (2005)
[8] Filipponi, L., Santini, S., Vitaletti, A.: Data Collection in Wireless Sensor Networks for Noise Pollution Monitoring. In: Proc. of the 4th IEEE International Conference on Distributed Computing in Sensor Systems, Santorini Island, Greece, pp. 492–497 (2008)
[9] Tandon, N., Choudhury, A.: A review of vibration and acoustic measurement methods for the detection of defects in rolling element bearings. Tribology International 32(8), 469–480 (1999)

Distribution of a Reasoning Engine over Wireless Sensor Networks

Carmen Iniesta, Juan A. Botia, and Pedro M. Ruiz

Abstract. In this paper we set out an approach to distribute the evaluation of a rule-based engine over a wireless sensor network. We propose the use of genetic algorithms to calculate a good distribution of the rule evaluation. This way, we pretend to avoid the need of having a base station connected to the WSN, where the reasoning is done in a centralized way.

1 Introduction

Nowadays there is a growing interest in wireless sensors networks [1][2][3]; systems based in them are being used in a very vast amount of fields. These networks are often integrated with expert systems since it is a very useful way to process the amount of data obtained from sensors readings[4]. In the great majority of these approaches the information processing task is done in a centralized way[5], making it mandatory to have a single base station connected to the sensor network in which all the mentioned information processing will be done. In other cases the sensors also perform small reasoning tasks, in order to reduce the power compsumption by doing data aggregation or other in network processing tasks. To avoid the need for that base station we set out an approach to distribute the reasoning over the network. With this we pretend to have more scalable networks, with more fault tolerance since they will not be tied to a single base station. This distribution of the reasoning implies not only to distribute the evaluation of the rules, so that every rule will be evaluated only by a concrete sensor. But also, to assign, in a similar fashion, the management of the variables that are working memory elements but are not linked to environment measures.

There are several constraints that need to be taken into account when designing an strategy for the mentioned distribution. Our solution not only

---

Carmen Iniesta · Juan A. Botia · Pedro M. Ruiz
Universidad de Murcia
e-mail: carmen.iniesta1@um.es

J.B. Pérez et al. (Eds.): Highlights in PAAMS, AISC 89, pp. 223–231.
springerlink.com

has to provide the same results that would be obtained in a distributed fashion, but also it has to be aware of the constraints associated to wireless sensor networks. The main problem that we face is the power consumption of the motes that make up the network. The most important issue concerning this energy restriction is the cost involved in the wireless communication. Therefore, the solution we are looking for must try to minimize these costs.

We can find numerous approaches to distributed reasoning in the literature, but practically none of them is referred to do this distribution over sensor networks, nevertheless there are some aspects of them that can be quite useful for our goals. And also there are many in-network processing (treatment of the data gathered by the sensors done within the network, such as data aggregation) examples that our problem can slightly relate to.

Based on the inherent computational complexity of our problem, we set out that a feasible solution could be to find an heuristic to assign the rule and variable set management over the sensor network by means of a genetic algorithm [6]. Since the search space of the possible solutions grows exponentially with the number of rules and sensors this approach seems a better idea than a greedy one.

Besides setting out the approach to solve the problem, along this paper we also expose the results obtained with simulations of the evaluation costs of distributed rule systems, comparing the results that we get with the genetic algorithm distribution, with the ones obtained doing the distribution randomly and with a simple heuristic approach.

The rest of the paper is organized as follows, in Section 2 we can find a description of the problem, followed by a description of the chosen solution. In Section 3 we set out the trials performed and the results that we obtained in them. And finally, in Section 4, we expose the conclusions and future research lines for these subjects.

2 Distributed Inference in Rule Based Systems over Sensor Networks

As we have already set out, sensor networks are a very active field in research nowadays. And also, they are very often combined with reasoning engines, as a way to obtain intelligence from the enormous amounts of data that are usually acquired by the sensors.

In a vast majority of these approaches the reasoning task is done in a centralized way, but there are several reasons that are encouraging to try to distribute it. The current trend in this field is to add intelligence to the devices so all the processing can be integrated in the environment and thus avoid all the inconveniences associated to the presence of a base station. Such as the lack of robustness and scalability that implies depending on a single unit to do all the processing, in whom all tha data has to be gathered.

There are not many approaches in the literature to this problem, but it is possible to use many ideas in the field of distributed rule evaluation, such as the representation of the relationship between rules and its variables in the form of a graph. This idea is interesting though, normally, that distribution usually refers to a more computational than physical view of the distribution. In this case, the physical view is crucial because it is directly related with power consumption in sensors at the time of communication. Moreover, some aspects and techniques of in-network processing over sensor networks can be really useful, such as, for example, hierarchical organization and the use of heterogeneous nodes.

The computational cost of rule processing increases as the size of the rule set grows. In consequence there are numerous works in which the reasoning is distributed, in order to handle these costs. Also, in some cases, this distribution is required because of the inherent geographic locations of the different knowledge bases involved. Two meaningful examples of this are [7] and [8]. In the first one the distribution of the rules it's done in an explicit way, grouping the rules in clusters. A relevant characteristic of this approach is the treatment of the access to remote variables to avoid incoherencies. The second one is another significant example, that our case can relate to, where the reasoning is represented as a RETE network [9] and the aim is to distribute the RETE nodes amongst different locations, to parallelize the reasoning.

As we have already mentioned some ideas from in-network processing techniques can be applied to the development of our solution, due to the numerous amount of shared characteristics. In [10], a solution to satisfy data requests over a sensor network with streams of continuous data appears. The aim there is to reduce communication costs and energy consumption, and attend the requests in a efficient way. To do so, the work proposes a solution based on an heterogeneous network in which the most powerful nodes are in charge of doing more data processing and aggregation.

Alternatively, in [11], it's set out the convenience of the use of clustering when the aim is to execute an algorithm in a distributed way over distributed nodes. Since it can reduce communication and computational costs in medium and large size networks, assuming a cluster works as an entity, making the solution more scalable and independent of the number of nodes.

Distributing a rule set over a sensor network has also associated problems that need to be solved. One concrete problem is the distribution of the control algorithm of a rule-based system.

The scenario we consider is a heterogeneous network in which we have two types of sensors, the most simple ones, with constrained capabilities and other ones with more communication and computation skills, longer lifetime, etc. The sensors are grouped forming clusters. In these clusters the complex motes take the role of cluster heads, and they are in charge of the coordination amongst the other members on the groups and the contact with the rest of cluster heads in the network. Assuming this topology, we pursue the design as a rule distribution approach and a decentralisation of the control algorithm.

Notice that the distribution of the control algorithm of the rule-based system not only involves distributing the rules among the sensors, but also to distribute the variables used in the rules. Some of these variables are bind to a specific sensor because they are the results of its measures. But there is another type of variables, the intermediate ones that are not linked with any sensor, instead, they are created or modified by the rules. That is, they are not sensed from the environment but generated from the rules. Thus, when we distribute the evaluation we also have to assign these variables to sensors. This distribution is not immediate since an intermediate variable can be linked to multiple rules.

We can see the set of rules and variables, that is, the production memory and the working memory, and the relationships between them as a reasoning graph, where variables and rules are the nodes and there is an arc directed from a variable to a node if that variable appears in any of the conditions of the antecedent of the rule. And there is a directed link from a rule to a variable if the variable appears in the consequent of the rule. This concept of reasoning graph is similar to the one set out by Robert Doorenbos in his Thesis.[9]

Once we have the reasoning graph we can focus on reducing the cost associated with the distribution of the nodes amongst the motes of the network, that way we can identify the links with the cost that implies accessing the variables from the rules. Also we take into account the frequency of activation of the rules and variables, so the only thing that matters is not the communication cost to access a variable but also how often it is accessed.

Since there are an exponential number of combinations to explore in relation to the number of rules and sensors to get the optimal solution, an exhaustive search is not affordable. Instead, we propose to use a genetic algorithm [12] to help us to solve this task. This algorithm searches for an optimal distribution of the rule set, in terms of the estimated communication cost incurred when working normally.

In a genetic algorithm, we have a population of individuals that represent solutions to our problem, and some of these individuals are selected to cross with each other to create new individuals that will be members of a new generation. The algorithm simulates basic evolutive mechanisms we may find in biology, and evolves getting better individuals each time. Such individuals are possible solutions to our problem.

In the codification of individuals proposed in this paper each one is formed by a number of genes equal to the sum of the number of rules and the number of intermediate variables. These genes represent the sensors to which the variables or rules are assigned. In what concerns to the selection of the individuals first we have to highlight that elitism is used, so the best individual is always selected to pass to the crossing phase. And the selection of the rest is performed using the roulette method. Thus, the better an individual is, the higher probability it has to be chosen for the following phase.

The fitness function encodes the idoneity of each individual solution. And in this case it represents the communication cost associated with the evaluation of the rules, providing that we consider the cost of a rule as the cost of accessing each one of the variables that appear either on its consequent or antecedent, multiplied by the activation frequency of the rule. The following function is used to calculate the fitness of each individual:

$$Fitness(i) = Max - \sum_{j=1}^{N}(EvaluationCost(j) * EvaluationFrequency(j))$$

Being N the number of rules of the production memory, Max an upper bound of the fitness used so the function can indicate the goodness of the individual, EvaluationCost the cost in hops of the access to all the variables that appear on the rule, and EvaluationFrequency the estimated frequency of evaluation of the rule. Note that this frequency is a simplification with simulation purposes, but in a real scenario there would be a frequency of activation of each variable, and a frequecy of triggering of each rule.

Once the selection phase is done, some of the individuals selected are crossed with each other to form new ones. The probability of being chosen to cross with another individual it's 0,1 for each one; if an individual is not selected then it goes straightly to the next phase.

After that, there comes the mutation phase, in which some of the genes of some of the individuals are modified in a random way, with a probability of one divided by the size of the individual.

3 Experimental Results

Two basic aproaches for distributing rule sets over sensor networks were defined. They are used here as baseline algorithms, to assess the effectiveness of the approach proposed in this paper. One of them consists on doing the distribution randomly. The other one is a simple heuristic approach. With the random one we want to represent a blind approach to solve the problem. It is a simple way to arrange the distribution of the rules. To implement this solution it is not necessary to have information about the relationships amongst the rules and the variables.

We can divide the Heuristic approach in two phases. In the first one, every rule is assigned to the cluster head in whose zone the bigger amount of non-intermediate variables that appear in the conditions of the antecedent of the rule are allocated. Once the rules are allocated, intermediate variables are allocated according to a similar criterion. In this case every one of them is assigned to the cluster head that "owns" the higher amount of rules in whose conditions the variable appears. Thus, this approach represents a reasonable trade-off between the complexity of the task and the quality of the solution.

Still it does not take into account neither the activation frequencies nor the distances amongst the motes.

We have accomplished numerous trials. The main goal of the performed experiments was to compare the performance of the allocation obtained with the genetic algorithm propossed here, and how it adapts to different scenarios. The evaluation metric is based on hops per rule and hops per variable access. These are the communication hops between sensors involved in the evaluation of the rule set, so they are the sum of the hops that are needed to evaluate every rule of the rule set, taking into account the different activation weights of the rules. These costs are caculated in a similar fashion as the fitness. Since the main goal that we pursue is to minimize the power comsumption of the devices, the best ranked solution will be the one that takes less hops to do the evaluation.

Moreover three different types of rules are considered, depending on the access to variables. When a condition access a variable, it means that this access not only implies a query to one concrete sensor, but to all of the sensors that measure variables of that certain kind, therefore the presence of non-linked variables increases the complexity of the distribution. An example of this is a rule that would state that all the sensors with temperature higher than 40 °C would set their state variable to "warm". When this rule is triggered it will have to interact with all the sensors measuring temperature instead of with a concrete one.

Thus, the chosen types of rules are the following:

- **Type 1,** rules in whose conditions there is no presence of variables.
- **Type 2,** rules in whose there are only presence of variables in the antecedent.
- **Type 3,** rules in whose conditions there is presence of variables both in the left and right part.

Nevertheless we'll only show the graphics corresponding to type 3 rules, because they are the most representative and the ones that relate more to real world ones.

In the first place, we have accomplished several trials with different rule sizes, so we can check how the solutions adapt to different problems.

Those trials have been made with 50, 100, 200, 300, 400 and 500 number of rules. The percentage between intermediate and not-intermediate variables is fixed to 50 percent of each type, having a total amount of 1050 variables. This way we can measure more accurately the impact of the change in the size of the knowledge base, without taking into account other factors.

We can notice in figure 1(a) and figure 1(b),that we obtain the best results with the genetic algorithm based distribution of rule sets. The worst results are always obtained with the random distribution, and we can appreciate that the heuristic approach always offers an intermediate solution, not as good as the genetic, but not as bad as the random one. But it is notorious that as

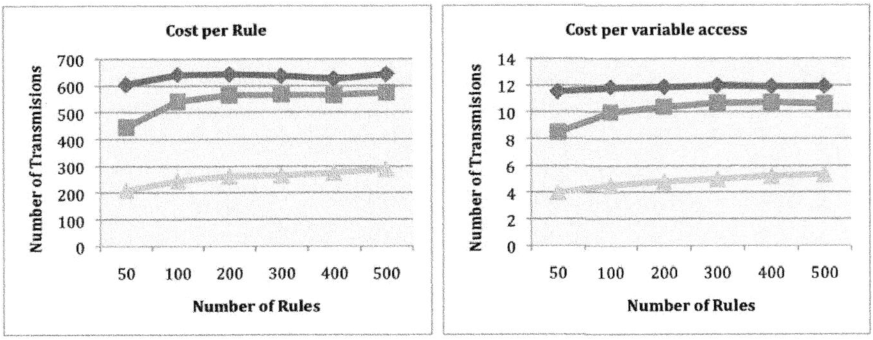

(a) (b)

Fig. 1 Graphics corresponding to the performance with different rule sizes. The triagles represent the random approach, the squares the heuristic one and the diamonds the genetic one.

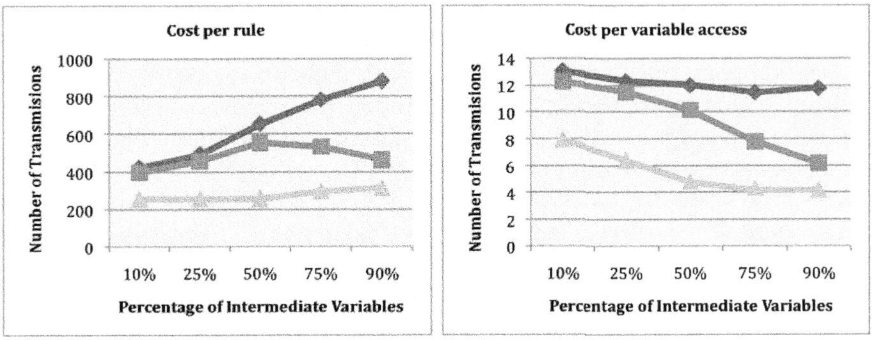

(a) (b)

Fig. 2 Graphics corresponding to the performance with different percentages of variables. The triagles represent the random approach, the squares the heuristic one and the diamonds the genetic one.

the complexity of the rules increases, the performance of this last algorithm decreases, making it closer to the performance of the random solution.

In the other set of trials that we have accomplished, we have focus in the variation of the percentage between non-intermediate and intermediate variables. The total amount of variables is set to 512 and the rules to 250. And the cases that we have studied have been with the following percentages of intermediate variables 10%, 25%, 50%, 75% y 90%. Even tough more than a fifty percent is not a real scenario; we have included these percentages to assess the behavior of the algorithm in extreme cases.

We can appreciate in figures 2(a) and 2(b) that for the three kind of rules the performance of the solution obtained with the genetic algorithm shows the best results. But, it's quite remarkable that when the percentage of intermediate variables goes above 50% percent, the performance of the heuristic approach gets closer to the genetic one. This sudden improvement is caused by the peculiar characteristics of theses extreme cases.

4 Conclusions and Future Works

With this research we were aiming to find a method to distribute the evaluation of a rule-based system over a sensor network in an efficient way. This way we pretend to obtain environments with the power and applicability that the combo of wireless sensor networks and rule-based reasoning provides. But without the inconveniences associated with the need for performing the reasoning in a centralized way. We have set out the use of a genetic algorithm to help us distribute the rules and variables among the sensors. And we have evaluated the efficiency in terms of the communication costs associated with the evaluation of the rules, comparing the results that we get with distributions calculated with the genetic algorithm with the ones calculated with the heuristic or generated randomly.

According to the results, generating the distribution with a genetic algorithm seems to be the best option among these three, in terms of communication costs associated with the evaluation of the rules. Thus, this one seems like a very promising choice. Besides, the additional cost linked to the distribution of the assignations doesn't seem very high, since this algorithm it's not supposed to be relaunched very often. But the computational cost is quite higher. In any case the results that we have obtained are positive, since they encourage us to keep on working on this subject. And to do further research on the diverse aspects associated with the deployment of the solution on a real environment. So we can find out if this one was indeed a proper solution or, otherwise, we should search in a different direction. Other relevant point about the solution is that it is generic enough to be adapted to different scenarios with different constraints.

Briefly, it seems quite interesting to keep on working on this line of work, that leaves the following questions opened:

- Perform the tests with real data about the activation frequencies of rules and variables over a long period of time. So we can check how the solution adapts to a real environment. It would be interesting to compare the results obtained depending on the criteria chosen to relaunch the distribution algorithm.
- Develop the described environment in a simulator and perform further studies about it. Such as the extra cost involved in the distribution of the allocations every time the algorithm is performed.
- Study how the method adapts to small networks without clusters.
- Try to find a totally distributed approach to solve the problem.

- Study how to improve the variables access management. This includes several different aspects, such as a previous study of the characteristics of the variables to measure, or a study of the convenience of using a pull model instead of a push for accessing the variables in the less active rules.

References

1. de Morais Cordeiro, C., Agrawal, D.P.: Ad hoc and sensor networks. World Scientific, Singapore (2006)
2. Karl, H., Willig, A.: Protocols And Architectures For Wireless Sensor Networks, 2nd edn. John Wiley 'I&' Sons, Chichester (2005)
3. Chong, C., Kumar, S., Hamilton, B.: Sensor networks: Evolution, opportunities, and challenges. Journal Article: Proceedings of the IEEE (2003)
4. Pauwels, E.J., Salah, A.A., Tavenard, R.: Sensor networks for ambient intelligence. In: IEEE 9th Workshop on Multimedia Signal Processing (2007)
5. Paganelli, F., Giuli, D.: An ontology-based context model for home health monitoring and alerting in chronic patient care networks. In: AINAW 2007 Proceedings of the 21st International Conference on Advanced Information Networking and Applications Workshops, vol. 02 (2007)
6. Holland, J.: Genetic algorithms. Scientific American (1992)
7. Hsu, C.C., Wu, S.M., Wu, J.J.: A distributed approach for inferring production systems. In: Proceedings of the Tenth International Joint Conference on Artificial Intelligence (1987)
8. Lin, P., Tabet, S.: System and method to distribute reasoning and pattern matching in forward and backward chaining rule engines - us patent application uspto 20050246301 (2005)
9. Doorenbos, R.B.: Production matching for large learning systems (1995)
10. Jain, N., Biswas, R., Nandiraju, N., Agrawal, D.P.: Energy aware routing for spatio-temporal queries in sensor networks. In: 2005 IEEE Wireless Communications and Networking Conference (2005)
11. Li, W., Dai, H.: Cluster-based fast distributed consensus. In: IEEE International Conference on Acoustics, Speech and Signal Processing (2007)
12. Mitchell, T.M.: Machine Learning. McGraw-Hill, New York (1997)

Efficient Use of Voice Activity Detector and Automatic Speech Recognition in Embedded Platforms for Natural Language Interaction

Marcos Santos-Pérez, Eva González-Parada, and José Manuel Cano-García

Abstract. Nowadays people are faced daily with the management of all types of technological devices that have embedded processors inside. The desire of users and the industry is to achieve a natural interaction with these devices. Researchers have spent many years working in spoken dialog systems which are now used in many applications. In these systems it is crucial to achieve correct speech recognition. Usually the largest research effort focuses on the robustness to noise in all kinds of adverse conditions but often the response time is ignored. This paper focuses on a new approach for the efficient use of voice activity detection and speech recognition in embedded devices for natural language interaction. We propose a new approach to adjust the response time of the recognition system to the requirements of the overall implementation without sacrificing too much accuracy.

1 Introduction

Owing to the ongoing advancement of electronics, embedded systems use has spread and it is common to find them in consumer electronic products, such as appliances, as well as mobile phones, avionics, automotive, and GPS navigation or air conditioning systems. This development announces that this trend will go higher in the near future, so that over time other emerging technologies like Ambient Intelligence and Robotics will be incorporated to the daily lives of people. Due to the enormous impact of all these technologies, there is a growing interest in facilitating access to such technologies to all types of users and minimize the technological barrier between people. Therefore, this will aim to interact as naturally and humanly as possible with new technology systems and applications. Considering that speech is the most natural way of interaction between humans, the aim of researchers and engineers should be to obtain usable spoken dialog systems.

Marcos Santos-Pérez · Eva González-Parada · José Manuel Cano-García
Electronic Technology Department, School of Telecommunications Engineering, University of Malaga, Teatinos Campus, 29071 Malaga, Spain
e-mail: {marcos_sape,gonzalez,jcgarcia}@uma.es

J.B. Pérez et al. (Eds.): Highlights in PAAMS, AISC 89, pp. 233–242.
springerlink.com

We say that a spoken dialog system or conversational agent is usable if it is robust and efficient. There are recent studies that address the subject of robustness with remarkable success [11]. Efficiency measures the success rate of the voice interface but also takes into account the computational resources and time spent.

The early stages of conversational agents are the voice activity detector (VAD) and the automatic speech recognizer (ASR). This paper focuses on a new approach for the efficient use of VAD and ASR in embedded platforms for natural language interaction. The work presented in this paper is part of a research project called AVATAR. Its main objective is the development of an embedded platform with low cost and low power consumption aim at the creation of embodied conversational agents (ECAs).

This paper is organized as follows: after this introduction, Section 2 describes briefly VAD and ASR elements. Section 3 presents in detail our proposal for the efficient use of VAD and ASR. Section 4 presents the platform were the tests are performed and Section 5 shows the tests executed to validate the design and their results. Finally, in Section 7 we summarize the conclusions drawn from the paper.

2 VAD and ASR Overview

2.1 Voice Activity Detector

The VAD role is to differentiate the audio segments where the user speaks from those containing noise. VAD are usually used in other scenarios, such as mobile communications and Voice over IP (VoIP). In these scenarios, the VAD function aims to achieve a reduction of network traffic in order to save bandwith. Although the operating principles are the same, the purpose of the VAD in conversational interfaces is to segment the user's speech into sentences.

The most common VAD algorithms are based on energy threshold and zero crossing rate of the signal. In our platform this role is performed by the SphinxBase library [3].

2.2 Automatic Speech Recognition

Voice recognition is a crucial part of the conversational system and its functionality is essential to allow voice communication between human beings and electronic systems. Although any task that is interacting with a computer can eventually use speech recognition, currently, the most common speech-driven applications are automatic dictation, command control, telephony, mobile systems, systems designed for people with disabilities, etc.

The complexity of such systems lies in the diversity of factors that includes human speech (acoustics, phonetics, phonology, lexicon, semantics). In many cases, the sense of naturalness of the conversational interface depends heavily on the robustness of speech recognition [7]. Despite these difficulties, some notable advances in this field has been achieved in recent years making possible automatic speech recognition with acceptable error levels for a large number of applications [17]. In addition to a low error rate, the construction of systems that are not prone to be rejected by the users requires the recognition task to be performed within a limited time. The progressive enhacement of speech recognition methods in addition to the improving performance of current computers have made it possible to meet these requirements in real-time desktop stations seamlessly, and even allow to execute them in embedded systems [12] [13].

Real Time Ratio (xRT) and Word Error Rate (WER) are two common metrics used in the literature on speech recognition to measure the system performance. Their respective formulas are 1 and 2.

In the xRT formula, the time to recognize the input is meant as the time it takes the ASR to recognize the user's sentence since he stops talking. By making the parallel use of VAD and ASR, part of the recognition process takes place while the user is still speaking and the value xRT value measured corresponds to the time spent on recognition once the user completes the sentence.

$$xRT = \frac{Time\ to\ recognize\ the\ input}{Duration\ of\ input\ speech} \tag{1}$$

$$WER = \frac{Substitutions + Deletions + Insertions}{Total\ Words} \tag{2}$$

Reducing the time spent by the recognizer usually means increasing the rate of wrong words. Thus, there must be a compromise between xRT and WER values.

There are currently several commercial solutions for speech recognition, such as: Nuance [9], IBM ViaVoice [5] or Loquendo [8]. The last two offer versions for embedded systems. As an alternative to commercial software, several universities have developed their own voice recognition systems. The vast majority are based on a free version of IBM ViaVoice libraries or the Sphinx platform from the Carnegie Mellon University [2], widely distributed since this code was released.

In our AVATAR platform, speech recognition is performed by PocketSphinx library, which belongs to the CMU Sphinx family. The reasons for this choice are that this library is free and open source, allows speaker-independent speech recognition and has been used in real-time applications in embedded systems [4].

3 Efficient Use of VAD and ASR

In a typical spoken dialog system the ASR processes the full utterances obtained by the VAD. In this case it could be said that the ASR and VAD systems work in series. PocketSphinx recognizer can process parts of each sentence separately, but the developers warn that it can result in loss of recognition accuracy. Moreover, the VAD reads the capture buffer of the microphone every few tens of milliseconds. If the ASR system is able to work in real time, it can process the frames from the VAD while still capturing the speaker's voice. In this case the VAD and ASR systems work in parallel. Thus, at the end of the utterance the ASR has already processed

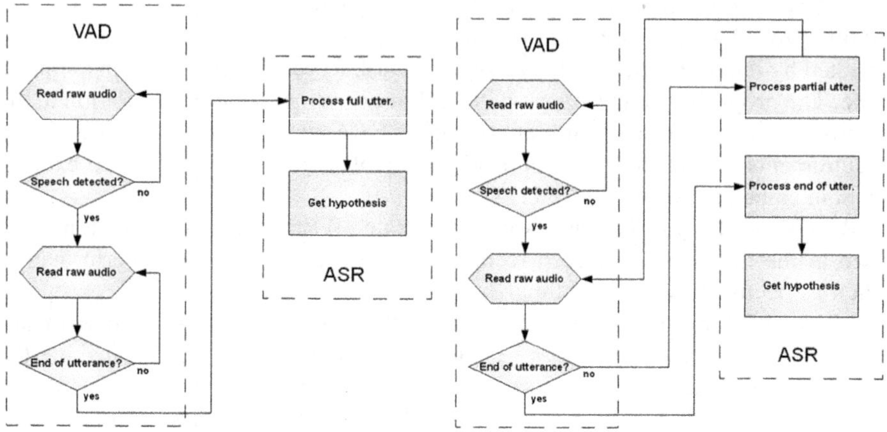

a) Serial scheme b) Parallel scheme

Fig. 1 Serial and parallel schemes of VAD and ASR.

a large part of the frames that make up the complete utterance and can offer its final hypothesis in much less time than the serial configuration. In figure 1 both schemes are shown.

3.1 Adding the Confidence Score to the Scheme

As noted above, it is assumed that the parallel use of VAD and ASR improves performance in terms of speed but may cause a loss of accuracy. Because of this we propose a hybrid approach that aims to make the most of the serial and parallel schemes. When the ASR matches the text, it also provides a confidence score [6]. We can take advantage of this if we set a threshold and, in case the conficence score for a given utterance is below it, force the ASR to perform a second recognition process with the complete utterance. This hybrid scheme is shown in Figure 2.

With the idea proposed above, we can establish a theoretical model for the response time of the recognizer in the hybrid scheme:

$$xRT_h(th) = SAR(th)\, xRT_s + xRT_p \tag{3}$$

where xRT_h, xRT_s and xRT_p are the average response times of the ASR in the hybrid, serial and parallel schemes respectively. We noted with th the variable confidence threshold and SAR is the Second Attempt Rate. It represents the percentage of the total utterances obtaining a confidence score below the threshold with which the ASR must make a second attempt.

In the case of the word error rate, its equation is as follows:

$$WER_h(th) = SAR(th)\, WER_s[cs < th] + (1 - SAR(th))\, WER_p[cs \geq th] \tag{4}$$

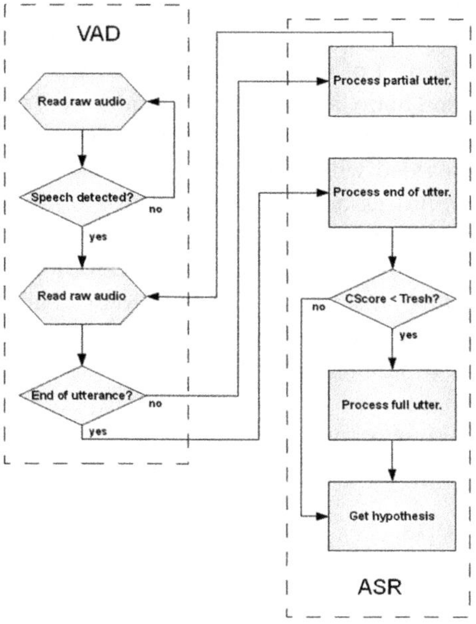

Fig. 2 Hybrid scheme of VAD and ASR.

where WER_h is the average WER in the hybrid scheme and SAR means the same as in equation 3. Now $WER_s[cs < th]$ and $WER_p[cs \geq th]$ are the average WER in the serial and paralel schemes too, but the first is measured on the subset of sentences that scores below the threshold and the second on the subset that scores equal or above.

The appropiate confidence threshold will be different depending on the application domain of the conversational agent. For example, if it is used to control home automation, the threshold should be high to achieve a reduced WER at the expense of a higher xRT. In this type of systems commands are usually short and simple and require little processing time. In contrast, the WER should remain low to maximize the understanding of the goal desired by the user. In another example where the conversational agent is used in a social robot, the threshold may be lower than before in order to reduce xRT and increase interactivity. In this scenario the sentences are longer and more complex and a greater WER is allowed as the conversation remains fluid.

4 Test Environment

The embedded platform used in this paper to test the different proposals discussed in previous sections has been selected so that the conclusions derived from the results can be considered as general as possible. All test were executed on a BeagleBoard rev B7 [1], a low power and low cost embedded computer based on Texas

Instruments OMAP 3530 system on chip. OMAP microprocessors are quite popular in current embedded computers, since they are the core of many handhelds and mobile phones such as most of the Nokia's N and E series, Samsung Omnia HD, Sony Eriksson Vivaz and Satio, and Motorola Droid serie. In addition, OMAP 3530 core is a variant of the also widely spread ARM Cortex A8 architecture. Revision B7 of the board is provided with 128 MB of RAM and 256MB of NAND Flash memory, and the microprocessor core is clocked at 600 MHz. These values can be considered to be representative since very similar ones can be found in millions of ordinary devices. Debian Linux 5.0 (Lennny) has been chosen as the supporting operating system for its flexibility, power and open source nature. Lenny kernel, core libraries and tools are up to date and similar versions can be found in many others contemporary linux distributions for embedded devices such as Android, Angstrom, Poky, Geekbox, Maemo, etc., so we can reasonably expect our application to behave similarly in these other environments.

5 Tests and Results

5.1 Test Configuration

Voice recognition needs both an acoustic and a language corpus to run. For this project we used the Voxforge Spanish corpus [16]. It is the first free, in spanish and speaker-independent acoustic model. The online available version is a continuous model with 1500 senones. Due to the increased computational complexity that continuous models require [14], it was necessary to perform a new training to obtain a semi-acoustic model with 1500 senones, 256 codebook Gaussians and 5 states per Hidden Markov Model (HMM) [15]. The language model is probabilistic and based on trigrams and the dictionary contains more than 22,000 words. They are the same as those used in acoustic model training.

5.2 Tests Results

The purpose of the tests is to compare the efficiency in terms of xRT, and WER for different confidence thresholds. The tests are performed with two different configurations of ASR. The first one holds the default parameters PocketSphinx. The second one is in accordance with the recommendations for embedded systems made by PocketSphinx library developers [10]. For both configurations 11 tests were performed, 1 for the serial scheme, 1 for the parallel scheme and the other for the hybrid scheme with 9 different confidence thresholds.

The value *S* of *THRESHOLD%* in Figures 3 and 4 corresponds to the serial scheme and the value *P* corresponds to the parallel scheme. The intermediate values *90-10* correspond to the hybrid scheme.

The first fact that can be seen in Figures 3 and 4 is that the empirical values *WER_e* and *xRT_e* fit almost perfectly to the theoretical values *WER_t* and *xRT_t* respectively.

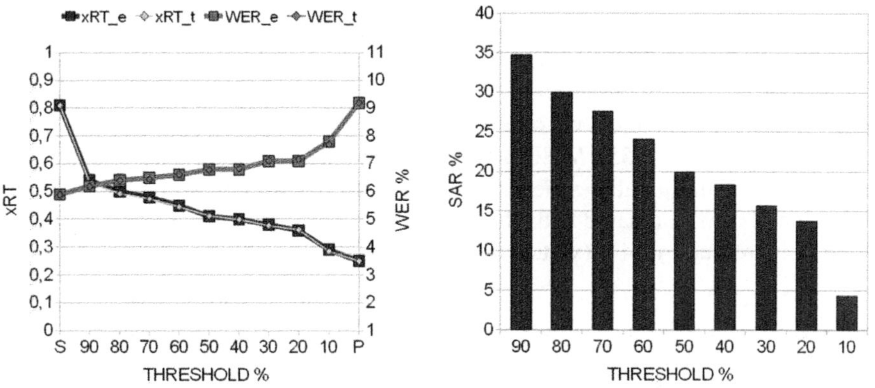

a) xRT and WER results

b) SAR results

Fig. 3 xRT, WER and SAR for baseline configuration.

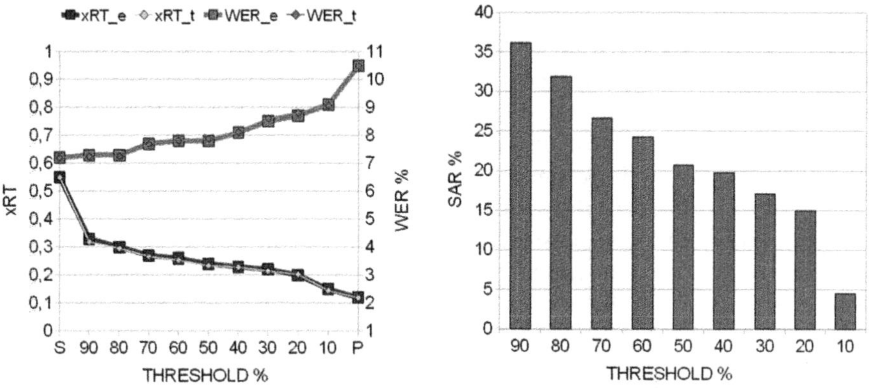

a) xRT and WER results

b) SAR results

Fig. 4 xRT, WER and SAR for optimized configuration.

As seen in Figure 3, the hybrid scheme achieves significantly lower values of xRT than the serial scheme. Even with a high confidence threshold of 90%, xRT is reduced from 0.81 to 0.54 (33%). Progressively the reduction continues until 0.29 xRT (64%) when the threshold is set to 10%. This is due to the SAR which goes from 35% to 4.3%. As expected, the WER increases from 5.9% in the serial scheme up to 9.2% in the parallel scheme.

By optimizing the system configuration parameters, we obtain in Figure 4 similar graphs to those in the Figure 3, but with xRT values lower than in the previous case and higher values of WER. xRT values compared with those of the serial approach go from 0.55 to 0.33 (40%) with confidence level of 90% and reach 0.15 xRT (73%) with the threshold set to 10%. SAR remains similar values to the previous case and WER ranges from 7.2% in the serial scheme set up to 10.5% in the parallel scheme.

Comparing the configurations to each other, we can see that in the serial scheme with optimized parameters we obtained xRT 32% lower than with the default ones (values from 0.81 to 0.55). WER is increased by 22% (values from 5.9% to 7.2%). In the case of the parallel scheme with the optimized configuration we obtain a reduction of 58% compared to the baseline configuration (values from 0.25 to 0.12). WER is increased by 14% (values from 9.2% to 10.5%).

6 Discussion

The results seen in Section 5 show that we are able to use the parallel scheme, since, as shown in Figure 3, the baseline system with serial scheme attains a xRT value below 1. It gets 0.81 xRT with a WER of 5.9%. But with this xRT value the user may experience a limited interactivity. With a typical input sentence of about 3 seconds long the system would take 2.43 seconds in average to respond. If the parameters are optimized to reduce the computational complexity, it can reach a value of 0.55 xRT at the expense of raising the WER up to 7.2%. Now the system would take 1.65 seconds to respond to the same request. If then we use the parallel scheme, 0.12 xRT values can be reached with a WER of 10.5%. We have drastically reduced the average response time of the system up to 0.36 seconds.

But at certain times the recognition process provides low confidence scores. Let's say we set the confidence threshold at 50%. If a sentence gets an equal or greater confidence score, the agent must attempt to obtain valid information and if necessary seek for clarification. But with a score less than 50% the ASR processes again the input sentence as a whole. SAR measures the percentage of utterances that need a second chance of recognition. Continuing the example, approximately 80% of the cases are processed by the parallel scheme obtaining xRT values of 0.12 but the remaining 20% is processed with xRT values of 0.67 due to the sum of serial and parallel processing times. If the response time for that 20% of sentences affect the interaction, the conversational agent can mitigate its impact by issuing grounding utterances while the second recognition is taking place.

Finally, the example system would achieve a xRT of 0.24 and a WER of 7.8%. This implies a reduction of 70% in terms of XRT compared to the initial baseline configuration. The downside is that the WER results in an increase of 32%, but it remains at acceptable levels.

SAR values are not too high considering that the system imposes no type of grammar to the user, who also has a vocabulary of more than 22,000 words. It can be assumed that by using medium-sized vocabularies (500-1,000 words) and a language model specific to a particular application domain, xRT and WER figures will improve significantly.

7 Conclusion

The main goal of this work was to propose a new approach for the efficient use of VAD and ASR in embedded platforms for natural language interaction. Thus, we describe the advantages of the parallel scheme and how we added the confidence score to get a hybrid scheme. We made an extra effort during the election of the components to be integrated so we could obtain a free and open source platform. This was followed by the presentation of tests and results. Finally, we discussed the results and extracted some considerations.

Our hybrid scheme offers a great capacity to adapt to the demands of the application domain of the final conversational agent developed in the embedded platform. By setting the confidence threshold, the conversational agent can fulfill the requirements of xRT or WER provided that SAR is maintained at reasonable values.

It is also worth noting that the characteristics of the hardware platform used in the tests are similar to those of the vast majority of mid-range mobile devices today, and, therefore, these results are fully transferable to everyday business systems.

Acknowledgments

This work was partially supported with public funds by the Spanish National Project TEC2009-13763-C02-01 and by the Andalusian Regional Project P08-TIC4198.

References

1. BeagleBoard website, `http://beagleboard.org/`
2. CMU Sphinx website, `http://cmusphinx.sourceforge.net/`
3. CMU Sphinxbase website, `http://sourceforge.net/projects/cmusphinx/`
4. Huggins-daines, D., Kumar, M., Chan, A., Black, A.W., Ravishankar, M., Rudnicky, A.I.: PocketSphinx: a free, real-time continuous speech recognition system for hand-held devices. In: Proc. of ICASSP, Touluse, France, May 2006, pp. 185–188 (2006)
5. Embedded ViaVoice website, `http://www-01.ibm.com/software/pervasive/embedded_viavoice/`
6. Jiang, H.: Confidence measures for speech recognition: A survey. Speech Communication 45(4), 455–470 (2005)
7. Jokinen, K.: Natural language and dialogue interfaces. In: The Universal Access Handbook, 1st edn., pp. 495–506. CRC Press Taylor & Francis Group (2009)
8. Loquendo ASR website, `http://www.loquendo.com/es/technology/asr.htm`
9. Nuance VoCon website, `http://www.nuance.es/vocon/`
10. Pocketsphinx optimizations for embedded devices, `http://cmusphinx.sourceforge.net/wiki/pocketsphinxhandhelds`
11. Sasou, A., Kojima, H.: Noise robust speech recognition applied to Voice-Driven wheelchair. EURASIP Journal on Advances in Signal Processing 2009, 1–10 (2009)
12. Schmitt, A., Zaykovskiy, D., Minker, W.: Speech recognition for mobile devices. International Journal of Speech Technology 11(2), 63–72 (2008), doi:10.1007/s10772-009-9036-6

13. Tan, Z.-H., Lindberg, B.: Speech recognition on mobile devices. In: Jiang, X., Ma, M.Y., Chen, C.W. (eds.) Mobile Multimedia Processing. LNCS, vol. 5960, pp. 221–237. Springer, Heidelberg (2010)
14. Vertanen, K.: Baseline WSJ acoustic models for HTK and sphinx: Training recipes and recognition experiments. Technical report, University of Cambridge, Cavendish Laboratory (2006)
15. Vertanen, K., Kristensson, P.O.: Parakeet: A continuous speech recognition system for mobile Touch-Screen devices. In: IUI 2009: Proceedings of the 14th International Conference on Intelligent User Interfaces, pp. 237–246. ACM Press, Sanibel Island (2009)
16. Voxforge Spanish Model website, `http://cmusphinx.sourceforge.net/2010/08/voxforge-spanish-model-released/`
17. Zhang, J., Ward, W., Pellom, B., Yu, X., Hacioglu, K.: Improvements in audio processing and language modeling in the CU communicator. In: Eurospeech 2001, Aalborg, Denmark (2001)

A Fuzzy Logic-Based and Distributed Gateway Selection for Wireless Sensor Networks

A. Triviño Cabrera, A.J. Yuste-Delgado, and D. Cintrano Macías

Wireless sensor networks are resource-constrained because of the difficulty of recharging the devices. In order to prolong the network lifetime, the nodes relaying packets should carefully be selected. In a scenario where multiple sinks (gateways) are available, this selection is extended to decide to which sink the sources should send their data. We propose a distributed algorithm in which the the path to the sink is dynamically decided using an anycast transmission scheme. In these schemes, multiple candidates are available to retransmit the frames. They determine their own goodness of being the next hop for every online transmission. The estimation is supported by a fuzzy-logic based system which takes into account the connectivity of the source, the connectivity of the candidate and the candidate's residual energy. The simulation results show that our proposal reduces the energy consumption of the wireless network and the end-to-end delay of the transmissions.

1 Introduction

Nowadays, there are numerous applications where sensors are used. In order to make their sensed information available for their processing, sensors communicate with a sink [1]. The sink may be also connected to a Gateway which allows the data to be accessed from the Internet.

In most applications, sensors are needed to be auto-sufficient as long as possible. In this sense, one requirement is prolonging the sensor lifetimes. Sensor lifetimes are limited because of their reduced battery and because of the difficulty of recharging them. Taking into account this restriction, several techniques are usually applied in the design of sensor applications. Firstly, the energy consumption derived in the transmissions is diminished by the support of multihop communications. In this scheme, sensors collaborate retransmitting others' packet. Thus, the

A. Triviño Cabrera
Dpto. Tecnología Electrónica, University of Málaga (Spain)

A.J. Yuste-Delgado
Dpto. Ingeniería de Telecomunicación, University of Jaén (Spain)

J.B. Pérez et al. (Eds.): Highlights in PAAMS, AISC 89, pp. 243–248.
springerlink.com

packet is forwarded by intermediate nodes which send it to the final destination through a route. This cooperation is easily achieved in a wireless environment. In contrast, a serious drawback is present: a node consumes some energy by receiving and analyzing data frames sent in their reception range, even when the node is not the destination of the data frame. To overcome this drawback, sensors may turn off and on their transceiver when needed [2]. The activation of the sensors can be synchronous or asynchronous. In the synchronous schemes, the activation of two nodes is triggered simultaneously. Although this kind of activation ensures the sending of the packets along the path, it is hard to ensure the synchrony in the wireless nodes. Alternatively, asynchronous nodes wait for their next hop to wake up. In this paper we will focus on asynchronous wireless sensor networks. In particular, we aim at defining a criterion to select the sink (gateway) when multiple ones are available. The selection is performed in a distributed way as every relaying node executes this criterion when retransmistting. The goal of our criterion is twofold: (i) to reduce the energy consumption of the nodes and (ii) to diminish the end-to-end delay in the transmissions. As multiple factors impact on these two parameters, a fuzzy logic-based algorithm is used [3].

The rest of the paper is structured as follows. Section 2 describes the anycast transmission. Section 3 describes our proposal. The algorithm is evaluated by means of simulations, as shown in Section 4. Finally, Section 5 draws the main conclusion of our work.

2 Anycast Transmission in Wireless Sensor Networks

In the multihop paradigm, nodes need to be provided of a routing protocol to become aware of the paths to the destination [5]. We assume that nodes execute a routing protocol which provides multiple routes to the sinks. In particular, the routing protocol offers the shortest routes to every sink. From this information, a node are able to construct its forwarding list, that is, the set of potential nodes (known as candidates) that could act as relay nodes to retransmit its packets to any sink [4]. Once the forwarding list is determined, an opportunistic/anycast routing paradigm could be applied. This kind of routing, based on the dynamic selection of the relaying nodes, is useful to increment the packet delivery ratio in wireless network [6]. The transmission of a frame works as follows. Firstly, the source transmits a preamble. In order to avoid the inclusion of the forwarding list in every preamble, the forwarding lists are periodically exchanged among the nodes. Upon reception of the frame, the neighbours check if they are candidates of the source (this can be done just by the awaken nodes). If so, they respond to the preamble with a specific frame. The time to respond to the frame depends on the priority assigned to the candidate. This is done to ensure that just one candidate will retransmit the frame. Once the source receives the frame, it triggers the data emission. The reception in the selected candidate is confirmed with an ACK message.

3 Fuzzy Logic- Based Criterion to Select the Relaying Node

In the anycast routing scheme, nodes should be assigned of a priority according to their convenience to act as the next relaying node of an on-line data frame. In particular, in the anycast routing, nodes delay its response to the preamble for a pause time. We propose this pause time to be inversely proportional to the goodness of the node to act as the next relay node. For us, the optimum relaying node is the node that forces the minimum energy consumption. The consumption of a transmission depends on the number of frames exchanged among the nodes. As the nodes need to wait for their next hop to wake up, the most convenient relaying node is the one whose next hop is active. Additionally, it is necessary to consider the energy of the nodes so that the network lifetime can be prolonged. In general, the best next hop is the one with the greatest number of neighbours connecting to the Gateway and with the biggest remaining energy. According to these conditions, we estimate the convenience of a node to act as the relaying node by means of a fuzzy logic-based system. The fuzzy-logic based system is a Mandani system with three inputs and one output. For the scenario to evaluate (described in the next Section), the crisp inputs are:

- Neighbours. It represents the number of neighbours that connect the candidate to any Gateway. This parameter is classified in three groups: FEW, MEDIUM and MANY. In order to ease the inclusion of fuzzy logic in sensors, the selected membership functions are lineal. In particular, for this input, the membership functions are reflected in Figure 1.

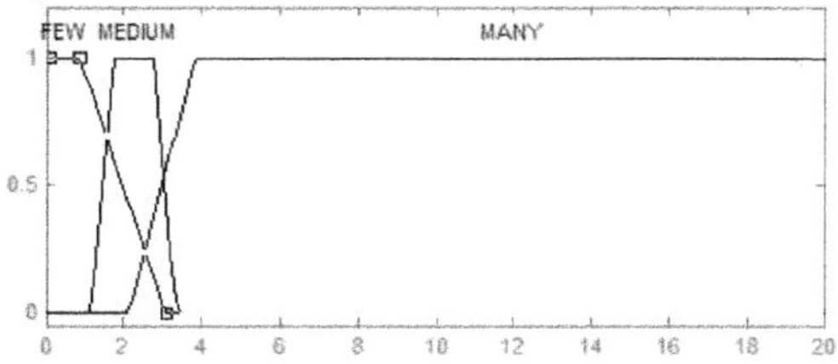

Fig. 1 Membership function for the variable Neighbours.

- Candidates. It corresponds to the number of candidates that the source has. According to this variable, the nodes are grouped into: FEW; MEDIUM and MANY. The membership functions for this input are depicted in Figure 2.

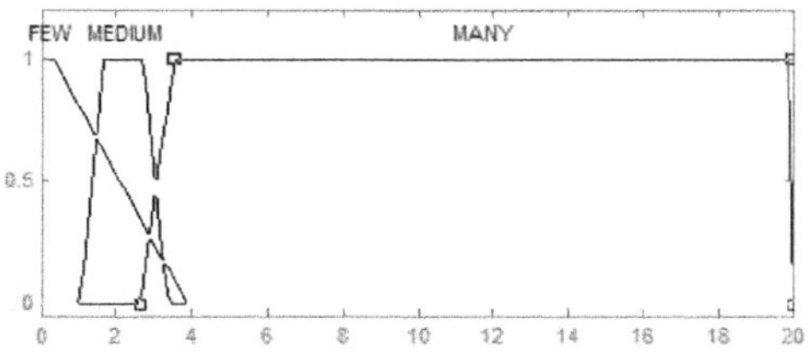

Fig. 2 Membership function for the variable Candidates.

- Energy. It is related to the percentage of remaining energy in the candidate's battery. The membership functions are shown in Figure 3.

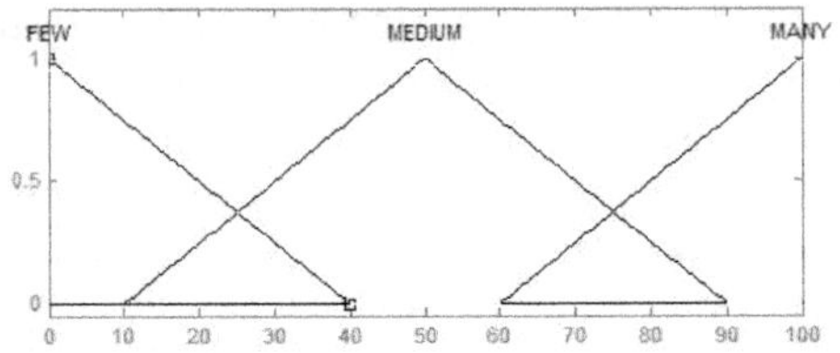

Fig. 3 Membership function for the variable Energy.

- In this system, the output corresponds to the suitability of the node to be the relaying node. The out is classified as: 'RETX' and 'NORETX' as shown in Figure 4. When the output is 'RETX', it means that it is an appropriate relaying node. Alternative, when the output is 'NORETX', it is preferable to select another node as the relaying node.

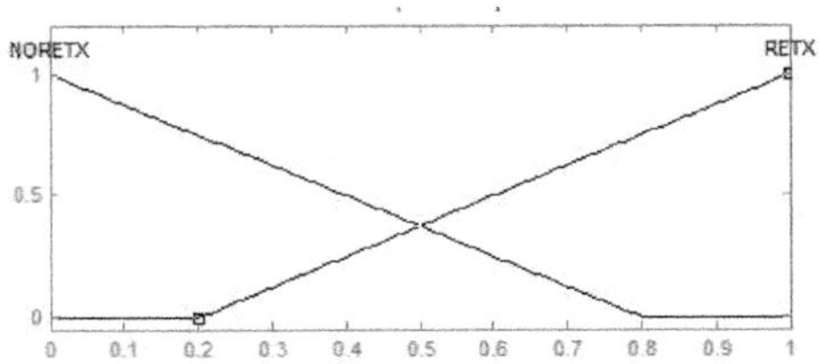

Fig. 4 Membership function of the Output.

Once the output is defuzzified, its value varies from 0 to 1. The defuzzyfication threshold is 0.5. The rules, summarized in Table 1, are and-type rules. An empty space in the table indicates that the input is not considered for the rule.

Table 1 Fuzzy rule set

Neighbours	Candidates	Energy	Output (Retransmit)
FEW	NOT(MANY)	NOT(MANY)	NO
MEDIUM	NOT(MANY)		YES
MANY			YES
FEW	FEW	MANY	YES
NOT(MANY)	MANY	NOT(MANY)	NO

4 Simulation Results

To evaluate the effectiveness of the proposed anycast technique, we have modelled the wireless sensor network in Matlab [7] considering a Free Space propagation with no losses in the transmission but with a delay of 0.03 seconds.

In this way, two nodes are connected if the distance between them is lower than the transmission range (which is set to 50 m). The source is always the same for all the transmissions. It is connected to 4 gateways by paths of 3-hop length which are depicted in Figure 5. These paths are not always available since nodes follow a sleep/awake cycle whose active time is 1 second and the mean sleep time is set to 5 or 10 seconds (two sets of experiments are executed for each mean sleep time). They follow a Rayleigh distribution for the sleep time.

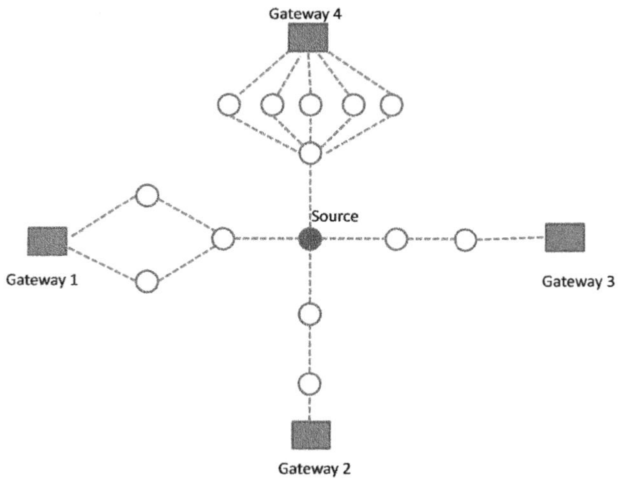

Fig. 5 Routing paths from the source to the Gateways.

Concerning the battery, in every simulation, nodes are assigned a random value for the remaining energy in their batteries. Assuming Mica2 nodes, nodes are powered by AA batteries whose maximum energy is 27MAs. For the cost of operation, we utilise the estimations derived in [8]. In this way, the sending of the preamble incurs in 0.17 mAs while its reception consumes 0.096 mAs. Transmitting the data frame leads to 1.7 mAs while receiving it leads to 0.96 mAs. We assume that the routing protocol has already been executed so their impact is not taken into account for our experiments. In our experiments, 1000 data frames are sent from the source to the any sink. We evaluate the performance of two anycast paradigms: our proposal based on a fuzzy logic based system and random anycast in which the candidate is selected randomly. For each mean sleep time (5 and 10 seconds), we run 10 simulations.

Using the fuzzy logic-based system, the energy consumption for each data frame is reduced 1.25% when the mean sleep time is 5 seconds and 3.25% when the mean sleep time corresponds to 10 seconds. Concerning the delay, the fuzzy logic-based system also improves the network performance. In particular, the end-to-end delay is reduced in 5.2% and 7.5% for a mean sleep time of 5 seconds and 10 seconds respectively. As we can see, the most important benefits are related to the end-to-end delay. This is due to the fact that using nodes with a large number of candidates increments the probability of having awaken nodes in the path but the receiving nodes not belonging to the path consumes part of their energy on the reception.

5 Conclusions

This paper has presented a novel scheme to dynamically select the path to any Gateway available in the scenario. The algorithm is supported by an anycast transmission technique. This mechanism to select the relaying node from the candidate list is supported by a fuzzy-based logic system which takes into account the connectivity of the candidate, its distance to the Gateway, its remaining energy and the number of available candidates. By means of simulations, we have demonstrated the effectiveness of our algorithm as our scheme is able to reduce the energy consumption per transmission. As future work, we intend to implement the protocol in a network simulation where realistic propagation conditions and realistic battery performance are taken into account.

Acknowledgments. This work has been partially supported by the National Project No. TEC2009-13763-C02-01.

References

[1] Akyildiz, I.F., Su, W., Sankarasubramaniam, Y., Cayirci, E.: Wireles sensor networks: a survey. Computer Networks 38(4), 393–422 (2002)

[2] Paruchuri, V., Basavaraju, S., Durresi, A., Kannan, R., Iyengar, S.S.: Random Asynchronous Wakeup Protocol for Sensor Networks. In: Proc. of the First International Conference on Broadband Networks (BROADNETS 2004) (October 2004)

[3] Bojadziev, G., Bojadziev, M.: Fuzzy Sets, Fuzzy Logic, Applications (Advances in Fuzzy Systems - Applications & Theory). World Scientific Publishing Company, Singapore (1996)

[4] El Rachkidy, N., Guitton, A., Misson, M.: Routing protocol for anycast communications in a wireless sensor network. In: Crovella, M., Feeney, L.M., Rubenstein, D., Raghavan, S.V. (eds.) NETWORKING 2010. LNCS, vol. 6091, pp. 291–302. Springer, Heidelberg (2010)

[5] García Villalba, L.J., Sandoval Orozco, A.L., Triviño Cabrera, A., Barenco Abbas, C.J.: Routing Protocols in Wireless Sensor Networks. Sensors 9(11) (2010)

[6] Jain, S., Das, S.R.: Exploiting Path Diversity in the Link Layer in Wireless Ad Hoc Networks. Ad Hoc Networks Journal 6(5) (July 2008)

[7] `http://www.mathworks.com/products/fuzzylogic/`

[8] Landsiedel, O., Werlhe, K., Götz, S.: Accurate Prediction of Power Consumption in Sensor Networks. In: Proc. of the 2nd IEEE Workshop on Embedded Networked Sensors (2005)

HERA: Hardware-Embedded Reactive Agents Platform

Dante I. Tapia, Ricardo S. Alonso, Óscar García, and Juan M. Corchado

Abstract. Wireless Sensor Networks is a key technology for gathering outstanding information from different sources. However, it is not easy to integrate devices from different technologies into a single network. Distributed architectures, such as Multi-Agent Systems, can facilitate the integration of heterogeneous sensor networks. In addition, Multi-Agent Systems expand the sensors' capabilities changing their behavior dynamically and personalizing their reactions. This paper presents the novel Hardware-Embedded Reactive Agents (HERA) platform. HERA allows developing applications where agents are directly embedded in heterogeneous wireless sensor nodes with reduced computational resources.

Keywords: Wireless Sensor Networks, Distributed Architectures, Multi-Agent Systems, Service-Oriented Architectures.

1 Introduction

Nowadays, there is a wide range of devices for gathering context information about both the environment and the users inside it [1]. However, the integration of such devices is not an easy task. Therefore, it is necessary to develop innovative solutions that integrate different approaches to create flexible and adaptable systems. In this sense, the implementation of distributed architectures is presented as a solution to such problems [2]. One of the most prevalent alternatives in distributed architectures is Multi-Agent Systems (MAS) which can help to distribute resources and reduce the central unit tasks [2]. A distributed agent-based architecture provides more flexible ways to move functions to where actions are needed,

Dante I. Tapia · Ricardo S. Alonso · Juan M. Corchado
Computers and Automation Department, University of Salamanca. Plaza de la Merced, s/n, 37008, Spain
e-mail: {dantetapia,ralorin,corchado}@usal.es

Óscar García
School of Telecommunications, University of Valladolid. Paseo de Belén 15, 47011 Valladolid, Spain
e-mail: oscgar@tel.uva.es

J.B. Pérez et al. (Eds.): Highlights in PAAMS, AISC 89, pp. 249–256.
springerlink.com

thus obtaining better responses at execution time, autonomy, services continuity, and superior levels of flexibility and scalability than centralized architectures [3].

Sensor networks are used to obtain context information about users and their environment. Wireless Sensor Networks (WSNs) are more flexible and require less infrastructural support than wired sensor networks. There are plenty of technologies for implementing WSNs, such as ZigBee, Wi-Fi or Bluetooth. ZigBee [4] is based on the IEEE 802.15.4 standard and operates in the 868/915MHz and 2.4GHz unlicensed bands. Unlike Wi-Fi or Bluetooth, ZigBee is designed to work with low-power nodes and allows up to 65,534 nodes to be connected in a star, tree or mesh topology network. Nonetheless, the lack of a common architecture may lead to additional costs due to the necessity of deploying non-transparent interconnection elements (*e.g.*, bridges or routers) between different networks.

This paper describes the HERA (*Hardware-Embedded Reactive Agents*) platform. In HERA, unlike other approaches, agents are directly embedded on the WSN nodes and their services can be invoked from other nodes in the same WSN or other WSN connected to the former one. HERA focuses specially on devices with reduced resources to save CPU time, memory size and power consumption.

Next section presents the problem description that essentially motivated the development of HERA. Then, it is depicted the main characteristics and components of HERA. After that, some experiments aimed at testing the HERA performance are described. Finally, it is presented the conclusions and the future lines of work.

2 Background and Problem Description

HERA have stemmed from the necessity to cover more efficiently the challenges produced by systems that use WSNs in their infrastructure. The fusion of the multi-agent technology and WSNs is not easy due to the difficulty in developing, debugging and testing distributed applications for devices with limited resources. The interfaces developed for these distributed applications are either too simple or, in some cases, do not even exist, which complicates even more their maintenance. Therefore, there are researches that develop methodologies for the systematic development of muti-agent systems for WSNs [5]. Some researches that relate multi-agent technologies with WSNs [6] insist that the combination of such technologies extends the life of wireless sensor nodes trough the reduction of the power consumption. ActorNet [7] is a study that describes a mobile agent platform for WSNs. However, each mobile agent is only centered on a sensor node. Baker *et al.* [8] present the integration of an agent-based WSN within an existing MAS focused on condition monitoring. In this research, it is used SubSense, a multi-agent middleware platform developed to allow condition monitoring agents to be deployed onto a WSN. SubSense platform is implemented over 512KB RAM Sun-SPOT sensor nodes using the Java Mobile Edition (J2ME). On the contrary, HERA platform can run on lightweight sensor nodes with just 8KB RAM. Likewise, SubSense platform is not focused on working with heterogeneous WSNs. Furthermore, most of the works that relate multi-agent systems and WSNs talk about Mobile Agents based on WSN (MAWSN). Fok *et al.* [9] describe their system, Agilla, as a mobile agent that facilitates the fast development of applications on WSNs and apply it to fire tracking. In a similar way, Zboril *et al.* [10] proposes WSageNt, a platform that is implemented through mobile agents running on wireless sensor nodes. This research poses that in WSN-based agent platforms the resources limitations of sensor motes do not allow to afford its development as an ordinary agent platform that should accomplish the FIPA specifications. Opposite to Agilla, WSageNt is supposed to be fault tolerant and not to be only focused on WSNs. However, it has not context-awareness features and, as Agilla, does not seem to contemplate the interconnection of heterogeneous WSNs.

The HERA platform proposes a model that can successfully solve the problems discussed above. HERA is specially designed to implement hardware agents. HERA allows devices from different radio and networks technologies to coexist in the same distributed virtual network. The platform facilitates and speeds up the integration between agents and sensors to reuse resources. This approach allows the development of multi-agent systems with increased scalability. It also expands the agents' capabilities to obtain information about the context and to automatically react over the environment. A totally distributed approach and the use of heterogeneous WSNs provide an architecture that is better capable of recovering from errors, and more flexible to adjust its behavior in execution time.

3 The HERA Platform

HERA platform facilitates the inclusion of agents into dynamic and self-adaptable heterogeneous WSNs. HERA is an evolution of the SYLPH (*Services laYers over Light PHysical devices*) platform [11] and is focused specifically on devices with reduced resources in order to save CPU time, memory size and energy consumption. The underlying layers of HERA, provided by SYLPH, follow a Service-Oriented Architecture (SOA) model [12]. There have been several attempts to integrate WSNs and MASs [5] or SOA [1]. Unlike those approaches, agents in HERA are directly embedded on the WSN nodes and their services can be invoked from other nodes in the same network or another network connected to the original one. Therefore, each wireless node is an agent and works as another software agent running on platforms such as RETSINA or JADE. This way, there is no difference between a software or hardware agent. HERA has been designed to enable an extensive integration of WSNs and optimize the distribution, management and reutilization of the available resources and functionalities in their networks. Unlike other approaches, the platform provides the possibility of connecting WSNs based on different radio and link technologies using SYLPH Gateways [11] (devices with several hardware network interfaces connected to distinct WSNs). Likewise, HERA can be executed over multiple wireless devices independently of their microcontroller or the programming language they use. In other words, HERA allows the interconnection of several networks from different wireless technologies, such as ZigBee or Bluetooth. Thus, a node designed over a specific technology can be virtually connected to a node from a different technology. This facilitates the inclusion of context-aware capabilities into systems because developers can dynamically integrate and remove nodes on demand.

HERA implements an organization based on a stack of layers. Each layer in one node communicates with its peer in another node through an established protocol. In addition, each layer offers specific functionalities to the immediately upper layer in the stack. These functionalities are usually called *interlayer services*, which must not be confused with the services invoked from node to node. These interlayer services are abstract functions and independent of the implementation of the platform. The HERA layers are added over the SYLPH layers [13] and also the existent application layer of each WSN stack, allowing the platform to be re-utilized over different technologies. The HERA layers must be adapted and recompiled over the programming language/platform provided by each WSN architecture (*e.g.*, SunSPOT or n-Core). However, only the lowest layer (*i.e.*, the *SYLPH Message Layer*) has to be integrated with the WSN stack (*i.e.*, the API that provides access to the WSN). This way, HERA comprises a modular and scalable platform. From lowest to highest, HERA/SYLPH layers are described as follows:

- **SYLPH Message Layer (SML).** SML offers the upper layers the possibility of sending asynchronous messages between two nodes through the *SYLPH Services Protocol* (SSP). These messages specify the source and destination nodes and the service invocation in a *SYLPH Services Definition Language* (SSDL) format. The SSDL describes the service itself and the parameters to be invoked. The SML not only transports services invocations over the network, but also services registration and search functions.
- **SYLPH Services Directory Sub-layer (SSDS).** SSDS creates dynamical services tables to locate and register services in the network. A node that stores and maintains services tables is called *SYLPH Directory Node* (SDN). These tables are made up of a list of service entries, each of which includes the description of a service in SSDL format and the SSP address of the node that offers the service. A node in the network can make a request to the SDN to know the location (*i.e.*, network address) of a certain service. Requests are packed in SML messages and must follow the SSP.
- **SYLPH Application Layer (SAL).** SAL allows different nodes to directly communicate with each other using SSDL requests and responses that will be delivered in encapsulated SML messages following the SSP. SAL implements service code (*i.e.*, firmware) from within each node, allowing each one to communicate with the SYLPH platform and invoke services located in other nodes. Moreover, there are other interlayer services for registering services or finding services offered by other nodes. In fact, these interlayer services for registering and searching services call other interlayer services offered by the SYLPH Services Directory Sub-layer (SSDS).
- **HERA Agents Layer (or just HERA).** HERA agents are specifically intended to run on devices with reduced resources, precisely what SYLPH was designed for. To communicate with each other, HERA agents use HERACLES, the agent communication language designed for being used under the HERA platform. Each HERA agent is an intelligent piece of code running over the SYLPH Application Layer. As explained below, there must be at least one facilitator agent in every agent platform. This agent is the first created in the platform and acts as a directory for searching agents. In HERA, the equivalent of these agents is the HERA-SDN (*HERA Spanned Directory Node*).

Similarly, the corresponding protocols used by HERA/SYLPH are:

- **SYLPH Services Protocol (SSP).** SSP is the internetworking protocol of the SYLPH platform. SSP has functionalities similar to those of the Internet Protocol (IP). That is, it allows sending packets of data from one node to another regardless of the WSN to which each one belongs. Every node has a unique SSP 32-bit address in the entire SYLPH network. Therefore, a SSP packet includes a header that describes the SSP addresses of the source and destination node, as well as information for managing transmissions that involve multiple SSP packets (*i.e.*, number of SSP packet and remaining bytes).
- **SYLPH Services Definition Language (SSDL).** SSDL is the IDL (Interface Definition Language) used by SYLPH. Unlike other IDLs such as WSDL (Web Services Definition Language) [12], SSDL does not use as many intermediate separating tags, and the order of its elements is fixed. SSDL has been specifically designed to work with limited computational resources nodes. SSDL [11] has two distinct representations: one that is human-readable, similar to C language and used for services development proposals, and one embedded on frames that SYLPH nodes understand. This is done in this way because in nodes with reduced resources it is not convenient to overload the microcontroller and the memory space with a heavy parsing method.
- **HERA Communication Language Emphasized to Simplicity (HERACLES).** HERACLES language is directly based on the SSDL language. As SSDL, HERACLES has two distinct representations. When developing a program, programmers use the human-readable representation to define agents' functionalities. However, HERA agents transmit the more compact representation of HERACLES as frames. This kind of compact frames is what HERA agents transmit in a heterogeneous WSN based on HERA/SYLPH over the SSDL/SSP protocols. When HERACLES is translated to HERACLES frames, the actual data transmitted among nodes, they are encapsulated into simple SSDL frames using "HERA" as their *service id* field.

Every agent platform needs some kind of facilitator agent that needs to be created before other agents are instantiated in the platform [2]. Facilitator agents act as agent directories. This way, every time an agent is created, it is registered on one of the existing facilitator agents. This allows other agents to request one of the facilitator agents to know where an agent with certain functionalities is and how to invoke such functionalities. As HERA is intended to run on machines that are not more complex than sensor nodes themselves are, it was necessary to design some hardware facilitator agents that do not need more CPU complexity and memory size than what a regular sensor node has. In order do this, HERA's facilitator agents, called HERA-SDNs (HERA Spanned Directory Nodes) are based on the SYLPH Directory Nodes (SDN). This way, any HERA node can perform as a HERA-SDN, just as SDNs do in the SYLPH platform. However, a HERA-SDN does not also have to be a SDN. HERA-SDN instances itself and starts the HERA platform by registering a special SYLPH service called "HERA" on a SDN stored on any node of the SYLPH network. When a new HERA Agent wants to instantiate itself through a HERA-SDN, it looks for the "HERA" service on the SYLPH network. When a HERA Agent is correctly instantiated, the HERA layer also registers a "HERA" service for the agent in a SDN. HERA Agents can send HERACLES messages to each other over SYLPH using the "HERA" service.

4 Experiments and Results

Several experiments were carried out to evaluate the performance of the HERA platform, mainly to test how it handled the instances of HERA-SDNs and HERA Agents and the exchange of HERACLES frames. In this sense, we deployed two distinct distributed WSN infrastructures with HERA running separately over them. Each infrastructure consisted of a ZigBee network with 31 devices, one acting as coordinator and the rest as routers. The first of these infrastructures was formed by nodes including each of them an 8-bit C51-based microcontroller with 8448B RAM, 128KB Flash memory and an IEEE 802.15.4/ZigBee transceiver. The second infrastructure was formed by n-Core Sirius-A devices belonging to the novel n-Core platform [14]. Each n-Core Sirius-A 2.4GHz device includes an ATmega1281 microcontroller with 8KB RAM, 128KB Flash memory, an AT86RF231 transceiver and several communication ports (GPIO, ADC, I^2C and USB/RS-232 UART) to connect to a wide range of sensors and actuators.

For each infrastructure, the ZigBee nodes were distributed in a short-range simple mesh, with less than 10 meters between any router and the coordinator. Each time the ZigBee network was formed, the nodes were powered on different random times, so that the mesh topology was different each time. The experiments consisted of trying to start a platform with HERA over each ZigBee network. Both networks were not tested simultaneously to avoid interferences. In each network, the ZigBee coordinator was acting as SDN and the 30 routers as SYLPH nodes. For each infrastructure, after the entire network was correctly created the SDN tried to instance a HERA-SDN. The HERA-SDN instanced itself and started the HERA platform registering a special service called "HERA" on the SDN stored on the same ZigBee coordinator node. Then, 10 nodes tried to instance one HERA Agent in the HERA platform. Once the HERA-SDN and the 10 HERA Agents were successfully instantiated, the HERA-SDN started to "ping" every of the ten HERA Agents with a *request* HERACLES frame including an *inform-if* command and waiting for an *inform* frame as a "pong" response. Each HERA Agent was pinged by the HERA-SDN one time every 5 seconds during one hour (7200 total pings tried). The experiment was run until both the platform and the agents were successfully started/instantiated 50 times. When the network could not be correctly created the run was discarded and not taken into account in the 50 runs. Furthermore, if the HERA platform could not be completely started and created (*i.e.*, all 10 HERA Agents correctly instantiated), these runs were also discarded and not taken into account as forming part of the 50 runs. The results are shown on Table 1, which indicates that it is necessary to improve both the SYLPH network creation and the instantiation of HERA Agents. A better ARQ (Automatic Repeat Request) mechanism could increase SSP-over-WSN transmissions.

Table 1 Results of the experiments to evaluate the performance of the HERA platform.

	C51-based	n-Core
Total runs	55	52
SYLPH not created correctly	2 (3.6%)	0 (0.0%)
HERA not started correctly	3 (5.6%)	2 (3.8%)
Ping-pongs not completed	15 (0.2%)	6 (0.1%)

5 Conclusions and Future Work

The HERA platform allows wireless devices from different technologies to work together in a distributed way. HERA Agents can communicate in a distributed way regardless of the technology or the programming language they use. Furthermore, HERA Agents are light enough to be run on WSN nodes with limited resources. The HERA Agents are reactive because they act on devices with critical response times.

Future work includes the improvement of the overall performance of the HERA platform. This way, the underlying SYLPH platform will be also improved, especially in the network formation and the SYLPH Gateways. In this sense, it will be evaluated other characteristics related to HERA/SYLPH nodes such as agent execution time, sleep mode intervals and power consumption. Likewise, the platform will be developed over additional WSN architectures to evaluate HERA and compare with existing implementations. In addition, it will be added a set of cross-layering services, so that HERA agents can modify the network parameters of each specific radio technology in a uniform way. Furthermore, we are working in the design of an efficient mechanism that allows HERA agents to move throughout different nodes, no matter the WSN technology they use. This way, we will get, for example, HERA agents to move from a ZigBee node to a Bluetooth node.

Acknowledgments. This project has been supported by the Spanish Ministry of Science and Technology project OVAMAH: TIN 2009-13839-C03-03.

References

1. Marin-Perianu, M., Meratnia, N., Havinga, P., de Souza, L., Muller, J., Spiess, P., et al.: Decentralized Enterprise Systems: A Multiplatform Wireless Sensor Network Approach. IEEE Wireless Communications 14(6), 57–66 (2007)
2. Wooldridge, M.: An Introduction to MultiAgent Systems, 2nd edn. Wiley, Chichester (2009)
3. Jennings, N.R., Sycara, K., Wooldridge, M.: A Roadmap of Agent Research and Development. Autonomous Agents and Multi-Agent Systems 1(1), 7–38 (1998)
4. ZigBee Alliance. ZigBee Specification Document 053474r13. ZigBee Standards Organization (2006)
5. Tynan, R., O'Hare, G., Ruzzelli, A.: Multi-Agent System Methodology for Wireless Sensor Networks. Multiagent and Grid Systems 2, 491–503 (2006)

6. Liu, Y., Zhou, C., Wang, K., Li, D., Guo, D.: Multi-agent ERA Model Based on Belief Interaction Solves Wireless Sensor Networks Routing Problem. In: Corchado, E., Abraham, A., Pedrycz, W. (eds.) HAIS 2008. LNCS (LNAI), vol. 5271, pp. 30–37. Springer, Heidelberg (2008)
7. Kwon, Y., Sundresh, S., Mechitov, K., Agha, G.: ActorNet: an actor platform for wireless sensor networks. In: Proceedings of the Fifth International Joint Conference on Autonomous Agents and Multiagent Systems, pp. 1297–1300. ACM, Hakodate (2006)
8. Baker, P., Catterson, V., McArthur, S.: Integrating an Agent-Based Wireless Sensor Network within an Existing Multi-Agent Condition Monitoring System. In: 15th International Conference on Intelligent System Applications to Power Systems, pp. 1–6 (2009)
9. Fok, C., Roman, G., Lu, C.: Mobile agent middleware for sensor networks: an application case study. In: Fourth International Symposium On Information Processing in Sensor Networks, IPSN 2005, pp. 382–387 (2005)
10. Zboril, F., Horacek, J., Spacil, P.: Intelligent Agent Platform and Control Language for Wireless Sensor Networks. In: Third UKSim European Symposium On Computer Modeling and Simulation, EMS 2009, pp. 482–487 (2009)
11. Corchado, J.M., Bajo, J., Tapia, D.I., Abraham, A.: Using Heterogeneous Wireless Sensor Networks in a Telemonitoring System for Healthcare. IEEE Transactions on Information Technology in Biomedicine 14(2), 234–240 (2010)
12. Cerami, E.: Web Services Essentials: Distributed Applications with XML-RPC, SOAP, UDDI & WSDL, 1st edn. O'Reilly Media, Inc., Sebastopol (2002)
13. Tapia, D.I., Alonso, R.S., De la Prieta, F., Zato, C., Rodríguez, S., Corchado, E., Bajo, J., Corchado, J.M.: SYLPH: An Ambient Intelligence based platform for integrating heterogeneous Wireless Sensor Networks. In: IEEE International Conference on Fuzzy Systems (FUZZ), pp. 1–8 (2010)
14. n-Core: A Faster and Easier Way to Create Wireless Sensor Networks (2011), `http://www.n-core.info` (retrieved January 12, 2011)

A Global Ant Colony Algorithm for Word Sense Disambiguation Based on Semantic Relatedness

Didier Schwab and Nathan Guillaume

Abstract. Brute-force WSD algorithms based on semantic relatedness are really time consuming. We study how to perform faster and better WSD. We focus here on an ant colony algorithm and evaluate it to exhibit some of its characteristics.

1 Introduction

Many textual applications such as multilingual information extraction, automatic summarisation, or machine translation can take advantage of Word Sense Disambiguation (WSD). Basically, the task consists in choosing the best sense among all possible senses for all words in a text. For example, in *"The mouse eats the cheese."*, the WSD task should choose for ‹*mouse*› the animal instead of the computer device. It exists many methods to perform WSD. Reader can read [10] for works before 1998 and [1] or [13] for a complete state of the art. Lot of these methods use semantic relatedness at local level (ie. between 2 words) then an algorithm to use them at global level.

In WSD, works as [15] use a brute-force (now BF) global algorithm which complexity is exponential time cost. In our works, a specific constant is that we aim at real time applications. When searching picture and moreover when calling someone who speak another language, one cannot wait a long time to get the answer. To obtain faster results in WSD, we currently study other methods than BF. In this paper, we focus on an adaptation of an ant colony algorithm. We present its characteristics and show that it permits to globally extend a local algorithm based on semantic relatedness to a whole text.

Didier Schwab · Nathan Guillaume
LIG-GETALP (Laboratory of Informatics of Grenoble - Study Group for Machine Translation and Automated Processing of Languages and Speech),
University of Grenoble, France
e-mail: didier.schwab@imag.fr

J.B. Pérez et al. (Eds.): Highlights in PAAMS, AISC 89, pp. 257–264.
springerlink.com

2 Local Algorithm

2.1 Semantic Relatedness Measures

These measures aim to give a score based on the likeness of the semantic content of compared linguistic items (often words or word senses). The score can be a mathematic similarity (value between 0 and 1), a distance (satisfying axioms of reflexivity, symmetry and triangle inequality) or is a positive value without higher bound.

Among them, we can cite *Hirst & Saint-Honge* based on graph distance between two senses in a lexical network ; *Rada et al.* and *Leacock & Chodorow* are similar but only consider hypernym links or distances between vectors (LSA, conceptual vectors [16]). One can consult [15], [4] ou [13] for a complete overview.

For WSD, these methods are applied locally for senses between two words and are then used at a global level. In this paper, we only focus on global algorithms.

2.2 Our Local Algorithm: A Variant of the Lesk Algorithm

Our local algorithm is a variant of the Lesk Algorithm [12]. This algorithm, invented more than 25 years ago, is really simple. It only needs a dictionary and no learning phase. The score given to a sense pair is the number of common words (considered as the text between to spaces) in the two definitions. It takes into account neither the word order (bag of word approach) nor the syntactic or morphologic information. Variants of this algorithm are still today among the best on English [15].

We use WordNet, a lexical database for English, where word senses (called synsets) are linked through relations (such as hypernymy, hyponymy or antonymy). In our local algorithm, we exploit these links. Our variant considers not only the definition of a sense but also the definitions of the linked senses as done in [2] [1].

Definitions are represented as vectors: a word for one component. Our ant algorithm incrementally builds such vectors to perform global WSD (see section 4.2.4).

3 Global Algorithms

Here, a global algorithm is a method which permits to extend a local algorithm to a whole text. The most direct algorithm is the BF method used, for example, in [2]. It is to consider combinations of all senses of the words in the context, to give a score to each combination and then to choose the combination with the best one. The main problem of this method is the combinatorial explosion. For example, in "*The pictures they painted were flat, not round as a figure should be, and very often the feet did not look as if they were standing on the ground at all, but pointed downwards as if they were hanging in the air.*"" which has 17 words, ‹*picture*› and ‹*air*› have 9 senses, ‹*paint*› 4, ‹*be*›, ‹*point*›, ‹*figure*› 13, ‹*flat*› 17, ‹*very*›, ‹*often*› 2, ‹*ground*›, ‹*foot*› 11, ‹*look*› 10, ‹*stand*› 12, ‹*at all*›, ‹*downwards*› 1, ‹*hang*› 15. Then, there are

[1] [2] introduce an idea of common word sequences but we don't use it there.

137,051,946,345,600 combinations to compute which is almost the number of operations (and a combination need many operations), 9000 Intel new 4 cores 3,8Ghz processors can do in one second. Hence, BF method is very difficult to apply in real conditions and, moreover, make it impossible to use a longer analysis context.

To circumvent this problem, several solutions have been proposed. For example, approaches using a corpus to reduce the number of dimensions to consider such as lexical chains [7] [17] or artificial intelligence approaches such as simulated annealing [3] or genetic algorithms [8].

Moreover, these methods share the problem to prevent exploitation of the linguistic structure as a graph (syntactic or morphological analysis), that is why we present here a method based on ant colony algorithm for semantic analysis inspired by [11].

4 A Global Ant Colony Algorithm

4.1 Ant Colony Algorithms

It has been demonstrated that cooperation inside an ant colony is self-organised and results from interactions between individuals. These interactions are often very simple and allow the colony to solve complex problems. This phenomenon is called swarm intelligence and is more and more used in computer science where centralized control systems are often successfully replaced by other types of control based on interactions between simple elements.

Artificial ants have first been used for resolving the Travelling Salesman Problem. In these algorithms, the environment is usually represented by a graph and virtual ants exploit pheromone deposed by others or pseudo-randomly explore the graph.

These algorithms are a good alternative for the resolution of problems modeled with graphs. They allow fast and efficient walkthrough close to other resolution methods. Their main interest is their important ability to adapt themselves to changing environment. Readers can consult [5] or [9] for a states of the art.

4.2 The Algorithm

4.2.1 Principle

Our algorithm (now ACA) starts with a graph. It is the environment of the ants. This graph can be a linguistic one as a morphological lattice [6] or a syntactic tree [11] or simply a graph organised following the text. For example, a word is linked in the graph to its predecessor in the sentence and to its successor in the sentence. Of course, results of the algorithm won't be the same. Research is currently undertaken on that point, this is not the focus of our article, we will just see interesting results with a simple graph (see fig. 1).

Following [16] or [9] experiments on French, each word sense is associated to a nest as a vector (see 2.2). A nest products ants which goal is to find energy and bring it back to their nests. The more energy deposed, the more nests produce ants.

Table 1 Parameters/notations for the ant algorithm

notation	description	value
F_A	Nest for the sense A	na
$V(X)$	Vector associated to X. X is a node or an ant	na
f_A	Ant born in the nest F_A	na
E_f	Energy used by a nest to produce an ant	na
$E(X)$	Energy on X. X is a node or an ant	na
E_{max}	Maximum of energy an ant can carry	5
$\varphi_c(A)$	Quantity of pheromone on the edge A at cycle c	na
θ	pheromone deposed when crossing an edge	1
δ	Evaporation of pheromone between each cycle	20%
$E_f(N)$	Evaluation of the node N for ant f	na
$\varphi_f(A)$	Evaluation of the pheromone on the edge A for ant f	na
	Number of cycles	100
	Initial quantity of energy on each node	20
	Ant life length	10
	Energy taken by an ant when it arrives on a node	1
	Odor vector length	50
	Amount of odor deposed by an ant when it arrives on a node	10%

An ant keeps a trace of its nest, its vector equals to the nest's. Ant displacements are function of scores given by the local algorithm, of the presence of energy and function of the other ants passage (pheromone). When an ant arrives on the nest from another term it can choose to come back directly to its mother nest. It builds a bridge that other ants will be able to cross and to reinforce with their pheromone. This reinforcement is done if lexical information (from the graph if it is a linguistics one and from local scores) lead other ants to take the bridge and disappears otherwise. Hence, ants make lots of links between compatible nets. Moreover, as an ant which come back to its nest leaves energy it carries, compatible nests have an emergent behave of meta-nest as they share resources and reinforce themselves comparatively to the ones who do not succeed in allying and impoverish their resources.

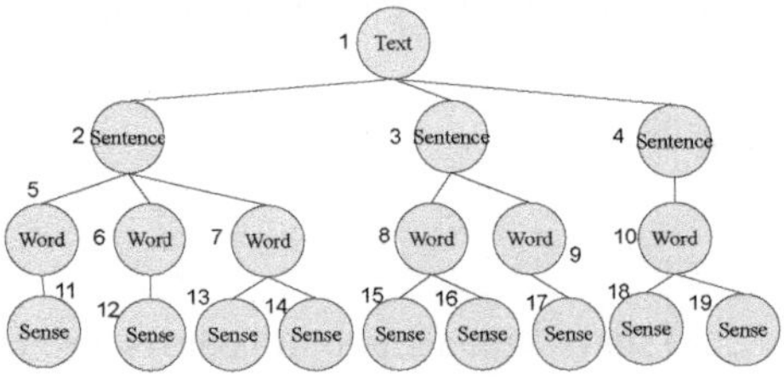

Fig. 1 Example of an environment:nests are grey-blue (1-10) and common nodes are grey (11-19).

4.2.2 Energy

Whatever the graph chosen (fig. 1 is just an example), we distinguish 2 types of nodes: the nests, one per word sense, and common nodes. At the initial stage, there is the same energy quantity on each node. Nests use it to produce ants with a probability function of its energy (sigmoid function). These ants will walk through the graph and bring back energy to their nest which will be able to produce new ants with it. When an ant dies, all energy it carries and all energy used by its nest to produce it is deposed on the node where it was. Hence there is nor increase nor decrease of energy in the system. This point is a fundamental element of the convergence of the algorithm to a solution. If the energy was not limited, all nests would receive energy and would be more and more activated and none would be inhibited.

4.2.3 Types of Nodes for an Ant

For an ant, a node can be : (1) *its mother nest*, the node where it was born; (2) *an enemy nest*, a nest which corresponds to another sense of the same word; (3) *a nest potentially friend*, a nest which is not an enemy; (4) a *common node* a node which is not a nest. For example, in fig. 1, for an ant born in the nest 19, it mother nest is 19, node 18 is an enemy as it has the same father (10), the potentially friends are nodes 11 to 17 and common ones are node 1 to 10.

4.2.4 Odor

Each node have a vector. For a nest, it is the sense vector (see 2.2) and it is never modified. When an ant arrives on a common node, it deposes some of its vector: some of its components, randomly chosen replace the same quantity, also randomly chosen, in the node vector. This is the vector odor. It is used by ants to find their nests when they bring back energy. Indeed, if a common node is near a nest their vectors share many components (and thus local score is strong) because lot of its ants go through it. Moreover, this phenomena permits to ants to make some errors and to go to potentially friend nests. This error can be beneficial as they can then build a bridge to their mother nest.

4.3 Simulation

It is a potentially infinite iteration of cycles. At the end of each cycle, it can be stopped and current state observed. During a cycle, the following tasks are done: (1) eliminate the older ants (ant life length is a parameter); (2) for each nest, request the production of an ant (an ant can or cannot born, in a probabilistic way); (3) For each edge, decrease the pheromone rate; (4) for each ant: determine its mode (seek energy, bring it back to its nest), move it ; (5) compute ant displacement consequences (interpretative bridge creation, energy and pheromone);

In an abstract way, one can summarize the ant displacements as follows. This displacement is random but influenced by environment. When an ant is on a node, it estimates all neighbor nodes N_i and all edges A_j they linked with $P(N_i,A_j) = max(\frac{Eval(N_i,A_j)}{\sum_{k=1,l=1}^{k=n,l=m} Eval(N_k,A_l)},\varepsilon)$. ε is there to permit to some ants to go to these destinations in the case the environment has behind them resources and lexical information.

The ant gives an evaluation of the energy on each node ($E_f(N) = \frac{E(N)}{\sum_0^m E(N_i)}$) and an evaluation of pheromone on each edge ($\varphi_f(A) = \frac{\varphi(A)}{\sum_0^n \varphi(N_j)}$).

A newborn ant seeks food. It is attracted by the nodes which carry much energy ($Eval_\varphi(N_i,A_j) = 1 - \varphi_f(A_j)$) but avoid to go through edges with lot of pheromone to permit to explore lot of solutions ($Eval_E(N_i,A_j) = E_f(N_i)$). It collects as much energy as it can carry, it transports more and more energy and the probability of return to the mother nest increases ($P(return) = \frac{E(f)}{E_{max}}$). Then, it moves while following (statistically) the ways which contain more pheromone ($Eval_\varphi(N_i,A_j) = \varphi_f(A_j)$) and to nodes which odor is close to their ones $Eval_{odor}(N_i,A_j) = \frac{Lesk(V(N_j),V(f_A))}{\sum_{i=1}^{i=k} Lesk(V(N_i),V(f_A))}$.

5 Experiments

We tested this algorithm with the SemEval 2007 coarse-grained English all-words task corpus [14]. It is composed of 5 texts of various domains (journalism, book review, travel, computer science, biography). The task consists in annotating 2269 words with one of their senses in WordNet. Evaluation of output is done considering coarse grained senses ie considering that close senses are equivalent (eg. snow/precipitation and snow/cover).

In the corpus, words are tagged with their part of speech. From that, we build the environment of the ants : a node for the text level, a node for each sentence, a node for each word, a node with a nest for each sense (see fig. 1). Then we launch our ants and at the end of the simulation, we choose for each word the nest with the highest quantity of sugar.

5.1 Comparison with the Brute-Force Algorithm

We compare results from the BF algorithm [2] and results of our algorithm with parameters shown in table 1. For BF, we have choosen as context the sentence as in [2]. That choice excludes sentences with only one word (four in the corpus). For calculatory reasons, we also exclude sentences with more than 10 billions combinations. We can see that only 77.3% of the corpus has been attempted with a time length incompatible with real-time applications[2]. On its side, ACA guarantees to make a choice between the different possibilities for a word. Hence, 100% of the corpus is attempted (22,7% more than the FB algorithm). First consequence is that precision, recall and F-measure are identical. The F-measure is clearly higher than the BF algorithm for a time length 2000 times lower. Note that on the subpart of the corpus attempted by the BF algorithm, ACA obtain around 76%.

[2] Experiments were done on Intel Xeon X5550, 4 core, 2.66Ghz. All time length are converted to mono-processor ones.

Global Algorithm	Attempted	Precision	Recall	F-measure	Length
Brute-force (BF)	77,30	77,82	60,16	67,86	≈ 300h
ACA	100,0	74,35	74,35	74,35	≈ 8m

5.2 *Executions of the Ant Colony Algorithm*

The algorithm does not exactly select the same senses at each execution nor at each cycle. On hundreds of simulation, we have noted that around 70-80 cycles, results remain globally constant. The diagram is an illustration of this phenomena. We have the same between two executions. To give an idea of the variability of results for several executions, we have repeated our experiment 100 times. The table shows a difference of only 2,16% between the better and the worst result (≈ 50 senses). In all cases, F-measure is higher than the BF one.

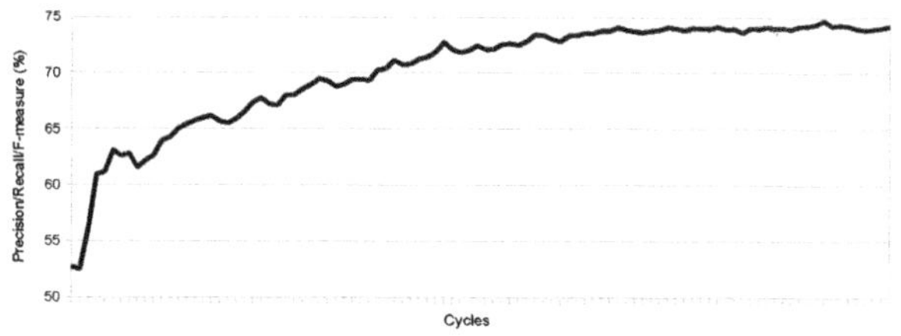

minimum	maximum	average	median	range	Standard deviation
73,20	75,36	74,25	74,24	2,16	0,51

6 Conclusions and Perspectives

We have presented an ant algorithm for word sense disambiguation based on semantic relatedness. This algorithm permits to choose for all words of a text a sense faster and better than BF algorithm. We presented interesting results that need to be extended. For example, how evolve results with linguistics graphs, how to use WordNet semantic graphs to improve results,is it possible to exploit several executions to get better results by votes. We also want to compare these ant algorithms to other global algorithm candidates as genetic algorithms or simulated annealing. Lots of interesting tracks of research we want to explore.

References

1. Agirre, E., Edmonds, P.: Word Sense Disambiguation: Algorithms and Applications (Text, Speech and LT). Springer-Verlag New York, Inc., Secaucus (2006)
2. Banerjee, S., Pedersen, T.: An adapted lesk algorithm for word sense disambiguation using wordnet. In: Gelbukh, A. (ed.) CICLing 2002. LNCS, vol. 2276, Springer, Heidelberg (2002)

3. Cowie, J., Guthrie, J., Guthrie, L.: Lexical disambiguation using simulated annealing. In: COLING 1992, Nantes, France, vol. 1, pp. 359–365 (1992)
4. Cramer, I., Wandmacher, T., Waltinger, U.: Modeling, Learning and Processing of Text Technological Data Structures. In: WordNet: An Electronic Lexical Database. Springer, Heidelberg (2010)
5. Dorigo, S.: Ant Colony Optimization. MIT Press, Cambridge (2004)
6. Falaise, A., Rouquet, D., Schwab, D., Blanchon, H., Boitet, C.: Extraction de contenu seímantique dans des masses de donneíes multilingues. le projet omnia. TAL 52(1), 11–19 (2010)
7. Gale, W., Church, K., Yarowsky, D.: One sense per discourse. In: Fifth DARPA Speech and Natural Language Workshop, Harriman, New-York, États-Unis, pp. 233–237 (1992)
8. Gelbukh, A., Sidorov, G., Han, S.Y.: Evolutionary approach to natural language wsd through global coherence optimization. WSEAS Transactions on Communications 2(1), 11–19 (2003)
9. Guinand, F., Lafourcade, M.: Artificial Ants. From Collective Intelligence to Real-life Optimization and Beyond. In: Artificial ants for NLP, Lavoisier, ch. 20, pp. 455–492 (2009)
10. Ide, N., Véronis, J.: Wsd: the state of the art. Computational Linguistics 28(1), 1–41 (1998)
11. Lafourcade, M., Schwab, D.: Multi-castes ants for holistic semantic text analysis. In: SNLP 2005: The 6th Symposium on NLP, Chiang Rai, Thailand (2005)
12. Lesk, M.: Automatic sense disambiguation using mrd: how to tell a pine cone from an ice cream cone. In: Proceedings of SIGDOC 1986, pp. 24–26. ACM, New York (1986)
13. Navigli, R.: Wsd: a survey. ACM Computing Surveys 41(2), 1–69 (2009)
14. Navigli, R., Litkowski, K.C., Hargraves, O.: Semeval-2007 task 07: Coarse-grained english all-words task. In: SemEval 2007, Prague, Czech Republic, pp. 30–35 (2007)
15. Pedersen, T., Banerjee, S., Patwardhan, S.: Maximizing Semantic Relatedness to Perform WSD. Research report, University of Minnesota Supercomputing Institute (2005)
16. Schwab, D.: Approche hybride pour la modélisation, la détection et l'exploitation des fonctions lexicales en vue de l'analyse sémantique de texte. Ph.D. thesis, U. Montpellier 2 (2005)
17. Vasilescu, F., Langlais, P., Lapalme, G.: Evaluating variants of the lesk approach for disambiguating words. In: Proceedings of LREC 2004, Lisbon, Portugal, pp. 633–636 (2004)

Modeling Context Information for Computational Semantics with the Language of Acyclic Recursion

Roussanka Loukanova

Abstract. The paper explores the expressiveness of the L_{ar}^{λ} language for richly informative lexical semantics. It introduces an extended version of the language L_{ar}^{λ} to represent semantic information associated with lexemes that have denotations dependent on context information such as space and time locations, "speaker" or "listener" agents, and other potential context components. The technique is demonstrated by rendering referential lexical items such as demonstratives, deictic pronouns, and definite descriptions. It is applicable to any language with richly informative inflection structure.

1 Background and Recent Developments

In a sequence of papers, Moschovakis introduced a class of languages of recursion as a new approach to the mathematical notion of algorithm for applications to computational semantics of languages, in similar ways as to semantics of programing languages, e.g., see [7] for FLR, and [8] for L_{ar}^{λ}. Brief introductions to the syntax and semantics of the higher-order type theory of recursion is also presented in other papers, see [4] and [6]. For the purpose of self-containment, we give a short introduction to the language L_{ar}^{λ}, with details that are relevant to the specifics of this paper.

The technical apparatus of the paper is an inspiration of Situation Semantics utterance types, see [3], into a version of Moschovakis' acyclic recursion language. This consideration does not induce direct relations between versions of Situation Semantics and the language L_{ar}^{λ}, primarily because L_{ar}^{λ} has functional types for total functions, while Situation Semantics uses relational, situated types that are complex information constructs representing partial objects. We extend L_{ar}^{λ} to a version with enriched type system, by adding "uttered" NL expressions inside formally defined objects, called utterances, that represent important components of context

Roussanka Loukanova
Uppsala, Sweden

J.B. Pérez et al. (Eds.): Highlights in PAAMS, AISC 89, pp. 265–274.
springerlink.com

information. The result is an extended version of L^{λ}_{ar}, with explicit representation of utterance information into formal terms.

2 Short Introduction to the Language of Acyclic Recursion

Types: The types of L^{λ}_{ar} is the smallest set *Types* defined recursively, by using a notational variant of Backus-Naur Form (BNF):

$$\sigma :\equiv e \mid t \mid s \mid (\tau_1 \rightarrow \tau_2) \qquad (Types)$$

Intuitively, the types represent: e the entities of the domain; t is for truth values; s is for states; $(\tau_1 \rightarrow \tau_2)$ is for unary functions from objects of type τ_1 to objects of type τ_2. A possibility for more detailed treatment of states is introduced further in this paper, to represent more detailed context information such as time, space location, speaker, etc.

Syntax of L^{λ}_{ar}: The *vocabulary of* L^{λ}_{ar} consists of pairwise disjoint sets, for each type τ: (1) a finite set of constants of type τ, $K_\tau = \{\mathsf{c}_0, \mathsf{c}_1, \ldots, \mathsf{c}_{k_\tau}\}$; (2) a set of pure variables of type τ, $PureVars_\tau = \{\mathsf{v}_0, \mathsf{v}_1, \ldots\}$; and (3) a set of recursion variables, called also locations, of type τ, $RecVars_\tau = \{\mathsf{p}_0, \mathsf{p}_1, \ldots\}$. *Terms of* L^{λ}_{ar}: The recursive rules for the set of L^{λ}_{ar} terms can be expressed by using a notational variant of BNF, which is loaded with additional notations for assigning types. We write $X : \xi$ just in case when X is a term of type ξ (which is not per se part of the term), and X^{ξ} when the type ξ is associated with X (X^{ξ} corresponds to type checking for X). Alternative rules are separated by "|" and are followed by brief explanations:

$$\begin{aligned}
A :\equiv\ & \mathsf{c}^{\tau} : \tau && \text{;; } A \text{ is a constant } c^{\tau} \text{ and of type } \tau; && \text{(const-term)}\\
& \mid x^{\tau} : \tau && \text{;; } A \text{ is a variable } x^{\tau} \text{ and of type } \tau; && \text{(var-term)}\\
& \mid B^{(\sigma \rightarrow \tau)}(C^{\sigma}) : \tau &&&& \text{(appl-term)}\\
& \mid \lambda(v^{\sigma})(B^{\tau}) : (\sigma \rightarrow \tau) &&&& (\lambda\text{-term})\\
& \mid A_0^{\sigma_0} \text{ where } \{p_1^{\sigma_1} := A_1^{\sigma_1}, \ldots, p_n^{\sigma_n} := A_n^{\sigma_n}\} : \sigma_0 &&&& \text{(rec-term)}
\end{aligned}$$

where: in (appl-term), A is a term of type τ, which is application of a term $B^{(\sigma \rightarrow \tau)}$ to a term C^{σ}; in (λ-term), A is a term of type $(\sigma \rightarrow \tau)$, which is λ-abstraction from a pure variable v^{σ} over a term B^{τ}; in (rec-term), A is a term (recursion term) of type σ_0, in case that, for each $0 \leq i \leq n$ ($n \geq 0$), $A_i^{\sigma_i}$, is a term of type σ_i, $\{p_i^{\sigma_i} : 1 \leq i \leq n\}$ are pairwise different locations of the corresponding types, and $\{p_1^{\sigma_1} := A_1^{\sigma_1}, \ldots, p_n^{\sigma_n} := A_n^{\sigma_n}\}$ is an acyclic system. And, where a set $\{p_1^{\sigma_1} := A_1^{\sigma_1}, \ldots, p_n^{\sigma_n} := A_n^{\sigma_n}\}$, for $n \geq 0$[1], is an *acyclic system of assignments* exactly when there is a ranking function $rank : \{p_1, \ldots, p_n\} \longrightarrow \mathbb{N}$ such that, for all $0 \leq i, j \leq n$, if p_j occurs free in A_i, then $rank(p_j) < rank(p_i)$.

[1] When $n = 0$, the system of assignments is empty, and $A :\equiv A_0^{\sigma_0}$ where $\{\} : \sigma_0 :\equiv A_0 : \sigma_0$.

Denotational Semantics of L_{ar}^{λ}: A frame of typed semantic objects $\mathbb{T}$ is a set of typed sets (or classes) $\mathbb{T} = \{\mathbb{T}_\sigma \mid \sigma \in Types\}$, where $\mathbb{T}_e \neq \varnothing$ are the entities, $\mathbb{T}_t = \{0, 1, er\}$ are *truth values*, $\mathbb{T}_s \neq \varnothing$ is a set of *states*. In this paper, we define the states as complex objects, with the purpose of modeling utterance information. Given the set G of all variable assignments, the *denotation function* den is:

(D1) $\mathsf{den}(x)(g) = g(x)$; $\mathsf{den}(c)(g) = \mathscr{I}(c)$;
(D2) $\mathsf{den}(A(B))(g) = \mathsf{den}(A)(g)(\mathsf{den}(B)(g))$;
(D3) $\mathsf{den}(\lambda x(B))(g) : \mathbb{T}_\tau \longrightarrow \mathbb{T}_\sigma$, where $x : \tau$ and $B : \sigma$, is the function defined as follows: for every $t \in \mathbb{T}_\tau$, $[\mathsf{den}(\lambda x(B))(g)](t) = \mathsf{den}(B)(g\{x := t\})$;
(D4) $\mathsf{den}(A_0 \text{ where } \{p_1 := A_1, \ldots, p_n := A_n\})(g) = \mathsf{den}(A_0)(g\{p_1 := \overline{p}_1, \ldots, p_n := \overline{p}_n\})$, where $\overline{p_i} = \mathsf{den}(A_i)(g\{p_{k_1} := \overline{p}_{k_1}, \ldots, p_{k_m} := \overline{p}_{k_m}\})$, with $p_{k_1}, \ldots, p_{k_m}$ all the recursion variables with ranks lower than $\mathsf{rank}(p_i)$, for $i \in \{1, \ldots, n\}$.

The Reduction Calculus of L_{ar}^{λ} provides a set of reduction rules (see, [8]) for simplifying the terms A into their canonical forms $\mathsf{cf}(A)$. The canonical form $\mathsf{cf}(A)$ represents the algorithm (when it exists) for calculating the denotation of A, in the simplest way: from the denotations of the basic facts that are computed incrementally and "stored" in locations. Intuitively, an acyclic system $\{p_1 := A_1, \ldots, p_n := A_n\}$ defines recursive "computations" of the values to be assigned to the locations $p_1, \ldots, p_n$, which end after a finite number of steps; $rank(p_j) < rank(p_i)$ means that the value A_i assigned to p_i, may depend on the values of the free p_j in A_i.

3 Richness of L_{ar}^{λ} Type-Theoretic Models

Expressiveness of the Language of L_{ar}^{λ}: As proved in [8], L_{ar}^{λ} is a proper extension of Gallin's TY_2, and, thus, of Montague's IL. That L_{ar}^{λ} strictly extends Montague's IL, is demonstrated by terms that are in L_{ar}^{λ}, but not in IL. Typed λ-calculus, with recursion terms, offers its own object language for modeling ambiguity and other semantic underspecification in NL. In addition, there is more subtle representation of different co-reference relations, i.e., co-denotation relations, and antecedent–anaphora relations, see [4] and [6]. In this paper, we explore further the expressiveness of the language L_{ar}^{λ}, with respect to representing semantic information that is associated with language expressions and contributed by their lexical components. In many cases, the denotation $\mathsf{den}(\varphi)$ of an expression strictly depends on the context of use, incl. on utterance components, and the concept of speaker's reference that vary over utterances. For example, among many other utterance dependent language units are the proper names, deictic pronominal expressions, demonstratives, etc. In this section, we initiate exploration of the rich type-theoretic semantic structures of L_{ar}^{λ}, with respect to capturing context dependent semantic information that is presented in NL utterances and. More specifically, we introduce a possibility to use model-theoretic states for modeling utterances of NL expressions and described possible-world indexes.

"Direct" Denotational Semantics of NL Expressions: In this subsection, we introduce the idea of using frames of typed semantic objects $\mathbb{T}$ for "direct" denotational interpretations of NL lexical and phrasal expressions, with the purpose of incorporating the notion of utterance into type-theoretic states. We consider a possibility for the objects in the domain of the states $\mathbb{T}_s$, which is a little bit more elaborated than suggested on p.9-10, of [8], and which borrows some ideas of state informativeness from Situation Semantics (e.g., see [1], [2]). Here, we consider a possibility to define states that represent language utterances (i.e., instances of language use) and speakers' references, as the elements of a set $\mathbb{T}_{refState} \subseteq (\mathbb{T}_u \times \mathbb{T}_{spRef} \times \mathbb{T}_d)$, where the elements of $\mathbb{T}_u$ and $\mathbb{T}_{spRef}$ (and, possibly, $\mathbb{T}_d$) have components that are systematically defined by a formal system of syntax and semantics of some object natural language (for example English, Bulgarian, Russian, etc.). Semantic frames include (as language modeling) representatives of language expressions as components of utterances $\mathbb{T}_u$. For the purposes of this paper, we set: $\mathbb{T}_u \subseteq \mathbb{T}_{speaker} \times \mathbb{T}_{time} \times \mathbb{T}_{space} \times \mathbb{T}_{expr}$ and $\mathbb{T}_d \subseteq \mathbb{T}_{world} \times \mathbb{T}_{time} \times \mathbb{T}_{space}$, where $\mathbb{T}_{time} \subseteq \mathbb{T}_e$ is a set of time moments; $\mathbb{T}_{space} \subseteq \mathbb{T}_e$ is a set of space locations; $\mathbb{T}_{expr}$ is a set of well-formed language expressions, defined by a formal (computational) system of lexicon and grammar; $\mathbb{T}_{speaker} \subseteq \mathbb{T}_{human} \subseteq \mathbb{T}_e$ is a set of people making utterances in the object NL; $\mathbb{T}_{world} \neq \varnothing$ is a nonempty set of possible worlds, taken as primitive objects of the type s; $\mathbb{T}_{spRef}$ is a set of functions, called *speakers' references*: $\delta : \mathbb{T}_{speaker} \times \mathbb{T}_{time} \times \mathbb{T}_{space} \times \mathbb{T}_{expr} \longrightarrow \mathbb{T}$.

Reference States: We assume that the formal grammar of NL can be assotiated, in a rigouros way, with a maping Υ, from the syntactic categories of the grammar, into the types of L^{λ}_{ar} (for example, see [9]). The translating procedure Υ should be such that, for every well-formed NL expression φ, of a syntactic category $\mathsf{SynCat}(\varphi)$, the type $\Upsilon(\mathsf{SynCat}(\varphi))$ has to reflect the intuitions about the objects that are (typically) denoted by uttarences of φ. We delimit the set $\mathbb{T}_{spRef}$ as follows: For every $u \in \mathbb{T}_u$, $\delta \in \mathbb{T}_{spRef}$ and $d \in \mathbb{T}_d$, the tuple $\langle u, \delta, d \rangle \in \mathbb{T}_{refState}$ iff for some $a \in \mathbb{T}_{speaker}$, $t_u \in \mathbb{T}_{time}$, $l_u \in \mathbb{T}_{space}$, $\varphi \in \mathbb{T}_{expr}$, $w_d \in \mathbb{T}_{world}$, $t_d \in \mathbb{T}_{time}$, $l_d \in \mathbb{T}_{space}$,

$$u = \langle a, t_u, l_u, \varphi \rangle,\ d = \langle w_d, t_d, l_d \rangle, \text{ and} \tag{2a}$$

$$\text{there is a function } \mathsf{evalindex} : \mathbb{T}_u \longrightarrow \mathbb{T}_d \text{ for which} \tag{2b}$$

$$d = \langle w_d, t_d, l_d \rangle = \mathsf{evalindex}(a, t_u, l_u, \varphi)$$

$$\delta(a, t_u, l_u, \varphi) \in \mathbb{T}_{\Upsilon(\mathsf{SynCat}(\varphi))} \tag{2c}$$

$$\text{For every } rs \in \mathbb{T}_{spRef}, \tag{2d}$$

$$[\mathsf{den}(\varphi)(g)](rs) = \begin{cases} \delta(a, t_u, l_u, \varphi), & \text{if } u = \langle a, t_u, l_u, \varphi \rangle \text{ and } rs = \langle u, \delta, d \rangle, \\ er, & \text{otherwise.} \end{cases}$$

Note that the states are not primitive objects: they are complex, structured objects in the semantic frame, represented by tuples of objects including agents (called speakers), time and space locations, and "uttered" language expressions, φ. Such tuples encode essential, context dependent information. While "uttered" expressions φ participate in the semantic frame, and they may have syntactic structure according to some grammar of the rendered language, the syntactical analyses of φ are not part of the semantic frame. In this paper, we do not make any assumptions for the rendered languages, neither for what kinds of computational syntax may be used. While we do not consider restrictions over what are the "uttered" expressions that may participate in the frame, except that they are potentially randed into L^{λ}_{ar}, primarily, we envisage lexical renderings.

We call the tuples $u = \langle a, t_u, l_u, \varphi \rangle \in \mathbb{T}_u$ *utterances* of φ, iff for some δ and d, $\langle u, \delta, d \rangle \in \mathbb{T}_{refState}$. We call the tuple $\langle u, \delta, d \rangle$ the *current state*, or the *reference state* for the utterance u iff $\langle u, \delta, d \rangle \in \mathbb{T}_{refState}$, i.e., iff (2a)-(2d) hold. In such a case, we also call the component d the *evaluation index* of the (current) state $\langle u, \delta, d \rangle$, which is determined by the following function for every $\langle a, t_u, l_u, \varphi \rangle$, $\mathsf{evalindex}(a, t_u, l_u, \varphi) = \langle w_d, t_d, l_d \rangle = d$. Each tuple $u = \langle a, t_u, l_u, \varphi \rangle$ represents an instance of usage of φ (which is an expression or lexical item, such as an word), by the speaker a, at the moment t_u and the location l_u. The values of the function $\delta \in \mathbb{T}_{spRef}$, $\delta(a, t_u, l_u, \varphi)$, will be called *a's references* in the utterance $u = \langle a, t_u, l_u, \varphi \rangle$. The clause in (2d) includes the possibility for errors, because the denotation of φ can be erroneous for some states rs, for example, where the utterance u component of rs is not of φ, i.e., $u = \langle a, t_u, l_u, \psi \rangle$ and $\varphi \not\equiv \psi$. The clauses (2a)-(2d) represents utterance dependent denotations, where the speaker refers to a state component d to evaluate the denotation of φ in d. In possible-world semantics, such as Montague's IL, the elements $d \in \mathbb{T}_d$ are called *indexes*. In Situation Semantics, the element $d \in \mathbb{T}_d$ corresponds to a *resource state* (i.e., more correctly, in Situation Semantics, d would be a *resource situation*) for evaluation of the expression φ. When the expression φ is a sentence, the index d, which is provision of the utterance, represents intuitions for a *described state*, because the tuple $\langle w_d, t_d, l_d \rangle$ represents a world w_d, and a space-time point $\langle t_d, l_d \rangle$, referred to by the speaker, as holding (satisfying) the propositional content denoted by φ. In possible-worlds semantics, e.g., as in Montague's IL or its variants, d represents the current, or the "actual" world, for evaluation of the denotation of an expression. Since, in this paper, we use the L^{λ}_{ar} calculus for computational semantics of NL, we use terminology, which (we believe) is in the spirit of L^{λ}_{ar}. The denotation $[\mathsf{den}(\varphi)(g)]$ is the Carnap intension of the proposition made by the speaker. Intuitively, the denotation $[\mathsf{den}(\varphi)(g)](rs) = \delta(a, t_u, l_u, \varphi)$ is the object to which the speaker's refers in the reference state rs. When φ is a sentence, this is the truth value, which the speaker associates with the statement made by the utterance u, for the reference index d. This is justified when the reference index d supports the factual content of the utterance of φ.

Deictic Use of Demonstratives in "Direct" Semantics: Demonstrative lexemes, e.g., "here" and "there", are strictly dependent on utterance components and speaker's references, for their strong deictic usage. Let $u \in \mathbb{T}_u$, $\delta \in \mathbb{T}_{spRef}$, and $d \in \mathbb{T}_d$, be such that for arbitrary, but fixed $a \in \mathbb{T}_{speaker}$, $t_u \in \mathbb{T}_{time}$, $l_u \in \mathbb{T}_{space}$, $\varphi \in \{\text{here}, \text{there}\} \subset \mathbb{T}_{expr}$, $w_d \in \mathbb{T}_{world}$, $t_d \in \mathbb{T}_{time}$, $l_d \in \mathbb{T}_{space}$, such that (2a)-(2d) hold, with $\tau \equiv e$, i.e., $\mathsf{den}(\varphi)(g)\colon rs \to e$. In particular, $u = \langle a, t_u, l_u, \varphi \rangle$, are:

$$[\mathsf{den}(\text{here})(g)](u, \delta, d) = \delta(a, t_u, l_u, \text{here}) = l_u = l_d \tag{3a}$$

$$[\mathsf{den}(\text{there})(g)](u, \delta, d) = \delta(a, t_u, l_u, \text{there}) = l_d \in \mathbb{T}_{space}, \text{ and } l_d \neq l_u \tag{3b}$$

Deictic Use of Pronouns in Direct Semantics: Assume that in an utterance u of a sentence like "She smiles", the pronoun "she" is used by the speaker to refer to a particular individual $c \in \mathbb{T}_e$, i.e., $\delta(a, t_u, l_u, \text{she}) = c$. Then

$$[\mathsf{den}(\text{she})(g)] = rs \mapsto \delta(a, t_u, l_u, \text{she}) = rs \mapsto c \tag{4a}$$

$$[\mathsf{den}(\text{she})(g)](u, \delta, d) = \delta(a, t_u, l_u, \text{she}) = c \tag{4b}$$

If $\langle u, \delta, d\rangle \in \mathbb{T}_{refState}$ is the *reference state*, the propositional denotation of the sentence "She smiles" is $smiles(rs \mapsto c) \in \mathbb{T}_{\tilde{t}}$. The denotation of this sentence in $\langle u, \delta, d\rangle$ is the truth value 1 just in case the individual c smiles in the evaluation index d; i.e., the proposition $smiles(rs \mapsto c) \in \mathbb{T}_{\tilde{t}}$ is true in d iff $smiles(rs \mapsto c)(u, \delta, d) = 1$. In neither of the above cases, the denotation carries the information that the individual c is female, i.e., the gender information that (depending on the specific language) is part of the semantics of the pronoun "she". The denotational value of the above sentence, with respect to the utterance, by itself, does not specify how it is obtained, i.e., "computed" from the denotations of the lexical components of the sentence. A canonical form, together with formalization of utterances, provides information for that. Pronouns like "she", "he", "it", etc., in their deictic use, resemble demonstratives like "here" and "there", for their strong denotational dependency on the speaker's reference δ and thus on the utterance u. However, the semantics of these pronouns differ from the semantics of the demonstrative "here", in an essential way: these pronouns can participate in anaphoric relations, which is part of their abstract linguistics meaning. This is not necessarily associated with the demonstrative "here", and is not possible for the pronoun "I". The denotation of "I", similarly to that of the demonstratives, is strictly dependent on its utterance, by denoting the speaker, but is independent on the speaker's reference.

4 "Indirect" Lexical Semantics of NL with L^{λ}_{ar}

Why "Indirect" Semantic Representation of NL? It is possible to present "direct" semantics of NL, by formulating direct correspondences, from formal syntax of well-formed NL constructs, to semantic structures, based on typed domain frames. Such attempts for "direct" semantics (exemplified above) would be less perspicacious. Indirect model-theoretic semantics of NL, by rendering it into type-theoretic language, is far more transparent and comes with many other superiorities. Type-theoretical languages are well-studied, have rich theoretical results and model theories. In addition, they offer possibilities for software developments and automatic processing. In this paper, we focus on rendering lexical items into L^{λ}_{ar}. Rendering NL expressions will be discussed only for expository purpose and is topic of other work. More specifically, we will use the language L^{λ}_{ar} to represent semantic information associated with some lexemes, the denotations of which depend on their utterances. Such information is "propagated" recursively into larger NL constructs, by grammar rules.

Rendering Utterance Information into Type-theoretic Terms: The specific denotations of various language units can be dependent on various components of specific utterances: however, the dependency on utterance components can be part of their abstract semantics, in a pattern-like way, as linguistic items out of any specific utterance. This is why, rendering NL items that depend on context of use, can be facilitated by representing utterance components as parts of type-theoretic terms, i.e., in this paper, L_{ar}^{λ}-terms. For this purpose, we take a version of L_{ar}^{λ} language with special constants. For every NL expression φ, let $\tau_{refState}(\varphi)$ be the type of the reference states in the domain $\mathbb{T}_{refState}$, essentially determined by utterances of φ. Note that $\tau_{refState}(\varphi)$ are not ground types. Let $\sigma_u(\varphi)$ be the type of utterances of φ, and let $\mathsf{Type}(\mathsf{render}(\varphi))$ be the type of the rendering of the expression φ. For the current expository purpose, we can take the type of the reference states to be the type of their characteristic function: $\tau_{refState}(\varphi) \equiv (\sigma_u(\varphi) \to ((\sigma_u(\varphi) \to \mathsf{Type}(\mathsf{render}(\varphi))) \to (s \to (\tilde{e} \to (\tilde{e} \to \tilde{t})))))$. We add constants to the language L_{ar}^{λ}, to represent context information: *speaker* for the utterance speaker, *uTime* for the utterance time, *uSpace* for the utterance space, *lWorld* for the local world of the index, *lTime* for the local time, and *lSpace* for the local space of the index, and *spRefs* for the speaker's reference function. The denotations of these constants give the respective components of the reference states, by being defined as follows. For every reference state $r \equiv \langle\langle a, t_u, l_u, \varphi\rangle, \delta, \langle w_d, t_d, l_d\rangle\rangle \in \mathbb{T}_{refState}$,

$$speaker\colon (\tau_{refState}(\varphi) \to e) \text{ and } [\mathsf{den}(speaker)](r) = a \tag{5a}$$

$$uTime\colon (\tau_{refState}(\varphi) \to e) \text{ and } [\mathsf{den}(uTime)](r) = t_u \tag{5b}$$

$$uSpace\colon (\tau_{refState}(\varphi) \to e) \text{ and } [\mathsf{den}(uSpace)](r) = l_u \tag{5c}$$

$$lWorld\colon (\tau_{refState}(\varphi) \to s) \text{ and } [\mathsf{den}(lWorld)](r) = w_d \tag{5d}$$

$$lTime\colon (\tau_{refState}(\varphi) \to e) \text{ and } [\mathsf{den}(lTime)](r) = t_d \tag{5e}$$

$$lSpace\colon (\tau_{refState}(\varphi) \to e) \text{ and } [\mathsf{den}(lSpace)](r) = l_d \tag{5f}$$

and, the constant *spRefs* for speaker's references is such that, for every expression φ, assumed as being a part of utterance types:

$$spRefs\colon (\varphi \to (\tau_{refState}(\varphi) \to \mathsf{Type}(\mathsf{render}(\varphi)))) \tag{6}$$

$$[\mathsf{den}(spRefs(\varphi))](r) = \begin{cases} \delta(a, t_u, l_u, \varphi), & \text{if } \langle a, t_u, l_u, \varphi\rangle \text{ is utterance component in } r \\ er, & \text{otherwise.} \end{cases} \tag{7}$$

5 Utterance Dependency in the Lexicon: Rendering Deictic Lexical Items

There are different kinds of lexical ambiguities, either by association to different part of speech lexical category, or due to "pure" semantic ambiguity of some lexemes. An overview of how lexical ambiguity propagates into phrasal ambiguity is presented in [5]. The semantic representation of many lexical items is tightly dependent on prominent context elements such as participating agents, speakers, listeners, space and time locations, some of which we consider in what follows. We demonstrate the technique by considering primarily deictic lexical items.

Deictic Use of Demonstratives in Indirect Semantics. Now, demonstrative lexical items, for example such as "here" and "there", can be *rendered* into L_{ar}^{λ} as follows:

$$\text{here} \xrightarrow{\text{render}} the(h) \,\mathsf{where}\, \{h := \lambda x(r(x) \wedge u(x) \wedge l(x)), \\ r := \lambda x(x(\text{here}) = spRefs(\text{here})), \\ u := \lambda x(x(\text{here}) = uSpace), \\ l := \lambda x(x(\text{here}) = lSpace)\}, \tag{8}$$

where $x\colon \varphi \to (\tau_{refState}(\varphi) \to e)$ is a pure variable.

Deictic Use of Pronouns in Indirect Semantics

$$\text{she} \xrightarrow{\text{render}} the(h) \,\mathsf{where}\, \{h := \lambda x(r(x) \wedge i(x)), \\ r := \lambda x(x = spRefs(\text{she})), \\ i := \lambda x\, female(x)\} \tag{9}$$

Anaphoric usages and underspecified ambiguities between deictic and anaphoric occurrences of pronouns in NL expressions is not a subject of this paper. Here is just a possible rendering:

$$\text{she} \xrightarrow{\text{render}} the(h) \,\mathsf{where}\, \{h := \lambda x(r(x) \wedge i(x)), \\ i := \lambda x\, female(x)\} \tag{10}$$

Note that, in (10), the location variable r is free, which gives the possibility for associating a specific reading (meaning and denotation) to the pronoun "she" at a later stage, when context prompts for a specific usage. I.e., an underspecified rendering of a pronominal, like in (10), can be bound, either by the speaker's references, or by an antecedent element in a larger expression. The rendering (10) of "she" carries the linguistic information, which is encoded explicitly by the lexical entry, about the gender of the individual denoted by it. Similar gender and number information that is encoded in some lexical items, and is proper semantic information, rather than merely syntactic agreement co-occurrence flag, needs to be included in the rendering type-theoretic terms, as part of their "logic form". Next we show how semantic contributions by inflections can be included in the L_{ar}^{λ} terms that serve as semantic representatives of inflected forms. Exhaustive treatment of inflection rules, such as verbal tenses and aspects, is not a subject of the paper.

Definite Determiner in English. Note that the reference states (via utterance expressions φ) provide "resource" states for evaluation of the denotations:

$$\text{the} \xrightarrow{\text{render}} \mathit{the}: \tag{11}$$
$$(\varphi \to (((\tau_{refState}(\varphi) \to e) \to (\tau_{refState}(\varphi) \to t)) \to (\tau_{refState}(\text{the } \varphi) \to e)))$$

Definite NPs in English. The speaker's references reflect the referential use of definite descriptions. This is achieved by a constraint over the denotation value via the assignment (12b), which requires the value *spRefs*(the homework) for the denoted object (via grammar rules for NPs):

$$\text{the homework} \xrightarrow{\text{render}} \mathit{the}(H) \,\mathsf{where}\, \{H := \lambda x(r(x) \wedge h(x)), \tag{12a}$$
$$r := \lambda x(x = \mathit{spRefs}(\text{the homework})), \tag{12b}$$
$$h := \mathit{homework}\} \tag{12c}$$
$$: (\tau_{refState}(\text{the homework}) \to e) \tag{12d}$$

Attributive use of definite NPs is represented by terms like (12a)-(12c), without the assignment (12b).

Definite NPs in Bulgarian

$$\text{домашното} \xrightarrow{\text{render}} \mathit{the}(H) \,\mathsf{where}\, \{H := \lambda x(r(x) \wedge h(x)), \tag{13a}$$
$$r := \lambda x(x = \mathit{spRefs}(\text{домашното})), \tag{13b}$$
$$h := \mathit{homework}\} \tag{13c}$$
$$: (\tau_{refState}(\text{домашното}) \to e) \tag{13d}$$

Indefinite NPs in Bulgarian. The semantic distinction between the part of speech of common nouns and indefinite NPs, in Bulgarian, is represented by the types of the rendering terms. For example:

$$\text{домашно} \xrightarrow{\text{render}} \mathit{homework} \qquad \text{common noun} \tag{14a}$$
$$: ((\tau_{refState}(\text{домашно}) \to e) \to (\tau_{refState}(\text{домашно}) \to t))$$
$$\text{домашно} \xrightarrow{\text{render}} \mathit{homework} \qquad \text{existential quantifier} \tag{14b}$$
$$: (\tilde{e} \to \tilde{t}) \to (\tau_{refState}(\text{домашно}) \to t)$$

6 Conclusion

Rendering NL lexicon and grammatical constructs into terms of a type-theoretical language, for the purpose of semantic "representations", has many other positive arguments, as demonstrated firstly by the original works of [9]. The continuing theoretical and computational advances of type theories bring more arguments, especially when targeting computational grammar that covers lexicon with morphological structure that propagates into the phrasal structure of sentences, and complex syntax-semantics inter-relations in languages. The language of acyclic recursion

L_{ar}^{λ} is one of the latest examples for such advances. L_{ar}^{λ} is an elegant mathematical system, which is also richly expressive, not simply by being a proper extension of a higher-order language such as Montague's IL. Language ambiguities and underspecified semantic representation are under major attention in computational semantics and, generally, in NLP, not only because complete resolution of ambiguities present NLP with computational intractability, but is also unnecessary in the absence of relevant context dependant information. The language and theory of L_{ar}^{λ} offers not only a novel algorithmic treatment of Frege's distinction between sense and denotation, it also has its own object language and calculi for representing semantic underspecification of NL expressions. Additionally, our major focus has been the exploration of the expressiveness of L_{ar}^{λ} for more adequate representation of semantic information that is carried by lexical entities, such as deictic language units and lexical inflection, which is typically interpreted depending on information provided by specific context. Such language phenomena, with Slavonic languages just as examples, are in need of computational theory that faithfully captures semantic information and incorporation into computational syntax-semantics interface. Morphology is a rich topic, which requires extensive exploration by development of computational syntax-semantics interface, and the research in this paper is in such directions.

References

1. Barwise, J., Perry, J.: Situations and Attitudes. MIT press, Cambridge (1983)
2. Devlin, K.: Situation theory and situation semantics. In: Gabbay, D., Woods, J. (eds.) Handbook of the History of Logic, vol. 7, pp. 601–664. Elsevier, Amsterdam (2008)
3. Loukanova, R.: Quantification and intensionality in situation semantics. In: Gelbukh, A. (ed.) CICLing 2002. LNCS, vol. 2276, pp. 32–45. Springer, Heidelberg (2002)
4. Loukanova, R.: Typed lambda language of acyclic recursion and scope underspecification. In: Muskens, R. (ed.) Workshop on New Directions in Type-theoretic Grammars, ESSLLI 2007, Dublin, Ireland, pp. 73–89 (2007)
5. Loukanova, R.: Computational syntax-semantics interface. In: Bel-Enguix, G., Jiménez-López, M.D. (eds.) Language as a Complex System: Interdisciplinary Challenges, pp. 111–150. Cambridge Scholars Publishing (2010)
6. Loukanova, R.: Reference, co-reference and antecedent-anaphora in the type theory of acyclic recursion. In: Bel-Enguix, G., Jiménez-López, M.D. (eds.) Linguistics, Biology and Computer Science: Interplays, Cambridge Scholars Publishing (to appear)
7. Moschovakis, Y.N.: Sense and denotation as algorithm and value. In: Oikkonen, J., Vaananen, J. (eds.) Lecture Notes in Logic, vol. 2, pp. 210–249. Springer, Heidelberg (1994)
8. Moschovakis, Y.N.: A logical calculus of meaning and synonymy. Linguistics and Philosophy 29, 27–89 (2006)
9. Thomason, R.H. (ed.): Formal Philosophy: Selected Papers of Richard Montague. Yale University Press, New Haven (1974); Edited, with an introduction, by Richmond H. Thomason

Complete Obligatory Hybrid Networks of Evolutionary Processors

Artiom Alhazov and Gemma Bel-Enguix*, Alexander Krassovitskiy*, and Yurii Rogozhin

Abstract. We consider obligatory hybrid networks of evolutionary processors (OHNEPs) introduced recently. It is a variant of hybrid network of evolutionary processors (HNEPs), differing from the main model as follows: in OHNEP a node discards the strings to which no operations are applicable, and underlying graph is directed. In this paper we consider OHNEPs with complete underlying graph. We show that such networks are computationally complete even using very simple evolutionary processors: one operation per node (obligatory left deletion, right insertion and obligatory substitution). In our proof, the total power of input/output filters does not exceed 3 symbols.

1 Introduction

Insertion, deletion, and substitution are fundamental in formal language theory. Networks of evolutionary processors (NEPs) [4] are systems built of agents connected in a graph, each one specializing in one of these operations. Such teams of limited agents have been shown to be very powerful and flexible, taking advantage of their connectivity. They operate on a set of words by rewriting them and redistributing them according to the system's protocol. The usual associated result are the words ever appearing in a specific node. NEPs are inspired by cell biology: each processor can represent a cell with DNA point mutations and a filtering mechanism. NEPs with a very small number of nodes are known [3] to be computationally complete.

Particularly interesting variants of these devices are the so-called *hybrid networks of evolutionary processors* (HNEPs) [8], where each language processor

Artiom Alhazov · Yurii Rogozhin
Institute of Mathematics and Computer Science, Academy of Sciences of Moldova
e-mail: {artiom,rogozhin}@math.md

Gemma Bel-Enguix · Alexander Krassovitskiy
Rovira i Virgili University, Research Group on Mathematical Linguistics
e-mail: {gemma.bel@,alexander.krassovitskiy@estudiants}urv.cat

* Supported by the Spanish Ministerio de Ciencia y Tecnología, project MTM2007-63422.

J.B. Pérez et al. (Eds.): Highlights in PAAMS, AISC 89, pp. 275–282.
springerlink.com

performs only one of the above operations in a certain position of the words in that node. Furthermore, the filters are random-context conditions, i.e., they check presence/absence of certain symbols in the words. It is known that HNEPs with 7 nodes are universal [2] and Accepting HNEPs with 6 nodes are universal [7], while HNEPs with 2 nodes are not computationally complete [2].

In this paper, we continue to study a new variant of HNEPs, called Obligatory Networks of Evolutionary Processors (OHNEP for short) introduced in [1]. The differences between HNEP and OHNEP are the following:

1. in deletion and substitution: a node discards a string if no operations in the node are applicable to the string (in HNEP case, this string remains in the node),
2. an underlying graph is a directed graph (in HNEP case, this graph is undirected).

These differences make OHNEPs universal [1] with 1 operation per node, no filters and only left insertion and right deletion. (The second point makes no difference for this paper, since the graph is complete).

In this paper we consider complete OHNEPs, i.e., OHNEPs with complete underlying graph. One may now regard complete OHNEP as a set of very simple evolutionary processors "swimming in the environment". One of the goals of this paper is to prove the complete OHNEPs with very simple evolutionary processors, i.e., evolutionary processors with only one operation (obligatory deletion, obligatory substitution and insertion) and filters containing total not more than 3 symbols are computationally complete. Open problems are whether one can decrease the total power of the filters keeping the computational completeness, and whether one can avoid substitutions using these very simple evolutionary processors.

2 Definitions

We recall some notions we shall use throughout the paper. An *alphabet* is a finite and nonempty set of symbols. The cardinality of a finite set A is written as $card(A)$. A sequence of symbols from an alphabet V is called a word over V. The set of all words over V is denoted by V^*, and the empty word is denoted by ε; we use $V^+ = V^* \setminus \{\varepsilon\}$. The length of a word x is denoted by $|x|$, while we denote the number of occurrences of a symbol a in a word x by $|x|_a$. For each nonempty word x, $alph(x)$ is the minimal alphabet W such that $x \in W^*$.

Circular Post Machines (CPMs) were introduced in [6], where it was shown that all introduced variants of CPMs are computationally complete, and moreover, the same statement holds for CPMs with two symbols. In this article we use the deterministic CPM0s.

A *Circular Post Machine* is a quintuple $(\Sigma, Q, \mathbf{q}_1, \mathbf{q}_f, R)$ with a finite alphabet Σ where 0 is the blank, a finite set of states Q, the initial state $\mathbf{q}_1 \in Q$, the final state $\mathbf{q}_f \in Q$, and a finite set of instructions R with all instructions having one of the forms $\mathbf{p}x \to \mathbf{q}$ (erasing the symbol read by deleting a cell), $\mathbf{p}x \to y\mathbf{q}$ (overwriting and moving to the right), $\mathbf{p}0 \to y\mathbf{q}0$ (overwriting and creating a blank cell), where $x, y \in \Sigma$ and $\mathbf{p}, \mathbf{q} \in Q$, $\mathbf{p} \neq \mathbf{q}_f$. We also refer to all instructions with $\mathbf{q}_f$ in the left hand side as halt instructions. The storage of this machine is a circular tape, the read and write head move only in one direction (to the right), and with the possibility to delete a cell or to create and insert a new cell with a blank.

We now summarize the necessary notions concerning *obligatory evolutionary operations*. For an alphabet V, we say that a rule $a \to b$, with $a,b \in V \cup \{\varepsilon\}$ is an *obligatory substitution operation* if $a \neq \varepsilon$ and $b \neq \varepsilon$; it is an *obligatory deletion operation* if $a \neq \varepsilon$ and $b = \varepsilon$; and it is an *(obligatory) insertion operation* if $a = \varepsilon$ and $b \neq \varepsilon$. The set of all obligatory substitution, deletion, and insertion operations over an alphabet V are denoted by Sub_V, Del_V, and Ins_V, respectively. Given such rules π, ρ, σ, and a word $w \in V^*$, we define the following *obligatory evolutionary actions* of π, ρ, σ on w if $\pi \equiv a \to b \in Sub_V$, $\rho \equiv a \to \varepsilon \in Del_V$, and $\sigma \equiv \varepsilon \to a \in Ins_V$:

$$\pi^*(w) = \{ubv \mid w = uav,\ u,v \in V^*\},\ \rho^*(w) = \{uv \mid w = uav,\ u,v \in V^*\} \quad (1)$$
$$\rho^r(w) = \{u \mid w = ua\},\ \rho^l(w) = \{v \mid w = av\} \quad (2)$$
$$\sigma^*(w) = \{uav \mid w = uv,\ u,v \in V^*\},\ \sigma^r(w) = \{wa\},\ \sigma^l(w) = \{aw\} \quad (3)$$

Notice that in (1) – (2) the result of obligatory evolution operation may be the empty set (this is the main difference between obligatory hybrid networks of evolutionary processors and hybrid networks of evolutionary processors).

Symbol $\alpha \in \{*, l, r\}$ denotes the way of applying an insertion or a deletion rule to a word, namely, at any position ($\alpha = *$), in the left end ($\alpha = l$), or in the right end ($\alpha = r$) of the word, respectively. Note that a substitution rule can be applied at any position. For a rule σ, an action $\alpha \in \{*, l, r\}$ and a language $L \subseteq V^*$, we define the α-action of σ on L by $\sigma^\alpha(L) = \bigcup_{w \in L} \sigma^\alpha(w)$. For a given finite set of rules M, we define the α – *action* of M on a word w and on a language L by $M^\alpha(w) = \bigcup_{\sigma \in M} \sigma^\alpha(w)$ and $M^\alpha(L) = \bigcup_{w \in L} M^\alpha(w)$, respectively.

Before defining an evolutionary processor, we define the filtering mechanism.

For disjoint subsets $P,F \subseteq V$ and a word $w \in V^*$, we define the predicate φ (see [5] and [2]) as $\varphi(w;P,F) \equiv (P = \emptyset \vee alph(w) \cap P \neq \emptyset) \wedge (F \cap alph(w) = \emptyset)$. This corresponds to predicate $\varphi^{(\beta)} = \varphi^{(2)}$ in the definitions from [5], i.e., when the filtering parameter is $\beta = 2$. Other values of β correspond to different filtering conditions, but we do not need them in this paper. The construction of this predicate is based on *random-context conditions* defined by the two sets P *(permitting contexts)* and F *(forbidding contexts)*. For every language $L \subseteq V^*$ we define $\varphi(L,P,F) = \{w \in L \mid \varphi(w;P,F)\}$.

An *obligatory evolutionary processor over alphabet* V is a 5-tuple $EP = (M, PI, FI, PO, FO)$, where:

- Either $M \subseteq Sub_V$ or $M \subseteq Del_V$ or $M \subseteq Ins_V$. The set M represents the set of obligatory evolutionary operations of the processor. Note that every processor is dedicated to only one type of the above obligatory evolutionary operations.
- $PI, FI \subseteq V$ are the *input* permitting/forbidding contexts of the processor, while $PO, FO \subseteq V$ are the *output* permitting/forbidding contexts of the processor.

We denote the set of obligatory evolutionary processors over V by OEP_V.

Definition 1. *An obligatory hybrid network of evolutionary processors (an OHNEP for short) is a 6-tuple* $\Gamma = (V, G, N, C_0, \alpha, i_0)$, *where the following conditions hold:*

- V *is a finite set (the alphabet).*
- $G = (X_G, E_G)$ *is a directed graph where the set of nodes is* X_G *and the set of edges is* E_G. G *is called the* underlying graph *of the network.*
- $N : X_G \longrightarrow OEP_V$ *is a mapping which associates with each node* $x \in X_G$ *the obligatory evolutionary processor* $N(x) = (M_x, PI_x, FI_x, PO_x, FO_x)$.
- $C_0 : X_G \longrightarrow 2^{V^*}$ *is a mapping which identifies the initial configuration of the network. It associates a finite set of words with each node of the graph* G.
- $\alpha : X_G \longrightarrow \{*, l, r\}$; $\alpha(x)$ *defines the action mode of the rules performed in node* x *on the words occurring in that node. We indicate* α *as a superscript of* M_x.
- *For every node,* $x \in X_G$, *we define the following filters: the input filter is given as* $\rho_x(\cdot) = \varphi(\cdot; PI_x, FI_x)$, *and the output filter is defined as* $\tau_x(\cdot) = \varphi(\cdot, PO_x, FO_x)$. *That is,* $\rho_x(w)$ *(resp.*τ_x*) indicates whether or not the word* w *can pass the input (resp. output) filter of* x. *More generally,* $\rho_x(L)$ *(resp.* $\tau_x(L)$*) is the set of words of* L *that pass the input (resp. output) filter of* x.
- $i_0 \in X_G$ *is the* output node *of the OHNEP.*

An OHNEP is said to be a *complete* OHNEP, if its underlying graph is a complete graph. A configuration of an OHNEP Γ, as above, is a mapping $C : X_G \longrightarrow 2^{V^*}$ which associates a set of words with each node of the graph. A component $C(x)$ of a configuration C is the set of words that can be found in the node x in this configuration, hence a configuration can be considered as the sets of words which are present in the nodes of the network at a given moment. A configuration can change either by an *evolutionary step* or by a *communication step*. When it changes by an evolutionary step, then each component $C(x)$ of the configuration C is changed in accordance with the set of evolutionary rules M_x associated with the node x and the way of applying these rules $\alpha(x)$. Formally, configuration C' is obtained in *one evolutionary step* from the configuration C, written as $C \Longrightarrow C'$, iff $C'(x) = M_x^{\alpha(x)}(C(x))$ for all $x \in X_G$.

When the system evolves by a communication step, each evolutionary processor $N(x)$, where $x \in X_G$, sends all its words passing the output filter of x to the processors in all nodes connected with x; from all the words sent by processors to $N(x)$, this processor $N(x)$ receives those words that pass the input filter of x. Formally said, configuration C' is obtained in *one communication step* from configuration C, denoted $C \vdash C'$, iff $C'(x) = (C(x) - \tau_x(C(x))) \cup \bigcup_{(y,x) \in E_G} (\tau_y(C(y) \cap \rho_x(C(y)))$ for all $x \in X_G$.

For an OHNEP Γ, a computation in Γ is a sequence of configurations $C_0, C_1, C_2, \ldots$, where C_0 is the initial configuration of Γ, $C_{2i} \Longrightarrow C_{2i+1}$ and $C_{2i+1} \vdash C_{2i+2}$, for all $i \geq 0$. If we use OHNEPs as language generating devices, then the generated language is the set of all words which appear in the output node at some step of the computation. Formally, the language generated by Γ is $L(\Gamma) = \bigcup_{s \geq 0} C_s(i_0)$.

3 Main Result

Theorem 1. *Any CPM0 P can be simulated by a complete OHNEP P' where evolutionary processors are with only one operation per node (obligatory deletion on the left end and insertion on the right end of the string and obligatory substitution).*

Proof. Let us consider a CPM0 P with symbols $a_j \in \Sigma$, $j \in J = \{0,1\ldots,n\}$, $a_0 = 0$ is the blank symbol, and states, $q_i \in Q$, $i \in I = \{1,2,\ldots,f\}$, where q_1 is the initial state and the only terminal state is $q_f \in Q$.

So, we consider CPM0 P with the set R of instructions of the forms $q_ia_j \longrightarrow q_l$, $q_ia_j \longrightarrow a_kq_l$, $q_i0 \longrightarrow a_kq_l0$, $q_fa_j \longrightarrow Halt$, where $q_i \in Q\setminus\{q_f\}$, $q_l \in Q$, $a_j, a_k \in \Sigma$.

A configuration $v = q_ia_jW$ of CPM0 P describes that P in state $q_i \in Q$ considers symbol $a_j \in \Sigma$ to the left of $W \in \Sigma^*$. To this configuration corresponds a string $v' = a_jWq_i$ in nodes of OHNEP P'. To the final configuration q_fa_jW of P corresponds a string a_jWq_f in the output node $\langle out\rangle$ of P'.

Now we construct a complete OHNEP P' that simulates P. Let $q'_{\langle q_ia_j\rangle} = q'_{\langle q_ia_j\rangle,1}$.

$$\begin{aligned}
P' &= (V,G,N,C_0,\alpha,i_0),\ V = \{S\}\cup Q\cup Q'\cup\Sigma\cup E', \text{ where}\\
E' &= \{\varepsilon'_{\langle q_ia_j\rangle}\},\ Q' = \{q'_{\langle q_ia_j\rangle,t}\}, q_i \in Q\setminus\{q_f\},\ a_j\in\Sigma,\ t=1,2,\\
G &= (X_G,E_G), \text{ is a complete graph}: E_G = X_G\times X_G,\\
X_G &= \{\langle init\rangle,\langle out\rangle = i_0\}\cup\{\langle q_ia_j\rangle_t^a\} \mid (q_ia_j\to q_l)\in R,\ 1\le t\le 4\}\\
&\cup\ \{\langle q_ia_j\rangle_t^b\} \mid (q_ia_j\to a_kq_l)\in R, 1\le t\le 6\}\\
&\cup\ \{\langle q_ia_j\rangle_t^c\} \mid (q_i0\to a_kq_l0)\in R,\ 1\le t\le 6\}.\\
C_0(x) &= \{SWq_1\},\ \text{if } x=\langle init\rangle,\ \text{ where } q_1W \text{ is the input of } P,\\
C_0(x) &= \emptyset,\ x\in X_G\setminus\{\langle init\rangle\},\ N(\langle init\rangle) = (\{S\to\varepsilon\}^l,\emptyset,\emptyset,\emptyset,\emptyset),\\
N(\langle out\rangle) &= (\emptyset,\{q_f\},\emptyset,\emptyset,\emptyset) \text{ and we define the other nodes below.}
\end{aligned}$$

CPM0 P starts a computation from a configuration q_1a_jW and OHNEP P' starts computation from a string Sa_jWq_1 in the input node $\langle init\rangle$ accordingly (other nodes of P' contain empty sets of strings): $Sa_jWq_1 \overset{N(\langle init\rangle)}{\longrightarrow} a_jWq_1$ (it means that evolutionary processor $N(\langle init\rangle$ is applied to string Sa_jWq_1 and resulting string a_jWq_1 is sent to all nodes of P', including node $\langle init\rangle$). Clearly, string a_jWq_1 will be rejected by $N(\langle init\rangle)$ as there is no operation applicable to this string. Further we continue to construct P' and describe how P' simulates three types of rules of CPM0 P. Notice that first three evolutionary processors of P' that correspond to types of operations of P, namely $N(\langle q_ia_j\rangle_t^a)$ (for rule $q_ia_j\to q_l$), $N(\langle q_ia_j\rangle_t^b)$ (for rule $q_ia_j\to a_kq_l$), and $N(\langle q_i0\rangle_t^c)$ (for rule $q_i0\to a_kq_l0$), $1\le t\le 3$, serve to determine the leftmost symbol a_j of string a_jWq_i, and they have the same forms.

Case (A). Rule $q_ia_j\to q_l$ is simulated by the following nodes.

$$N(\langle q_ia_j\rangle_1^a) = (\{a_j \to \varepsilon'_{\langle q_ia_j\rangle}\}, \{q_i\}, E', \emptyset, \emptyset),$$
$$N(\langle q_ia_j\rangle_2^a) = (\{q_i \to q'_{\langle q_ia_j\rangle}\}, \{\varepsilon'_{\langle q_ia_j\rangle}\}, \emptyset, \emptyset, \emptyset),$$
$$N(\langle q_ia_j\rangle_3^a) = (\{\varepsilon'_{\langle q_ia_j\rangle} \to \varepsilon\}^l, \{q'_{\langle q_ia_j\rangle}\}, \emptyset, \emptyset, \{\varepsilon'_{\langle q_ia_j\rangle}\}),$$
$$N(\langle q_ia_j\rangle_4^a) = (\{q'_{\langle q_ia_j\rangle} \to q_l\}, \emptyset, \emptyset, \emptyset, \{\varepsilon'_{\langle q_ia_j\rangle}\}).$$

Let $q_ia_jW \overset{q_ia_j \to q_l}{\longrightarrow} q_lW$ be a computation step in P, i.e., rule $q_ia_j \to q_l$ is applied to configuration q_ia_jW and q_lW is the next configuration.

In P' the string a_jWq_i (copies of this string) is distributed among all nodes $\langle q_ia_t\rangle$, $a_t \in \Sigma$ (it is clear that this word will be rejected by other evolutionary processors $N(\langle q_sa_r\rangle)$, $q_s \neq q_i$, $a_r \in \Sigma$ of P'). Thus for nodes $\langle q_ia_j\rangle_s^a$, $s = 1, \ldots, 4$ evolution of string a_jWq_i leads to the correct result Wq_l:

$$a_jWq_i \overset{N(\langle q_ia_j\rangle_1^a)}{\longrightarrow} \varepsilon'_{\langle q_ia_j\rangle}Wq_i \overset{N(\langle q_ia_j\rangle_2^a)}{\longrightarrow} \varepsilon'_{\langle q_ia_j\rangle}Wq'_{\langle q_ia_j\rangle} \overset{N(\langle q_ia_j\rangle_3^a)}{\longrightarrow} Wq'_{\langle q_ia_j\rangle} \overset{N(\langle q_ia_j\rangle_4^a)}{\longrightarrow} Wq_l.$$

Assume that evolutionary processor $N(\langle q_ia_j\rangle_1^a)$ is applied in an "improper position". More precisely, let $W = W'a_jW''$. String $a_jW'a_jW''q_i$ evolves in P' as follows: $a_jW'a_jW''q_i \overset{N(\langle q_ia_j\rangle_1^a)}{\longrightarrow} a_jW'\varepsilon'_{\langle q_ia_j\rangle}W''q_i$. It is easy to check that during an evolution of string $a_jW'\varepsilon'_{\langle q_ia_j\rangle}W''q_i$ symbol $\varepsilon'_{\langle q_ia_j\rangle}$ cannot be deleted and the resulting string cannot enter at node $N(\langle q_ia_j\rangle_1^a)$ and will be lost. The same arguments are valid in the case when "improper evolutionary processor" $N(\langle q_ia_t\rangle)$, $t \neq j$ is applied.

Thus OHNEP P' correctly simulates rule $q_ia_j \to q_l$ of CPM0 P.

Case (B). Rule $q_ia_j \to a_kq_l$ is simulated by the following nodes.

$$N(\langle q_ia_j\rangle_t^b) = N(\langle q_ia_j\rangle_t^a),\ 1 \le t \le 3, \text{(here } N(\langle q_ia_j\rangle_t^b) \text{ have the same forms as } N(\langle q_ia_j\rangle_t^a)\text{)},$$
$$N(\langle q_ia_j\rangle_4^b) = (\{\varepsilon \to q'_{\langle q_ia_j\rangle,2}\}^r, \{q'_{\langle q_ia_j\rangle}\}, \{q'_{\langle q_ia_j\rangle,2}\}, \emptyset, \{\varepsilon'_{\langle q_ia_j\rangle}\}),$$
$$N(\langle q_ia_j\rangle_5^b) = (\{q'_{\langle q_ia_j\rangle} \to a_k\}, \{q'_{\langle q_ia_j\rangle,2}\}, \emptyset, \emptyset, \emptyset\}),$$
$$N(\langle q_ia_j\rangle_6^b) = (\{q'_{\langle q_ia_j\rangle,2} \to q_l\}, \emptyset, \emptyset, \emptyset, \{q'_{\langle q_ia_j\rangle}\}).$$

We use the same techniques as above and show that OHNEP P' correctly simulates rule $q_ia_j \to a_kq_l$ of CPM0 P.

Case (C). Rule $q_i0 \to a_kq_l0$. We consider 3 subcases, depending on behavior of P in state q_l under symbol 0: (i) $q_l0 \to q_s$; (ii) $q_l0 \to a_pq_s$; (iii) $q_l0 \to a_pq_s0$.

Clearly, case (i) is equivalent to applying instruction $q_i0 \to a_kq_s$, so we let P' simulate that instruction (see above) instead. Consider a computation fragment in P in case (ii): $q_i0W \overset{q_i0 \to a_kq_l0}{\longrightarrow} q_l0Wa_k \overset{q_l0 \to a_pq_s}{\longrightarrow} q_sWa_ka_p$. In P' we have the following evolutionary processors to simulate this computation:

$$N(\langle q_i 0\rangle_t^c) = N(\langle q_i 0\rangle_t^a),\ 1 \le t \le 3, (\text{here } N(\langle q_i 0\rangle_t^c) \text{ have the same forms as } N(\langle q_i 0\rangle_t^a)),$$
$$N(\langle q_i 0\rangle_4^c) = (\{\varepsilon \to q'_{\langle q_i 0\rangle,2}\}^r, \{q'_{\langle q_i 0\rangle}\}, \{q'_{\langle q_i 0\rangle,2}\}, \emptyset, \{\varepsilon'_{\langle q_i 0\rangle}\}),$$
$$N(\langle q_i 0\rangle_5^c) = (\{q'_{\langle q_i 0\rangle} \to a_k\}, \{q'_{\langle q_i 0\rangle,2}\}, \emptyset, \emptyset, \emptyset),$$
$$N(\langle q_i 0\rangle_6^c) = (\{q'_{\langle q_i 0\rangle,2} \to q'_{\langle q_l 0\rangle}\}, \emptyset, \emptyset, \emptyset, \{q'_{\langle q_i 0\rangle}\}).$$

Now consider computation in P in case (iii):

$$q_i 0W \overset{q_i 0 \to a_k q_l 0}{\longrightarrow} q_l 0W a_k \overset{q_l 0 \to a_p q_s 0}{\longrightarrow} q_s 0W a_k a_p.$$

In P' the evolutionary processors to simulate this computation are the same as above, i.e., $N(\langle q_i 0\rangle_1^c), \ldots, N(\langle q_i 0\rangle_6^c)$. Consider evolution of string $0Wq_i$ in P':

$$0Wq_i \overset{N(\langle q_i 0\rangle_1^c)}{\longrightarrow} \varepsilon'_{\langle q_i 0\rangle} W q_i \overset{N(\langle q_i 0\rangle_2^c)}{\longrightarrow} \varepsilon'_{\langle q_i 0\rangle} W q'_{\langle q_i 0\rangle} \overset{N(\langle q_i 0\rangle_3^c)}{\longrightarrow} W q'_{\langle q_i 0\rangle} \overset{N(\langle q_i 0\rangle_4^c)}{\longrightarrow}$$
$$W q'_{\langle q_i 0\rangle} q'_{\langle q_i 0\rangle,2} \overset{N(\langle q_i 0\rangle_5^c)}{\longrightarrow} W a_k q'_{\langle q_i 0\rangle,2} \overset{N(\langle q_i 0\rangle_6^c)}{\longrightarrow} W a_k q'_{\langle q_l 0\rangle} \overset{N(\langle q_l 0\rangle_4^c)}{\longrightarrow}$$
$$W a_k q'_{\langle q_l 0\rangle} q'_{\langle q_l 0\rangle,2} \overset{N(\langle q_l 0\rangle_5^c)}{\longrightarrow} W a_k a_p q'_{\langle q_l 0\rangle,2} \overset{N(\langle q_l 0\rangle_6^c)}{\longrightarrow} W a_k a_p q'_{\langle q_s 0\rangle}.$$

So, the string $W a_k a_p q'_{\langle q_s 0\rangle}$ in P' corresponds the configuration $q_s 0 W a_k a_p$ of P then symbol 0 "is deleted". So, P' correctly simulates P.

Now we can conclude that if a computation in CPM0 P starts with configuration $q_1 W$ and ends with the final configuration $q_f W'$ then OHNEP P' starting with string SWq_1 in the node $\langle init\rangle$ will obtain string $W'q_f$ in the output node $\langle out\rangle$ and if P does not stop then in the node $\langle out\rangle$ of P' never appears some nonempty string. Thus P' correctly simulates P.

Corollary 1. *Any CPM0 P can be simulated by a complete OHNEP P' where evolutionary processors are with only one operation per node (obligatory deletion on the left end and insertion on the right end of the string and obligatory substitution) and every obligatory evolutionary processors also possesses an input and output filters with total power not exceed than 3 symbols.*

Proof. (Outline). We reduce the power of filters FI in evolutionary processors $N(\langle q_i a_j\rangle_1^a), N(\langle q_i a_j\rangle_1^b)$ and $N(\langle q_i a_j\rangle_1^c)$ to one symbol, i.e., $N(\langle q_i a_j\rangle_1^a) = N(\langle q_i a_j\rangle_1^b)$ $= N(\langle q_i a_j\rangle_1^c) = N(\langle q_i a_j\rangle_1^d) = (\{a_j \to \varepsilon'_{\langle q_i a_j\rangle}\}, \{q_i\}, \{\varepsilon'_{\langle q_i a_j\rangle}\}, \emptyset, \emptyset)$.

After applications of evolutionary processors of this type in the developing string at least one symbol from E' is present. After that P' is checking whether this symbol (symbols) is in a proper form, i.e., does it correspond to only one $\langle q_i a_j\rangle$? If the "checking" is successful, the process of simulation continues as in the cases (A), (B), (C) above.

Let us add to alphabet of P' new symbols Q'' as follows:

$$V = \{S\} \cup Q \cup Q' \cup Q'' \cup \Sigma \cup E', \text{ where } E' = \{\varepsilon'_{\langle q_i a_j\rangle}\},\ Q' = \{q'_{\langle q_i a_j\rangle,t}\},$$
$$Q'' = \{q''_{\langle q_i a_j,p\rangle}\}, q_i \in Q \setminus \{q_f\},\ a_j \in \Sigma,\ 1 \le t \le |Q| - 1, 1 \le p \le |\Sigma| - 1.$$

New evolutionary processors below serve for "checking".

$$
\begin{aligned}
N(\langle q_ia_j\rangle_1^d) &= (\{a_j \to \varepsilon'_{\langle q_ia_j\rangle}\},\{q_i\},\{\varepsilon'_{\langle q_ia_j\rangle}\},\emptyset,\emptyset),\\
N(\langle q_ia_j\rangle_2^d) &= (\{q_i \to q''_{\langle q_ia_j\rangle,1}\},\{\varepsilon'_{\langle q_ia_j\rangle}\},\emptyset,\emptyset,\{\varepsilon'_{\langle q_ia_1\rangle}\}),\\
N(\langle q_ia_j\rangle_s^d) &= (\{q''_{\langle q_ia_j\rangle,s-2} \to q''_{\langle q_ia_j\rangle,s-1}\},\emptyset,\emptyset,\emptyset,\{\varepsilon'_{\langle q_ia_{s-1}\rangle}\}),\ s\in\{3,\dots,j\},\\
N(\langle q_ia_j\rangle_t^d) &= (\{q''_{\langle q_ia_j\rangle,t-1} \to q''_{\langle q_ia_j\rangle,t}\},\emptyset,\emptyset,\emptyset,\{\varepsilon'_{\langle q_ia_{t+1}\rangle}\}),\ t\in\{j,\dots,|\Sigma|-1\},\\
N(\langle q_ia_j\rangle_{|\Sigma|+1}^d) &= (\{q''_{\langle q_ia_j\rangle,|\Sigma|-1} \to q'_{\langle q_ia_j\rangle}\},\emptyset,\emptyset,\emptyset,\emptyset).
\end{aligned}
$$

4 Conclusions

In a previous work [1] we introduced a new variant of HNEPs, Obligatory HNEPs. OHNEPs have a directed underlying graph and discard a string if operations at the node are not applicable to such string. In the present paper, we consider a special case of OHNEPs with complete graphs. We demonstrate that very simple COHNEPs are able to simulate any deterministic Circular Post Machine CPM0. Therefore, paper proves, once more, the flexibility of OHNEPs, and their suitability to simulate different systems. Not only Circular Post Machines, but other complex mechanisms can be successfully approached by OHNEPs. The efficiency of the model in simulations demonstrates that OHNEPs can have their own role in the field of artificial intelligence, building complexity from their extremely simple agents.

References

1. Alhazov, A., Bel-Enguix, G., Rogozhin, Y.: Obligatory Hybrid Networks of Evolutionary Processors. In: Proc. of the First International Conference on Agents and Artificial Intelligents, ICAART 2009, Porto, Portugal, January 19-21, pp. 613–618 (2009)
2. Alhazov, A., Csuhaj-Varjú, E., Martín-Vide, C., Rogozhin, Y.: On the Size of Computationally Complete Hybrid Networks of Evolutionary Processors. Theoretical Computer Science 410, 3188–3197 (2009)
3. Alhazov, A., Dassow, J., Martín-Vide, C., Rogozhin, Y., Truthe, B.: On Networks of Evolutionary Processors with Nodes of Two Types. Fundamenta Informaticae 91(1), 1–15 (2009)
4. Castellanos, J., Martín-Vide, C., Mitrana, V., Sempere, J.: Solving NP-complete Problems with Networks of Evolutionary Processors. In: Mira, J., Prieto, A.G. (eds.) IWANN 2001. LNCS, vol. 2084, pp. 521–528. Springer, Heidelberg (2001)
5. Csuhaj-Varjú, E., Martín-Vide, C., Mitrana, V.: Hybrid Networks of Evolutionary Processors are Computationally Complete. Acta Informatica 41(4-5), 257–272 (2005)
6. Kudlek, M., Rogozhin, Y.: Small Universal Circular Post Machines. Computer Science Journal of Moldova 9(1), 34–52 (2001)
7. Loos, R., Manea, F., Mitrana, V.: Small universal accepting hybrid networks of evolutionary processors. Acta Informatica 47, 133–146 (2010)
8. Martín-Vide, C., Mitrana, V., Pérez-Jiménez, M., Sancho-Caparrini, F.: Hybrid Networks of Evolutionary Processors. In: Cantú-Paz, E., Foster, J.A., Deb, K., Davis, L., Roy, R., O'Reilly, U.-M., Beyer, H.-G., Kendall, G., Wilson, S.W., Harman, M., Wegener, J., Dasgupta, D., Potter, M.A., Schultz, A., Dowsland, K.A., Jonoska, N., Miller, J., Standish, R.K. (eds.) GECCO 2003. LNCS, vol. 2723, pp. 401–412. Springer, Heidelberg (2003)

Agents in Formal Language Theory: An Overview

M. Dolores Jiménez-López

Abstract. Agent-based systems are one of the most important areas of research and development that have emerged in information technology in the 1990s. The concept of agent can be found in a range of disciplines. The metaphor of autonomous problem solving entities cooperating and coordinating to achieve their objectives is a natural way of conceptualizing many problems. An important topic in the field of multi-agent systems concerns *formalization*. To develop formal models for agency that can be used both as the basis of an implementation and also as a precise but general frameworks is necessary. In this paper we present three agent-based models that have been defined in the field of formal language theory: *colonies, eco-grammar systems* and *networks of evolutionary processors*. We claim that those agent models may contribute to the field of agent technologies by offering a highly formalized framework that could be applied to many different issues.

1 Introduction

Agent technology is one of the fastest growing areas of information technology. Researches agree on the fact that the apparatus of agent technology provides a powerful and useful set of structures and processes for designing and building complex applications. In general, multi-agent systems offer strong models for representing complex and dynamic real-world environments. The metaphor of autonomous problem solving entities cooperating and coordinating to achieve their objectives is a natural way of conceptualizing many problems. In fact, the multi-agent system literature spans a wide range of fields including robotics, mathematics, linguistics, psychology, and sociology, as well as computer science.

In this paper we want to show how agent technologies can offer good solutions and alternative frameworks to classic models in the area of processing and

M. Dolores Jiménez-López
Research Group on Mathematical Linguistics, Universitat Rovira i Virgili,
43002 Tarragona, Spain
e-mail: mariadolores.jimenez@urv.cat

J.B. Pérez et al. (Eds.): Highlights in PAAMS, AISC 89, pp. 283–290.
springerlink.com

computing languages. Since, languages, either natural or artificial, are particular cases of a symbol system and the manipulation of symbols is the stem of formal language theory, we focus on non-standard formal language models that can be considered grammatical models of agent systems.

The theory of formal languages mainly originated from mathematics and generative linguistics. It was born in the middle of the 20th century as a tool for modeling and investigating syntax of natural languages. After 1964, it developed as a separate branch with specific problems, techniques and results and since then it has had an important role in the field of computer science. The first generation of formal languages, fitted into the Chomskian hierarchy, were based on rewriting, and caused the generalization of the tree-like models for describing and explaining natural language. A further step in the development of formal languages has been the idea of several devices collaborating for achieving a common goal.

Multi-agent systems promote the interaction and cooperation of autonomous agents to deal with complex tasks. Taking into account that computing languages is a complex task, formal language theory has taken advantage of the idea of formalizing architectures where a hard task is distributed among several task-specific agents that collaborate in the solution of the problem: the generation/recognition of language. Those non-standard agent-based models present the same abstractness that has facilitated the application of classical formal language models to many issues, and, in addition, present several advantageous features: natural inspiration, parallelism, distribution, cooperation, etc.

Many non-stardard models in formal languages offer grammatical models of agent systems. However, taking into account the advantage of bio-inspired methods, in this paper we focus on three frameworks that can be defined as bio-inspired multi-agent systems: *colonies, eco-grammar systems* and *networks of evolutionary processors*. The main advantage of those bio-models is to account for languages in a more natural way than classical models.

2 Colonies

Colonies as well-formalized language generating devices have been proposed in [8], and developed during the nineties in several directions in many papers. Colonies can be thought of as grammatical models of multi-agent systems motivated by Brooks' subsumption architectures [1]. They describe language classes in terms of behavior of collections of very simple, purely reactive, situated agents with emergent behavior. Roughly, a colony consists of a finite number of simple agents which generate finite languages and operate on a shared string of symbols –the *environment*– without any explicitly predefined strategy of cooperation. Each component has its own reactive behavior which consists in: 1) sensing some aspects of the *context* and 2) performing elementary tasks in it in order to achieve some local changes. The environment is quite passive, its state changes only as result of acts agents perform on its string. Because of the lack of any predefined strategy of cooperation, each component participates in the rewriting of current strings whenever it can participate in it. The behavior of a colony is defined as the set of all the strings which can be generated by the colony from a given starting string. Formally, a colony is defined as follows.

Definition 1. A *colony* C is a 3-tuple: $C = (R,V,T)$, where $R = \{R_i | 1 \leq i \leq n\}$ is a finite set of regular grammars $R_i = (N_i,T_i,P_i,S_i)$ producing finite languages $L(R_i) = F_i$ for each i. R_i will be referred to as a component of C; $V = \bigcup_{i=1}^{n}(T_i \cup N_i)$ is the alphabet of the colony; $T \subseteq V$ is the terminal alphabet of the colony.

Components $R_i \in R$ $(1 \leq i \leq n)$ of a colony C are *regular grammars* generating finite languages and operating on a shared string of symbols –the *environment*– without any explicitly predefined strategy of cooperation of the components.

An *environment* of a colony is formed by strings of symbols from V. Strings are modified only by sequential activities of components of a colony. Because of the lack of any predefined strategy of cooperation between components, each component participates in the rewriting of the current string whenever its start symbol is present in the actual string. Conflicts are solved non-deterministically, as it is usual in the classical theory of formal grammars. The activity of components in a colony is realized by string transformation on a common tape. Elementary changes of strings are determined by a basic derivation step:

Definition 2. For $x,y \in V^*$ we define $x \Longrightarrow y$ iff $x = x_1 S_i x_2$, $y = x_1 z x_2$, where $z \in F_i$ for some $i, 1 \leq i \leq n$.

The behavior of a colony is defined as the set of all strings which can be generated by the colony from a given starting string. A terminal symbol of one component can occur as a non-terminal symbol of another one, so that the possibility of co-operation of components of the colony allows to generate substantially more than finite languages. The global behavior of the whole colony emerges from the strictly individual behaviors of components.

Definition 3. *The language determined by a colony* C *starting with the word* $w_0 \in V^*$ *is given by:* $L(C,w_0) = \{v | w_o \Longrightarrow^* v, v \in T^*\}$

A colony is outlined in Figure 1.

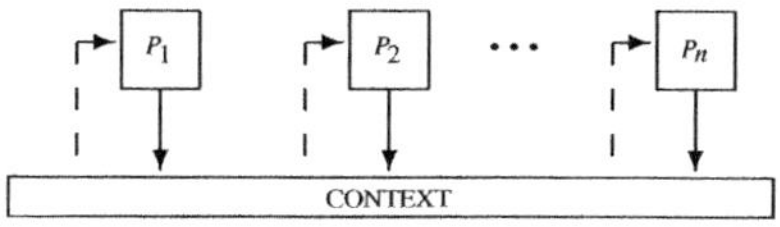

Fig. 1 A colony.

3 Eco-grammar Systems

The theory of eco-grammar systems is a subfield of grammar systems theory, a consolidated and active branch in the field of formal languages [3]. Eco-grammar systems have been introduced in [4] and provide a syntactical framework for eco-systems, this is, for communities of evolving agents and their interrelated environment. Briefly, an eco-grammar system is defined as a multi-agent system where different components, apart from interacting among themselves, interact with a special component called 'environment'. So, within an eco-grammar system we can distinguish two types of components *environment* and *agents*. Both are represented at any moment by a string of symbols that identifies current state of the component. These strings change according to sets of evolution rules. Interaction among agents and environment is carried out through agents' actions performed on the environmental state by the application of some productions from the set of action rules of agents. The concept of eco-grammar system is based on six postulates formulated according to properties of artificial life: (1) An ecosystem consists of an *environment* and a *set of agents*; (2) In an ecosystem there is a *universal clock* which marks time units, the same for all the agents and for the environment, according to which agents and environment evolution is considered; (3) Both *environment* and agents have characteristic *evolution rules* which are in fact L systems, hence are applied in a parallel manner to all the symbols describing agents and environment; such a (rewriting) step is done in each time unit; (4) *Evolution rules of environment* are *independent* on agents and on the state of the environment itself. Evolution rules of agents *depend on* the state of the environment; (5) Agents act on the environment according to *action rules*, which are pure rewriting rules used sequentially. In each time unit, each agent uses one action rule which is chosen from a set depending on current state of the agent; (6) *Action has priority over evolution* of the environment. At a given time unit exactly the symbols which are not affected by action are rewritten by evolution rules. Main features of eco-grammar systems are captured in Figure 2.

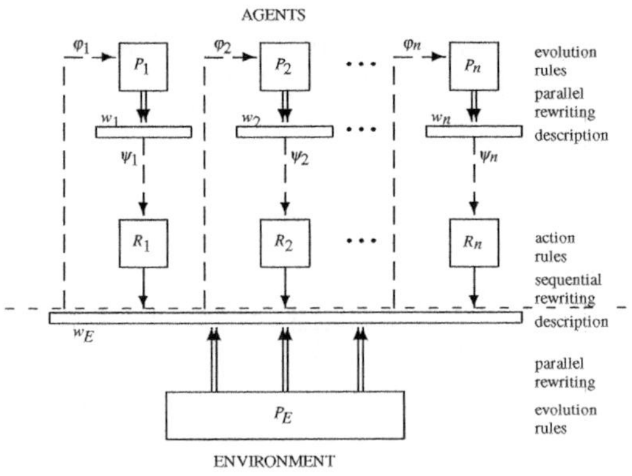

Fig. 2 An Eco-Grammar System.

Definition 4. An Eco-Grammar System is an $n+1$-tuple: $\Sigma = (E, A_1, \ldots, A_n)$, where:

- $E = (V_E, P_E)$, V_E is a finite alphabet; P_E is a finite set of 0L rules over V_E.
- $A_i = (V_i, P_i, R_i, \varphi_i, \psi_i)$ for i, $1 \leq i \leq n$, where: V_i is a finite alphabet; P_i is a finite set of 0L rules over V_i; R_i is a finite set of rewriting rules of the form $x \rightarrow y$ with $x \in V_E^+, y \in V_E^*$;
- $\varphi_i : V_E^* \longrightarrow 2^{P_i}$;
- $\psi_i : V_i^+ \longrightarrow 2^{R_i}$.

Definition 5. A state of an Eco-Grammar System $\Sigma = (E, A_1, \ldots, A_n)$ is an $(n+1)$-tuple: $\sigma = (w_E, w_1, w_2, \ldots, w_n)$, where $w_E \in V_E^*$ and $w_i \in V_i^*$, $1 \leq i \leq n$; w_E is the state of the environment, and w_i is the state of i-th agent, $1 \leq i \leq n$.

Definition 6. Let $\sigma = (w_E, w_1, w_2, \ldots, w_n)$ be a state of EG-system $\Sigma = (E, A_1, \ldots, A_n)$. Agent A_i is said to be *active* in state σ if the set of its current action rules, this is $\psi_i(w_i)$, is nonempty. By an *action* of an active agent A_i in state σ we mean an application of an action rule r, $r \in \psi_i(w_i)$, to the environmental state w_E. A *simultaneous action* of agents $A_{i_1}, \ldots, A_{i_r}$, $\{i_1, \ldots, i_r\} \in \{1, \ldots, n\}$ being active in state σ, onto the environment is a parallel derivation step $w_E \Longrightarrow w'_E$ such that $w_E = x_1 u_1 x_2 u_2 \ldots u_r x_{r+1}$, and $w'_E = x_1 v_1 x_2 v_2 \ldots v_r x_{r+1}$, where $u_j \rightarrow v_j \in \psi_{i_j}(w_{i_j})$, $1 \leq j \leq r$, and $x_i \in V_E^*$, $1 \leq i \leq r+1$.

Definition 7. Let $\sigma = (w_E, w_1, w_2, \ldots, w_n)$ be a state of EG-system $\Sigma = (E, A_1, \ldots, A_n)$. We say that w'_i is a *current evolution* of agent A_i in state w_i, if w'_i can be derived from w_i by productions of $\varphi_i(w_E)$, in 0L-manner, $1 \leq i \leq n$. For two states w_E and w'_E of the environment we say that w'_E is an evolution of w_E if w'_E can be derived from w_E by productions of P_E in 0L-manner. A change of a state of an Eco-Grammar System means an evolution of the state of every agent and an evolution of the environment at each place except some distinguished ones where currently active agents perform simultaneously an action.

Definition 8. Let $\sigma = (w_E, w_1, w_2, \ldots, w_n)$ and $\sigma' = (w'_E, w'_1, w'_2, \ldots, w'_n)$ be two states of EG-system $\Sigma = (E, A_1, \ldots, A_n)$. We say that σ changes into σ', written as:$\sigma \Longrightarrow_\Sigma \sigma'$, iff the following conditions hold: (i) w'_E arises from w_E by an evolution affected by all the active agents in state σ: $w_E = z_1 x_1 z_2 x_2 \ldots z_m x_m z_{m+1}$ and $w'_E = z'_1 y_1 z'_2 y_2 \ldots z'_m y_m z'_{m+1}$ such that: $z_1 x_1 z_2 x_2 \ldots x_m z_{m+1} \Longrightarrow z_1 y_1 z_2 y_2 \ldots y_m z_{m+1}$ is a simultaneous action of all the agents $A_{i_1}, \ldots, A_{i_m}$, $\{i_1, \ldots, i_m\} \subseteq \{1, \ldots, n\}$, that are active in state σ and, $z'_1 z'_2 \ldots z'_{m+1}$ is an evolution of $z_1 z_2; \ldots z_{m+1}.$; (ii) w'_i is an evolution of A_i in state w_i, $1 \leq i \leq n$.

4 Networks of Evolutionary Processors

Networks of Evolutionary Processors (NEPs) [2] are inspired in both, in bio cellular models and in basic structures for parallel and distributed symbolic processing. The cellular basis of NEPs relates them with P systems, especially with *Tissue P systems* [10], a theory in the area of membrane computing whose biological referent is the structure and behavior of multicellular organisms. From the computational point of view, NEPs are related to the Connection Machine [7] and the Logic Flow paradigm [5]. NEPs can be described as *bio-inspired context-sensitive multi-agent systems.* They take the multiagent and modular features from classical theoretical computer science, while the environmental and evolutionary characteristics are typical of bio-inspired models.

Definition 9. A Network of Evolutionary Processors of size n is a construct: $\Gamma = (V, N_1, N_2, ..., N_n, G)$, where:

- V is an alphabet and for each $1 \leq i \leq n$,
- $N_i = (M_i, A_i, PI_i, PO_i)$ is the i-th evolutionary node processor of the network. The components of every processor are (we denote by e the empty word): M_i is a finite set of evolution rules of one of the following forms only (i) $a \to b, a, b \in V$ (substitution rules), (ii) $a \to e, a \in V$ (deletion rules), (iii) $e \to a, a \in V$ (insertion rules); A_i is a finite set of strings over V. The set A_i is the set of initial strings in the i-th node; PI_i and PO_i are subsets of V^* representing the input and the output filter, respectively. These filters are defined by the membership condition, namely a string $w \in V^*$ can pass the input filter (the output filter) if $w \in PI_i (w \in PO_i)$.
- $G = (\{N_1, N_2, \ldots, N_n\}, E)$ is an undirected graph called the underlying graph of the network. The edges of G, that is the elements of E, are given in the form of sets of two nodes. The complete graph with n vertices is denoted by K_n.

Definition 10. Configuration of a NEP is an n-tuple $C = (L_1, L_2, \ldots, L_n)$, with $L_i \subseteq V^*$ for all $1 \leq i \leq n$. It represents the sets of strings which are present in all the nodes at a given moment.

A given configuration of a NEP can change either by an evolutionary step or by a communicating step. When changing by an evolutionary step, each component L_i of the configuration is changed in accordance with the evolutionary rules associated with the node i. The change in the configuration by an evolutionary step is written as $C_1 \Rightarrow C_2$. When changing by a communication step, each node processor N_i sends all copies of the strings it has which are able to pass its output filter to all the node processors connected to N_i and receives all copies of the strings sent by any node processor connected with N_i, if they can pass its input filter. The change in the configuration by a communication step is written as $C_1 \vdash C_2$.

A scheme of the basic architecture of NEPs is shown in Figure 3.

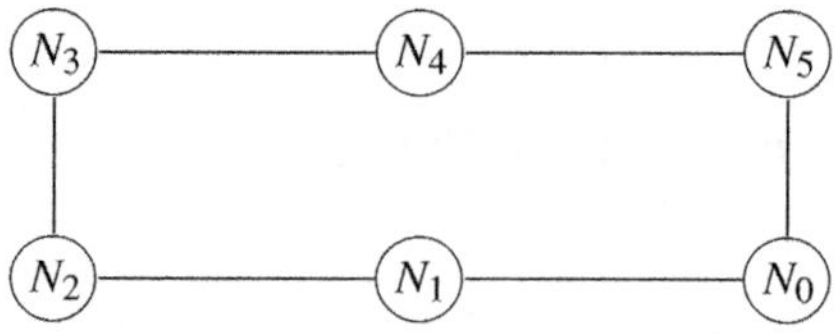

Fig. 3 Basic Architecture of a NEP.

We describe the main traits of NEPs as a multi-agent system in the following way:

- *Multi-level modularity*: NEPs are distributed systems of contributing nodes, each one of them carrying out just one type of operation. Every specialized processor can be modular as well.
- *Evolutionary*: agents/nodes can change their definition during the computation. Such modifications can be caused by the context, by the elements inside or by the learning of the modules.
- *Environmental*: By this feature we explain that NEPs are systems-in-context. Context has an important influence in the behavior of nodes/agents. By context of one node we understand: a) the other processors in the same system, and b) the environment in which the system is placed.

Inside of the construct, every agent is *autonomous, specialized,* and *context-interactive*:

- *Autonomous*: Autonomy and specialization are strongly related in NEPs. Even if every agent depends on the other to accomplish the final objective, in order to improve the final result they can change the strategy and the own sub-goals.
- *Specialized*: Every processor is designed for a specific task, in a way that the final success of the system depends on the correct working of every one of the agents.
- *Context-Interactive*: There is a constant *contextualization* and *redefinition* of agent capabilities during the computation. The interaction with context implies three different abilities of agents. They are: 1) *Context-sensitive*: Having a broad knowledge of the configuration of the environment and the other agents of the system; 2) *Context-reactive*: Changing their goal and behavior depending on the configuration of the environment; 3) *Context-modifier*: The productions and behavior of a given cell are able to modify the general context.

Finally, between the processors, there are social competences, specified by *communication* -agents can interact in the consecution of the same goal, work for their own interests or even isolate- and *coordination* -it is necessary to define protocols for synchronizing the communication.

5 Conclusions

Ferber [6] defines a multiagent system as having: an *environment*, a *set of objects* that exist in the environment, a *set of agents*, a *set of operations* that agents can use to sense and affect objects, a set of *universal laws* that define the reaction of the environment to agent operations. And Wooldridge [11] point out three important characteristics that agents in a multi-agent systems must have: 1) *Autonomy*: the agents are at least partially autonomous; 2) *Local views*, this is, no agent has a full global view of the system; 3) and *Decentralization*, this is, there is no designated controlling agent. All those characteristics and basic components can be found in the formal language frameworks we have presented here. So, colonies, eco-grammar systems and networks of evolutionary processors can be considered grammatical models of agent systems.

Agent technology provides a powerful a useful set of structures and processes for designing and building complex systems. Language is a complex system. Therefore, if we want to describe, explain and process language (natural or artificial) we must consider its complex nature. The metaphor of autonomous problem solving entities cooperating and coordinating to achieve their objectives is a natural way of conceptualizing many problems, among them, the processing of languages. Therefore, the multi-agent capabilities of the models we have presented make them suitable tools for simulating the processes of generation and recognition of languages. These models improve the features of classical formal language theory and may contribute to the field of agent technologies by offering a highly formalized framework that could be applied to many different issues.

One of the main advantages of the models presented here is that they are defined in formal and non ambiguous terms. According to [9], formalization provides clarity in characterizing the nature of concepts. Colonies, eco-grammar systems and NEPs are based on a consolidated and active research area -formal language theory- so they offer highly formalized frameworks easy to be implemented, due to the simplicity of the formalisms and the computational background of the theory they are placed in.

References

1. Brooks, R.A.: Elephants don't play chess. Robotics and Autonomous Systems 6, 3–15 (1990)
2. Castellanos, J., Martín-Vide, C., Mitrana, V., Sempere, J.M.: Networks of Evolutionary Processors. Acta Informatica 39, 517–529 (2003)
3. Csuhaj-Varjú, E., Dassow, J., Kelemen, J., Păun, G.: Grammar Systems. Gordon and Breach, London (1993)
4. Csuhaj-Varjú, E., Kelemen, J., Kelemenová, A., Păun, G.: Eco-Grammar Systems: A Grammatical Framework for Life-Like Interactions. Artificial Life 3(1), 1–28 (1996)
5. Errico, L., Jesshope, C.: Towards a New Architecture for Symbolic Processing. In: Plander, I. (ed.) Artificial Intelligence and Information-Control Systems of Robots 1994, pp. 31–40. World Science, Singapore (1994)
6. Ferber, J.: Multi-Agent Systems. An Introduction to Distributed Artificial Intelligence. Addison-Wesley, Reading (1999)
7. Hillis, W.D.: The Connection Machine. MIT Press, Cambridge (1985)
8. Kelemen, J., Kelemenová, A.: Grammar-Theoretic Treatment of Multiagent Systems. Cybernetics and Systems 23, 621–633 (1992)
9. Luck, M., d'Inverno, M.: A formal framework for agency and autonomy. In: Proceedings of the First International Conference on Multi-Agent Systems, pp. 254–260. AAAI Press/MIT Press (1995)
10. Martín-Vide, C., Pazos, J., Păun, G., Rodríguez-Patón, A.: A new class of symbolic abstract neural nets: Tissue P systems. In: Ibarra, O.H., Zhang, L. (eds.) COCOON 2002. LNCS, vol. 2387, pp. 290–299. Springer, Heidelberg (2002)
11. Wooldridge, M.: An Introduction to MultiAgent Systems. John Wiley, New York (2002)

Self Cross-Over Array Languages

A. Roslin Sagaya Mary and K.G. Subramanian

Abstract. Self cross over array language consists of arrays that are formed by applying the operation of array splicing in H-array splicing systems to identical copies of picture arrays. This paper compares the family of self cross-over array languages with other two-dimensional language classes. Various properties analogous to H-array splicing systems are proved.

1 Introduction

Head [4] introduced the concept of splicing system as an effective mathematical model for representing the reactionary behaviors of DNA molecules under the influence of certain restriction enzymes and ligases in the test tube environment. Paun et al [8] extended the definition of Head and defined Extended H systems which are computationally universal. Since then theoretical investigation of splicing on strings has been extensively done by different researchers.

Splicing operation for arrays was studied by Kirthivasan et al [7]. A different approach to the operation on images that constitute of rectangular arrays (*picture languages*) was introduced in [6] where splicing rules that involve dominoes are considered. Two rectangular arrays (also called *pictures*) are column spliced or row spliced by using the domino splicing rules in parallel. The resulting model is called as H array splicing system. Generative power and closure properties were discussed in [6]. Splicing on images in the context of grammar systems is considered in [10].

A. Roslin Sagaya Mary
GRLMC, Universitat Rovira I Virgili, 1 Plaa Imperial Tarraco,
Tarragona - 43005, Spain
e-mail: anthonath.roslinsagayamary@estudiants.urv.cat

K.G. Subramanian
Universiti Sains Malaysia, 11800 Penang, Malaysia
e-mail: kgsmani1948@yahoo.com

J.B. Pérez et al. (Eds.): Highlights in PAAMS, AISC 89, pp. 291–298.
springerlink.com

Dassow and Mitrana [1] investigated a very simple and natural restriction on the splicing operation, namely cross-over rule applicable only on identical strings in trying to capture features of the recombination of genes in a chromosome. The self cross-over operation on arrays is also introduced in [6] and the resulting language is called *self cross-over array languages* where it is defined by the H-array splicing on identical arrays. This family is referred by $\ell(A_{sco})$ in this paper. Closure properties on some operations like union, row and column concatenations are stated in [6].

In syntactic approaches to generation and recognition of images or pictures considered as digitized arrays, several two-dimensional grammars have been proposed and studied. In this paper we compare $\ell(A_{sco})$ with the languages of 2dimensional right linear grammars $\ell(2RLG)$ and with an elegant generalization of the concept of local and recognizable string languages to two-dimensional picture languages done by Giammarresi and Restivo [3].

We consider the self cross-over languages with respect to H-array splicing [6] where some closure properties have been studied, we compare it with the class of languages $LOC, 2RLG$. And also prove that $\ell(HAS), \ell(A_{SCO})$ are incomparable.

We now recall the basic definitions and results used in this paper on two-dimensional languages, H-array splicing systems and self cross-over array languages in [6, 3].

2 Preliminaries

Let Σ be a finite alphabet. Σ^* is the set of all words over Σ including the empty word λ. An image or a picture A over Σ is a rectangular $m \times n$ array of elements of Σ of the form

$$A = \begin{matrix} a_{11} & \cdots & a_{1n} \\ \vdots & \ddots & \vdots \\ a_{m1} & \cdots & a_{mn} \end{matrix}$$

The array A is also compactly written as $[a_{ij}]_{m \times n}$. The set of all images is denoted by Σ^{**}. A *picture language* or a *two dimensional language over* Σ is a subset of Σ^{**}. We denote bordered picture of p by $\hat{p}$ and $B_{2,2}(\hat{p})$ denotes the 2×2 subpicture arrays of $\hat{p}$, called tiles.

For an array A of dimension $m \times n$ and an array B of dimension $m' \times n'$, the column concatenation $A \oplus B$ (*B juxtaposed to the right of A*) is defined only when $m = m'$ and the row concatenation $A \ominus B$ (*B juxtaposed to the bottom of A*) is defined only when $n = n'$.

A *picture language* L is *local* if there exists a finite set Θ of tiles over alphabet $\{\Sigma \cup \#\}$ where $\#$ is a special border symbol such that $L = \{p \in \Sigma^{**} | B_{2,2}(\hat{p}) \subseteq \Theta$ where $\hat{p}$ is over $\{\Sigma \cup \#\}\}$. Then L is the local language defined by Θ.

Let us now recall the notion of a tiling system and language generated by such a system [3].

Definition 1 (Tiling systems). *A tiling system is a 4-tuple $\tau = (\Sigma, \Pi, \Theta, \pi)$, where Σ and Π are two finite alphabets, Θ is a finite set of tiles over the alphabet $\Pi \cup \{\#\}$ and $\pi : \Pi \to \Sigma$ is a projection.*

Given system τ, the language defined by the system, denoted as $L(\tau)$, is the projection π of the local language defined by Θ. We denote the class of languages of tiling systems by TS.

We now recall as stated in [4], the grammar model that consists of two sets of rewriting rules or grammars defined by Siromoney et al as stated in [3].

Definition 2 (Two-Dimensional Right Linear Grammar 2RLG). *is a 2-tuple (G_1, G_2) where $G_1 = (H_1, I_1, P_1, S)$ is a right linear grammar with H_1 a finite set of horizontal nonterminals, $I_1 = \{S_1, S_2, \cdots, S_k\}$ a finite set of intermediates and $H_1 \cap I_1 = \emptyset$. P_1 is a finite set of right linear production rules called horizontal production rules, $S \in H_1$ is the start symbol.*

And $G_2 = (G_{21}, G_{22}, \cdots, G_{2k})$ where $G_{2i} = (V_{2i}, T, P_{2i}, S_i)$ for $1 \leq i \leq k$ are right linear grammars with V_{2i} a finite set of vertical nonterminals and $V_{2i} \cap V_{2j} = \emptyset$ for $i \neq j$, T is a finite set of terminals, P_{2i} is a finite set of right linear production rules of the form $X \longrightarrow aY$ or $X \longrightarrow a$ where $X, Y \in V_{2i}, a \in T$, $S_i \in V_{2i}$ is the start symbol of G_{2i}.

The set $L(G)$ of all matrices generated by G consists of all $m \times n$ arrays $[a_{ij}]$ such that $1 \leq i \leq m$, $1 \leq j \leq n$ and $S \Rightarrow^*_{G_1} S_{i1} S_{i2} \cdots S_{in} \Rightarrow^*_{G_2} [a_{ij}]$. $\ell(2RLG)$ refers to the class of 2dimensional right linear languages

We next recall the notion of self cross-over systems on strings defined by Dassaow and Mitrana in [1].

Definition 3 (A self cross-over system). *is a triple (V, A, R) where V is an alphabet, A is a finite subset of V^*, R is a finite commutative relation, $R \subset (V^* \times V^*)^2$; With respect to a self cross-over system, we define for $x \in V^+$*

$x \bowtie y$ if and only if ($x = x_1\alpha\beta x_2 = x_3\gamma\delta x_4$; $y = x_1\alpha\delta x_4$; $(\alpha, \beta)R(\gamma, \delta)$.

$\bowtie^*$ is the reflexive and transitive closure of the relation $\bowtie$. The language generated by a self cross-over system is $L(SCO) = \{x \in V^* \;/\; w \bowtie^* x, w \in A\}$.

We now recall the notions of domino splicing rules and splicing of arrays using these rules [6].

Definition 4. *Let V be an alphabet. $\S, \$ \notin V$ are two special symbols. A domino splicing rule over V is of the form $r = \alpha_1 \,\S\, \alpha_2 \,\$\, \alpha_3 \,\S\, \alpha_4$; when each α_i is a column domino, it is called domino column splicing rule and when each α_i is a row domino it is called domino row splicing rule. A column domino and row domino over V is of the form $\boxed{\begin{matrix} a \\ b \end{matrix}}$ and $\boxed{a \mid b}$ respectively, for some $a, b \in V \cup \{\#\}$.*

Given two arrays X and Y of sizes $m \times p$ and $m \times q$ respectively,

$$\overset{X}{\begin{matrix} a_{11} & \cdots & a_{1,j} & a_{1,j+1} & \cdots & a_{1p} \\ a_{21} & \cdots & a_{2,j} & a_{2,j+1} & \cdots & a_{2p} \\ \vdots & \ddots & \vdots & \vdots & \ddots & \vdots \\ a_{m1} & \cdots & a_{m,j} & a_{m,j+1} & \cdots & a_{mp} \end{matrix}} \; , \; \overset{Y}{\begin{matrix} b_{11} & \cdots & b_{1,k} & b_{1,k+1} & \cdots & b_{1q} \\ b_{21} & \cdots & b_{2,k} & b_{2,k+1} & \cdots & b_{2q} \\ \vdots & \ddots \cdots & \vdots & \vdots & \ddots & \vdots \\ b_{m1} & \cdots & b_{m,k} & b_{m,k+1} & \cdots & b_{mq} \end{matrix}} \quad \overset{Z}{\begin{matrix} a_{11} & \cdots & a_{1,j} & b_{1,k+1} & \cdots & b_{1q} \\ a_{21} & \cdots & a_{2,j} & b_{2,k+1} & \cdots & b_{2q} \\ \vdots & \ddots & \vdots & \vdots & \ddots & \vdots \\ a_{m1} & \cdots & a_{m,j} & b_{m,k+1} & \cdots & b_{mq} \end{matrix}}$$

a_{ir}, $b_{is} \in V$, for $1 \le i \le m$, $1 \le r \le p$, $1 \le s \le q$. Let X, Y be surrounded by border symbol #, we write $(X,Y) \;\; |\!\stackrel{\textcircled{\tiny |}}{-} \;\; Z$ if there is a sequence of column splicing rules $r_1, r_2, ...r_m$ (not necessarily all different) such that

$$r_i \;=\; \boxed{\begin{array}{c} a_{i,j} \\ \hline a_{i+1,j} \end{array}} \;\; \S \;\; \boxed{\begin{array}{c} a_{i,j+1} \\ \hline a_{i+1,j+1} \end{array}} \;\; \$ \;\; \boxed{\begin{array}{c} b_{i,k} \\ \hline b_{i+1,k} \end{array}} \;\; \S \;\; \boxed{\begin{array}{c} b_{i,k+1} \\ \hline b_{i+1,k+1} \end{array}}$$

for all i, $1 \le i \le m-1$ and for some j, k $1 \le j \le p-1$, $1 \le k \le q-1$.

We can similarly define row splicing operation of two arrays U and V of sizes $p \times n$ and $q \times n$, using row splicing rules to yield an array W. c_{rj}, $d_{sj} \in V$, for $1 \le j \le n$, $1 \le r \le p$, $1 \le s \le q$, where the bordered arrays are of sizes $(p+2) \times (n+2)$, $(q+2) \times (n+2)$ obtained by surrounding U and V with # symbols.

We write $(U,V) \;\; |\!\stackrel{\ominus}{-} \;\; W$ if there is a sequence of row splicing rules $r_0, r_1, r_2, \cdots r_n$ (not necessarily all different) such that

$$r_j = \boxed{c_{r,j} \,|\, c_{r,j+1}} \;\; \S \;\; \boxed{c_{r+1,j} \,|\, c_{r+1,j+1}} \;\; \$ \;\; \boxed{d_{s,j} \,|\, d_{s,j+1}} \;\; \S \;\; \boxed{d_{s+1,j} \,|\, d_{s+1,j+1}}$$

for all $j, (0 \le j \le n)$ and for some $r, s(0 \le r \le p+1)$ and $(0 \le s \le q+1)$

Definition 5. *An H-array splicing scheme is a triple $\Gamma = (V, R_r, R_c)$, where V is an alphabet, R_c a finite set of domino column splicing rules, R_r a finite set of domino row splicing rules. For a given H-array splicing scheme $\Gamma = (V, R_r, R_c)$ and a language $L \subseteq V^{**}$, we define*

$$\Gamma = \{Z \in V^{**} | (X,Y) \vdash^{R_c} Z \text{ or } (X,Y) \vdash^{R_r} Z \text{ for some } X, Y \in L, \{R_c, R_r\}\}$$

*Iteratively, we define $\Gamma^0(L) = L, \Gamma^{i+1}(L) = \Gamma(L) \cup \Gamma(\Gamma^i(L)), i \ge 0$,
$\Gamma^*(L) = \bigcup_{i \ge 0} \Gamma^i(L)$. H-array splicing system is defined by $S = (\Gamma, I)$ where $\Gamma = (V, R_c, R_r)$ and I is a finite subset of V^{**}. The language of S is defined by $L(S) = \Gamma^*(I)$ and the class of languages by $\ell(HAS)$*

Based on the study of [1] on strings, we now introduce self cross-over array systems based on H-array splicing systems and state in theorem 2.1 the results in [6].

Definition 6. *A self cross-over array system (A_{SCO}) is defined by $S = (\Delta, I)$ where $\Delta = (V, R_c, R_r)$ is a H-array splicing scheme and I is a finite subset of V^{**}. A set of domino splicing rules is applied to two identical copies of* *the same array. A self cross-over array language is defined as in the case of linear strings and we denote the class of self cross-over array languages by* $\ell(A_{SCO})$

Theorem 2.1. *The family* $\ell(A_{SCO})$ *is*

(i) incomparable with the families of regular and context-free two-dimensional matrix languages.
(ii) not closed under union and column (row) concatenation.
(iii) closed under reflections on the base and right leg and rotations by $90°, 180°$ *and* $270°$.

We now discuss the various other properties of the self cross-over array languages by giving examples of the language classes generated by them.

3 Self Cross-Over Array Languages

By the following results it can be seen that various properties of self cross-over array languages are analogous to the properties of H-array splicing systems.

3.1 Examples and Counter-Examples

We now prove a comparison of string case H-splicing languages and self cross-over languages extended to family of H-array splicing languages $\ell(HAS)$ and the family of self cross-over array languages $\ell(A_{SCO})$.

Theorem 3.1. *The family* $\ell(HAS)$ *is incomparable with the language family* $\ell(A_{SCO})$.

Proof. It is known [6] that the language $L = \{(a^p)_m(b^q)_m(a^r)_m(b^s)_m / m \geq 2, p, q, r, s \geq 0\}$ cannot be generated by any A_{SCO} where $(x^n)_m$ is an array consisting of m rows with each row having x^n elements for a given n.

Now, to prove that L can be generated by a H-array splicing system we write the column(row) domino rules generating the language :

$$\text{Let } V = \{a, b\}, \qquad I = \left\{ \begin{array}{ccccc} a & a & b & a & b \\ a & a & b & a & b \end{array}, \begin{array}{cccc} a & b & a & b \\ a & b & a & b \end{array}, \begin{array}{cccccc} a & b & a & a & b & b \\ a & b & a & a & b & b \end{array} \right\},$$

$$R_c = \{p_1 : \begin{array}{|c|}\hline \# \\ \hline b \\ \hline\end{array} \ \S \ \begin{array}{|c|}\hline \# \\ \hline a \\ \hline\end{array} \ \$ \ \begin{array}{|c|}\hline \# \\ \hline b \\ \hline\end{array} \ \S \ \begin{array}{|c|}\hline \# \\ \hline a \\ \hline\end{array}\ , p_2 : \begin{array}{|c|}\hline b \\ \hline b \\ \hline\end{array} \ \S \ \begin{array}{|c|}\hline a \\ \hline a \\ \hline\end{array} \ \$ \ \begin{array}{|c|}\hline b \\ \hline b \\ \hline\end{array} \ \S \ \begin{array}{|c|}\hline a \\ \hline a \\ \hline\end{array}\ ,$$

$$p_3 : \begin{array}{|c|}\hline b \\ \hline \# \\ \hline\end{array} \ \S \ \begin{array}{|c|}\hline a \\ \hline \# \\ \hline\end{array} \ \$ \ \begin{array}{|c|}\hline b \\ \hline \# \\ \hline\end{array} \ \S \ \begin{array}{|c|}\hline a \\ \hline \# \\ \hline\end{array} \ \}$$

$$R_r = \{q_1 : \begin{array}{|c|c|}\hline a & a \\ \hline\end{array} \ \S \ \begin{array}{|c|c|}\hline \# & \# \\ \hline\end{array} \ \$ \ \begin{array}{|c|c|}\hline \# & \# \\ \hline\end{array} \ \S \ \begin{array}{|c|c|}\hline a & a \\ \hline\end{array}\ ,$$

$$q_2 : \begin{array}{|c|c|}\hline a & b \\ \hline\end{array} \ \S \ \begin{array}{|c|c|}\hline \# & \# \\ \hline\end{array} \ \$ \ \begin{array}{|c|c|}\hline \# & \# \\ \hline\end{array} \ \S \ \begin{array}{|c|c|}\hline a & b \\ \hline\end{array}$$

$$q_2 : \begin{array}{|c|c|}\hline b & a \\ \hline\end{array} \ \S \ \begin{array}{|c|c|}\hline \# & \# \\ \hline\end{array} \ \$ \ \begin{array}{|c|c|}\hline \# & \# \\ \hline\end{array} \ \S \ \begin{array}{|c|c|}\hline b & a \\ \hline\end{array} \ \}$$

we illustrate the column and row splicing rules given above:

$$\begin{array}{ccc|cc} a&a&b&a&b\\ a&a&b&a&b \end{array} \quad \begin{array}{cc|cccc} a&b&a&a&b&b\\ a&b&a&a&b&b \end{array} \;\;\vdash_{R_c}\;\; \begin{array}{ccccccc} a&a&b&a&a&b&b\\ a&a&b&a&a&b&b \end{array}\text{'}$$

$$\begin{array}{ccccccc} a&a&b&a&a&b&b\\ a&a&b&a&a&b&b\\ \hline \end{array} \quad \begin{array}{ccccccc} \hline a&a&b&a&a&b&b\\ a&a&b&a&a&b&b \end{array} \;\;\vdash_{R_r}\;\; \begin{array}{ccccccc} a&a&b&a&a&b&b\\ a&a&b&a&a&b&b\\ a&a&b&a&a&b&b\\ a&a&b&a&a&b&b \end{array}$$

A vertical bar '| ' or a horizontal bar '—' indicates the place where splicing is done. Thus, we know that iterating the above rules proves $L \in \ell(HAS)$.

Now we write the rules of A_{SCO} language $L' = \{(b)_m(a^{2n})_m(b)_m \,/\, m \geq 2, n \geq 0\}$ where m is the number of rows and the power counts the column by denoting the number of that element in that row. It can be seen that this language cannot be generated by a H-array splicing system. Now we construct the rules of A_{SCO};

Let $V = \{a, b\}$, $\quad I = \left\{ \begin{array}{ccc} b&a&b\\ b&a&b \end{array} \right\}$,

$$R_c = \{p_1 : \begin{array}{|c|}\hline a\\ \hline a\\ \hline\end{array} \;\S\; \begin{array}{|c|}\hline b\\ \hline b\\ \hline\end{array} \;\$\; \begin{array}{|c|}\hline b\\ \hline b\\ \hline\end{array} \;\S\; \begin{array}{|c|}\hline a\\ \hline a\\ \hline\end{array}\}$$

$$R_r = \{q_1 : \begin{array}{|c|c|}\hline b&a\\ \hline\end{array} \;\S\; \begin{array}{|c|c|}\hline \#&\#\\ \hline\end{array} \;\$\; \begin{array}{|c|c|}\hline \#&\#\\ \hline\end{array} \;\S\; \begin{array}{|c|c|}\hline b&a\\ \hline\end{array},$$
$$q_2 : \begin{array}{|c|c|}\hline a&b\\ \hline\end{array} \;\S\; \begin{array}{|c|c|}\hline \#&\#\\ \hline\end{array} \;\$\; \begin{array}{|c|c|}\hline \#&\#\\ \hline\end{array} \;\S\; \begin{array}{|c|c|}\hline a&b\\ \hline\end{array}\;\}$$

we can illustrate the column and row splicing rules similar to the above illustration. Thus we prove that the family of H-array splicing system is incomparable with the family of self cross-over array system.

We now compare $\ell(A_{SCO})$ with two-dimensional classes of languages LOC, $2RLG$.

Theorem 3.2. *The family $\ell(A_{SCO})$ is incomparable with the class of local languages (LOC) but is not disjoint with it.*

Proof. The picture language L' consisting of all $m \times n$ arrays for $(m \geq 2, n \geq 2)$ describing shape(pattern) L of x's is in LOC. Now we give an A_{SCO}, $S = (V, R_c, R_r, I)$ to describe L'.

Let $V = \{0, x\}$, $\quad I = \left\{ \begin{array}{cc} x&0\\ x&x \end{array} \right\}$,

$$R_c = \{p_1 : \begin{array}{|c|}\hline 0\\ \hline x\\ \hline\end{array} \;\S\; \begin{array}{|c|}\hline \#\\ \hline \#\\ \hline\end{array} \;\$\; \begin{array}{|c|}\hline x\\ \hline x\\ \hline\end{array} \;\S\; \begin{array}{|c|}\hline 0\\ \hline x\\ \hline\end{array},\; p_2 : \begin{array}{|c|}\hline 0\\ \hline 0\\ \hline\end{array} \;\S\; \begin{array}{|c|}\hline \#\\ \hline \#\\ \hline\end{array} \;\$\; \begin{array}{|c|}\hline x\\ \hline x\\ \hline\end{array} \;\S\; \begin{array}{|c|}\hline 0\\ \hline 0\\ \hline\end{array},$$
$$p_3 : \begin{array}{|c|}\hline \#\\ \hline 0\\ \hline\end{array} \;\S\; \begin{array}{|c|}\hline \#\\ \hline \#\\ \hline\end{array} \;\$\; \begin{array}{|c|}\hline \#\\ \hline x\\ \hline\end{array} \;\S\; \begin{array}{|c|}\hline 0\\ \hline 0\\ \hline\end{array},\; p_4 : \begin{array}{|c|}\hline x\\ \hline \#\\ \hline\end{array} \;\S\; \begin{array}{|c|}\hline \#\\ \hline \#\\ \hline\end{array} \;\$\; \begin{array}{|c|}\hline x\\ \hline \#\\ \hline\end{array} \;\S\; \begin{array}{|c|}\hline 0\\ \hline 0\\ \hline\end{array}\Bigg\} \text{ and}$$

$$R_r = \{q_1 : \begin{array}{|c|c|}\hline x&0\\ \hline\end{array} \;\S\; \begin{array}{|c|c|}\hline x&x\\ \hline\end{array} \;\$\; \begin{array}{|c|c|}\hline \#&\#\\ \hline\end{array} \;\S\; \begin{array}{|c|c|}\hline x&0\\ \hline\end{array}$$

$$q_2 : \boxed{0|0} \quad \S \quad \boxed{x|x} \quad \$ \quad \boxed{\#|\#} \quad \S \quad \boxed{0|0}$$

$$q_3 : \boxed{\#|x} \quad \S \quad \boxed{\#|x} \quad \$ \quad \boxed{\#|\#} \quad \S \quad \boxed{\#|x}$$

$$q_4 : \boxed{0|\#} \quad \S \quad \boxed{x|\#} \quad \$ \quad \boxed{\#|\#} \quad \S \quad \boxed{0|\#} \}$$

The picture language L'' of all images over $V = \{a\}$ with three columns is not in LOC. In fact, it is not possible to fix the number of columns using only one symbol. i.e. we can be move without restriction on the number of columns the block $\begin{bmatrix} a & a \\ a & a \end{bmatrix}$. But it is obtained by an A_{SCO} where $I = \{a \quad a \quad a\}$,

$$R_r = \{p_1 : \boxed{a|a} \quad \S \quad \boxed{\#|\#} \quad \$ \quad \boxed{\#|\#} \quad \S \quad \boxed{a|a}$$

$$p_2 : \boxed{\#|a} \quad \S \quad \boxed{\#|\#} \quad \$ \quad \boxed{\#|\#} \quad \S \quad \boxed{\#|a}$$

$$p_3 : \boxed{a|\#} \quad \S \quad \boxed{\#|\#} \quad \$ \quad \boxed{\#|\#} \quad \S \quad \boxed{a|\#} \} \quad \text{and} \quad R_c = \emptyset$$

It is known [3] that the picture language of square images in which diagonal positions carry symbol 1 but the remaining positions carry symbol 0 is in LOC. But it is not in A_{SCO}. Since row and column splicing are independently done, it is clear that arrays with a proposition between rows and columns and in particular pictures with only square size cannot be generated.

Theorem 3.3. *The family $\ell(A_{SCO})$ is incomparable with the language family of two-dimensional right linear grammars ($2RLG$) but is not disjoint with it.*

Proof. The picture language of chessboards with even side-length is generated by a A_{SCO} and is known to be generated by a $2RLG$ [3].

The picture language $L_2 = \{((ab)^p \cup (ba)^q)_m \,/\, p, q, m \geq 1\}$ cannot be described by any A_{SCO}. This is due to the fact that the column splicing of any two arrays $((ab)^p)_m$ and $((ba)^q)_m$ will yield an array which is not in L_2. But it is generated by the following $2RLG$.

$\Sigma = \{a, b\}; \Sigma_I = \{A_1, A_2, A_3, A_4\}; V_h = \{S, X\};$
$R_h = \{S \to A_1A_2X;\ S \to A_3A_4Y;\ X \to A_1A_2X;\ Y \to A_3A_4Y; X \to A_1A_2;\ Y \to A_3A_4\}; \quad V_v = \{A_1, A_2, A_3, A_4\}; R_v = \{A_1 \to aA_1;\ A_1 \to a; A_2 \to bA_2;\ A_2 \to b; A_3 \to bA_3;\ A_3 \to b; A_4 \to aA_4;\ A_4 \to a\}$.

The picture language L_1 consisting of arrays describing shape(pattern) H cannot be generated by any $2RLG$, as the horizontal row of x's cannot be maintained by any $2RLG$. But it is described by the A_{SCO} where $V = \{x, 0\}$, $I = \left\{\begin{matrix} x & 0 & x \\ x & x & x \\ x & 0 & x \end{matrix}\right\}$, and R_c is such that the domino rules are written to cut the last column and the first column of the two identical copies of I to paste and make the pattern of more number of columns. Similarly R_r is such that the cutting and pasting is at the bordered row to enlarge the pattern of I. Thus the incomparability results.

4 Conclusion

In this paper we have studied the classes of languages $\ell(A_{SCO})$ to prove its properties, comparing with other two-dimensional language classes like LOC, $2RLG$. Also the comparison with the class of languages $\ell(HAS)$ is proved analogous to the string case self cross-over systems and H-splicing systems. Several questions concerning the notion of $\ell(A_{SCO})$ has to be investigated, as well as the comparison of this approach with other array language classes like Tiling system [3] has to be studied.

References

1. Dassow, J., Mitrana, V.: Self cross-over systems. In: Paun, G. (ed.) Computing with Bio - Molecules: Theory and Experiments. Discrete Mathematics and Theoretical Computer Science, pp. 283–294. Springer, Heidelberg (1998)
2. Gatterdam, R.W.: Splicing systems and regularity. Int. J. Comp. Math. 31, 63–67 (1989)
3. Giammarresi, D., Restivo, A.: Two-dimensional languages. In: Rozenberg, G., Salomaa, A. (eds.) Handbook of Formal Languages, vol. 3, pp. 215–267. Springer, Heidelberg (1997)
4. Head, T.: Formal language theory and DNA: an analysis of the generative capacity of specific recombinant behaviours. Bull. Math. Biology 49, 735–759 (1987)
5. Head, T., Paun, G., Pixton, D.: Language theory and molecular genetics: Generative mechanisms suggested by DNA recombination. In: Rozenberg, G., Salomaa, A. (eds.) Handbook of Formal Languages, ch. 7, vol. 2, pp. 295–358. Springer, Heidelberg (1997)
6. Helen Chandra, P., Subramanian, K.G., Thomas, D.G.: Parallel Splicing On Images. IJPRAI 18(6), 1071–1091 (2004)
7. Krithivasan, K., Chakaravarthy, V.T., Rama, R.: Array splicing systems. In: Păun, G., Salomaa, A. (eds.) New Trends in Formal Languages. LNCS, vol. 1218, pp. 346–365. Springer, Heidelberg (1997)
8. Paun, G., Rozenberg, G., Salomaa, A.: Computing by splicing. Theoretical Computer Science 168, 321–336 (1996)
9. Siromoney, R., Subramanian, K.G., Rangarajan, K.: Parallel/sequential rectangular arrays with tables. Int. J. Comput. Math. 6A, 143–158 (1977)
10. Subramanian, K.G., Roslin Sagaya Mary, A., Dersanambika, K.S.: Splicing array grammar systems. In: Van Hung, D., Wirsing, M. (eds.) ICTAC 2005. LNCS, vol. 3722, pp. 125–135. Springer, Heidelberg (2005)

Multi-Agent System (GerMAS) Used to Identify Medicines in Geriatric Residences

Jose M. Pérez, F. Fernández, Juan A. Fraile, M. Mateos, and Miguel A. Sánchez

Abstract. This article proposes a multi-agent system (MAS) that personalizes medical information in geriatric residences. The personal medical can personalize each medication with the instructions for each patient by using a contactless card reader or a mobile with NFC technology. The use of RFID and NFC technology allows the identification of the patient and the application of preventative medical protocols to the patient. The proposed multi-agent system has reasoning capabilities in order to analyze the situation and achieve a high level of interaction with the patients and the personal medical. A distributed prototype system has been developed to test this multi-agent system. This system is aimed to improve health care and assistance to dependent persons in geriatric residences. Preliminary results are presented in this paper.

Keywords: NFC, Multi-Agent Systems, Health Care, Geriatric residences.

1 Introduction

There is an ever growing need to supply constant care and support to dependents [3] and the drive to find more effective ways to provide such care has become a major challenge for Europe and its scientific community [3]. During the last three decades the number of Europeans over 60 years old has risen by about 50%. Today they represent more than 25% of the population and it is estimated that in 20 years this percentage will rise to one third of the population, meaning 100 millions of citizens [1]. The importance of developing new and more reliable ways to provide care and support to the elderly is underlined by this trend [3].

On the other hand, the incorporation of Information and Communication Technology (ICT) into our daily lives has, undoubtedly, many positive consequences. The mobile telephone is no longer considered to be a simple device limited to making calls, rather it incorporates an array of features for use in daily life, such as messaging, connecting to internet, playing music, video, etc. As mobile technology advances and these devices begin to function as increasingly complex

Jose M. Pérez · F. Fernández · Juan A. Fraile · M. Mateos · Miguel A. Sánchez
Pontifical University of Salamanca, c/ Compañía 5, 37002 Salamanca, Spain
e-mail: {elcorreodechema,hiperfifo}@gmail.com,
{jafraileni,mmateossa,masanchezvi}@upsa.es

J.B. Pérez et al. (Eds.): Highlights in PAAMS, AISC 89, pp. 299–306.
springerlink.com

systems, the effort to allow disabled individuals to access these technologies is growing more and more. The main objective of these efforts is to provide access to different types of information from any place and at any time, which is evident in recent attempts to integrate mobile information devices into web-based business processes and e-commerce situations. In this sense, agents and multi-agent systems have become basic elements in the development of dynamic and distributed systems, and have been successfully applied in areas such as e-commerce, medicine, home care, robotics, etc [7] [9]. One of the advantages of the agents is their ability to easily adapt to any type of patient and his or her pathologies [6]. Additionally, agents are frequently used with mobile technology such as GPRS (General Packet Radio Service), NFC [7] (Near Field Communication) or WiFi or Bluetooth [5], to obtain patient information in novel ways.

The main objective of this study is to present a multi-agent GerMAS (Geriatric residences Multi-Agent System) architecture that has been specifically designed for geriatric residences. The architecture provides innovative mechanisms for integrating multi-agent systems and intelligent interfaces to obtain patient information. The GerMAS architecture utilizes the patient information it obtains to personalize the information that a patient should receive in a geriatric residence, while taking into account the patient's individual pathologies. The architecture incorporates Wireless, Wi-Fi, NFC and RFID (Radio Frequency Identification) technologies [12] to automatically identify patients and medications. Another important contribution of the GerMAS architecture is the incorporation of deliberative BDI agents [10]. GerMAS was used to develop a prototype (FabulaGer) to personalize the information in the medication that patients receive in the geriatric residences.

The remainder of the article is structured as follows: the next section describes the proposed multi-agent system, focusing on the description of how the agents interact with the contactless devices. Section 4 presents the FabulaGer prototype that was developed for this study. Finally, section 5 presents the results and conclusions obtained after installing the prototype in a care facility, and suggests future areas of investigation for improving the system.

2 GerMAS Multi-Agent System

Multi-agent systems are distributed systems based on the cooperation of multiple autonomous agents [10]. GerMAS is a multi-agent system that facilitates the integration of agents and mobile NFC devices using a contactless RFID technology. The GerMAS multi-agent system emerged in response to the need for improving techniques utilized for generating personalized information associated with medication and patient data processing. GerMAS also improves the assignment of medical diagnostics and monitoring for all types of patients. The GerMAS multi-agent system is a distributed agent platform that uses NFC technology to enable patients and personal medical to maintain secure intercommunication for pathologies and treatments.

Each of the devices used in GerMAS is based on NFC and RFID technology. These devices not only identify and contain information about the patients, but also identify and contain personalized information for the patient in the medication. Contactless readers and mobile devices with NFC technology are used to read the RFID tags. NFC is a protocol that is based on short range wireless interface. The communication is made between two entities. The protocol establishes a wireless connection between the network applications and electronic devices. NFC works on a 13.56 MHz band, which eliminates the need for any license required for its use. With NFC there is always a device that initiates and then monitors a conversation. NFC technology merges the usability of RFID chip technology with the portability and market penetration of the mobile telephone. It uses an electromagnetic field to communicate securely with a mobile telephone, PDA or laptop with an intelligent RFID tag, or another device, to perform micro payments, exchange of information and access control. Patients are assigned a mobile with NFC technology, with which they can identify objects and play multimedia content associated with RFID tags. The patients can also process or provide personal information or attributes using a mobile terminal [8]. The GerMAS system not only gathers and stores patient information, but it also serves as a proactive computing system since it takes the information associated with the medication and personalizes it according to the needs of the patient. The proactive computation, along with the use of devices utilizing NFC and RFID technology, allows GerMAS to generate a degree of independence and enrich the quality of life for the patients.

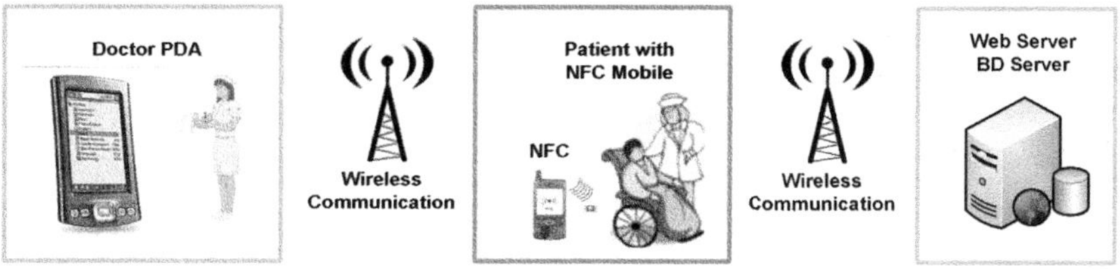

Fig. 1 Devices used in the GerMAS system.

Figure 1 illustrates the different devices that can be connected to the GerMAS system. Starting at the top there are three levels:

- The first level clearly shows a PDA that medical personnel can use to identify the patient and update his or her medical history.
- This middle level includes various devices (NFC mobiles, RFID readers, RFID tags, computers and portable equipment such as laptops) that provide the system with identification capabilities.
- The bottom level stores the data. This level provides computational capability that allows the data to be stored and processed prior to being used by the middle level.

As shown in figure 2, there are different types of agents that function on each of the levels. The GerMAS multi-agent system is based on the BDI (Belief, Desire and Intention) model in which the agents act as controllers and coordinators for the various tasks associated with medical care. The agents can initialize services based on demand, planned actions or user requests. The behavior of the agents can change according to any modification made in the patient's diagnosis. The same agents can modify the patient's treatment and respond to the information associated to the object. Additionally the information for the state of the system is continually stored, allowing easy recovery in the event of system failure.

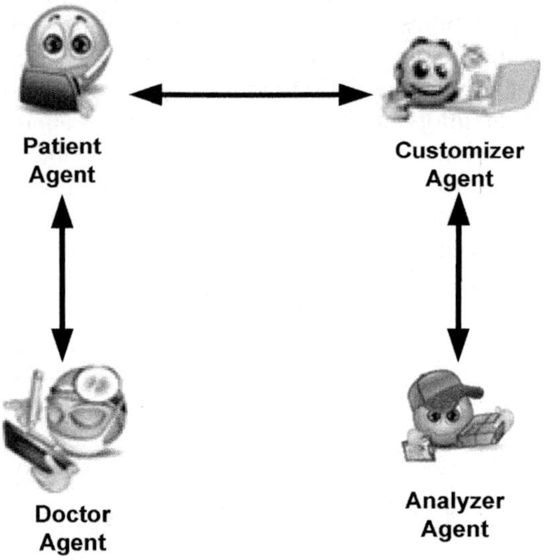

Fig. 2 Agents in the GerMAS system.

The types of agents shown in figure 2 are:

- Patient Agent: Manages all communication with medical personnel. Receives information from the Doctor agent on incidents and data regarding patient supervision. All of this information can be sent via a wireless or GPRS connection. In the event of any failure in communication, this agent informs the Doctor agent to resend the information at a later time. The Patient agent is also responsible for copying the information associated with the medication, taking into account the type of format in which the information is stored. The Patient agent also communicates with the Customizer agent by identifying itself and updating medical history.
- Customizer Agent: Automatically generates personalized information for the patient by processing the information managed by the Analyzer agent. The Customizer agent uses the Text-To-Speech Loquendo system [2] to convert personalized information into an audio file. The Customizer agent assigns an identifier to the audio file and the Analyzer agent stores it in the system. The Customizer agent records the identifying information on the RFID tags, using a URI (Uniform Resource Identifier). This agent is also responsible for identifying the patient and sending the information to the

Analyzer agent in order to obtain all of the information available that is associated with the patient.

- Doctor Agent: Identifies the patient and communicates with the Patient data to update the medical history of the patient, taking into account the state and symptoms of the patient pathologies.
- Analyzer Agent: Logically orders and stores the information that is provided by the Patient and Customizer agents in the information. This task facilitates the work of the Customizer agent. The stored information can be accessed via Internet.

The Customizer agent in the GerMAS system integrates reasoning (CBR) and planning (CBP) mechanisms that permit the agents to utilize past experience to create better results and achieve their objectives. Case-Based Reasoning (CBR) [10] is a type of reasoning based on past experiences. CBR systems solve new problems by adapting the solutions that have already been used to solve similar problems in the past, and learning from the new experiences. Case-Based Planning (CBP) stems from CBR, but is specifically designed to generate plans (a sequence of actions) [10]. With CBP, the proposed solution for a problem is a plan. This solution is generated after taking into account the plans applied to solve similar problems in the past. The problems and their corresponding plans are stored in a memory of plans. The reasoning mechanism generates the plans and the planning strategies used in past experiences, which is what essentially constitutes the concept of Case-Based Planning [6].

3 Case Study: FabulaGer

A prototype of the GerMAS multi-agent system (FabulaGer) was developed for this case study. The system improves the patients' quality of life by simplifying, improving and automating the information provided by the personal medical to their patients. FabulaGer identifies patients at the geriatric residence and automatically personalizes the information that the patients receive regarding the product they are taking. The system gathers information about the patient, medical personnel and diagnostics, which it then stores, handles and processes to generate information at a later date that can be useful for each medication and patient. The system interacts with users by means of mobile devices (PDA, NFC mobile phones) and wireless communication technologies (Wi-Fi, GPRS, RFID and NFC). These technologies and devices work in a distributed way, providing the users with flexibility and easy access to the system services.

This specific RFID technology allows us to automatically capture data that is used for the electronic identification of patients and for associating multimedia content to the medication. A RFID system is essentially comprised of 2 components: tags and readers. The tags can be passive (not battery operated) or active (battery operated). The active tags can be read by a reader from a greater distance than the passive tags. The RFID tag reader, which can be a mobile device with NFC technology, gathers information it reads within the reading area and transmits it to the system. This information consists of the identification taken from the

RFID tag that has been read, the identification of the reader used, and the information contained in the RFID tag. The NFC mobiles use Wi-Fi or GPRS to transmit this information to the system, while the RFID readers use Wi-Fi to transmit the information through an internal network installed within the context.

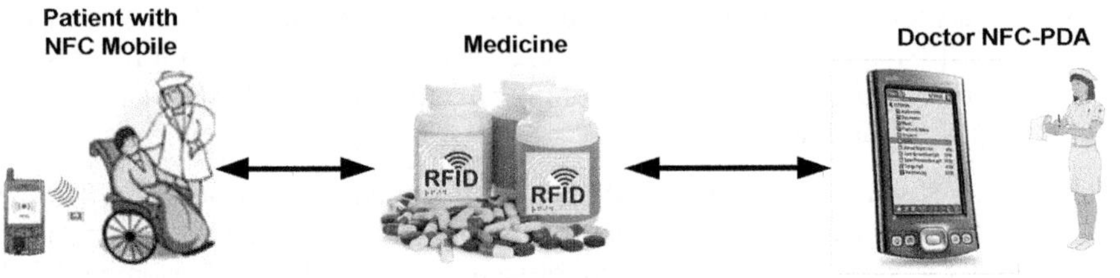

Fig. 3 FabulaGer user context process description.

As shown in figure 3 and figure 4, there are two distinct processes presented in the case study:

- User context: in this scenario, the patients use FabulaGer to receive an audio file description of the medication. To achieve this, it is necessary to use a mobile NFC terminal, such as a Nokia 6131 NFC, and a medicine box that contains an RFID tag inside. This tag must have been previously configured, as explained in the description for Admin context, with a specific code associated with the medication. The patients need only move the mobile NFC terminal close to the medicine box containing the RFID tag inside. When the mobile is close enough, the Patient agent reads the information code associated with the tag. The Patient agent communicates with the Analyzer agent, sending the identification information for the medicine. The Analyzer agent recognizes the medication identifier and sends the corresponding audio file with the description of the medication, personalized for the patient, to the patient's mobile phone. The user can play the audio file on the mobile phone and listen to the information for the medication.
- Admin Context: this scenario configures all of the components described in the User context. The personal medical use the Customizer agent to identify the patient and communicate with the Patient agent. The Patient agent contains the medical diagnosis received by the Doctor agent. At the same time, the Customizer agent communicates with the Analyzer agent to know what information is associated with the patient and the medication that the patient wishes to purchase. At this point, the Customizer agent generates personalized and useful indications for the patient regarding the medication and converts the text into an audio file using the Text-To-Speech Loquendo system. The Customizer agent then assigns a unique identifier to the audio file and the personal medical uses an application installed in the NFC mobile or a personal computer with a contactless reader to record a URI on the RFID tag for the medication. The URI is generated by the Analyzer agent. It also identifies the sound file stored by the Analyzer agent.

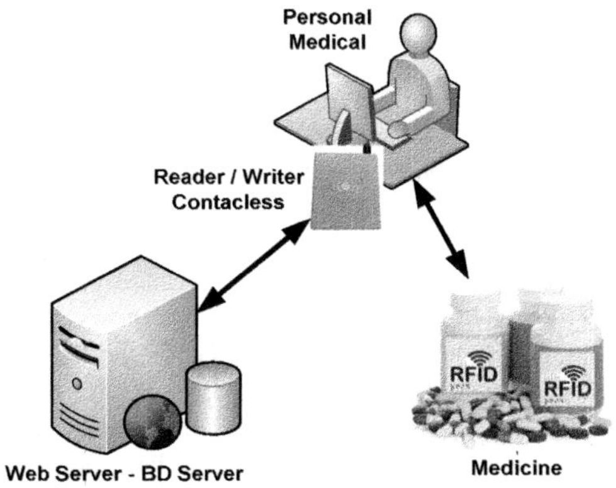

Fig. 4 FabulaGer admin context process description.

4 Results and Conclusions

Multi-agent systems have been shown to be valid in resolving health problems [4] [10]. They have also been applied to control wireless devices in geriatric care facilities [6]. The GerMAS multi-agent system is very appropriate for solving problems related to the coordination of medical specialists and pharmacists in geriatric residence. The agents can communicate and cooperate among themselves to build a personalized solution. Many times the solutions are complex and difficult to find. A multi-agent system searches for a solution to the problem by breaking it down into sub-problems.

NFC and RFID technology are particularly promising as a support tool for health care environments. Tags and communication devices facilitate the process of sending and receiving information in a ubiquitous and non-intrusive manner. The FabulaGer system presented in this paper proposes an innovative technological solution consisting of an intelligent environment based on the use of NFC mobile devices, RFID tags and multi-agent systems. The FabulaGer system was tested in a real environment, and the results obtained are promising.

The main innovative aspect of the FabulaGer software lies in the combination of NFC mobile devices and RFID to improve the classical object recognition, providing practical information in multimedia format. It can also provide personal information for each patient. It is simple, powerful, flexible and easy to use. Furthermore, it does not require a complex technical infrastructure and it is inexpensive to deploy.

Additionally, FabulaGer has been well received by the official ONCE (Spanish National Organization for the Blind) in Salamanca due to the value add that it offers to the blinds. In hopes of improving the project even more and making it useful for the greatest possible number of users, we are working in collaboration with the ONCE of Salamanca to develop a prototype that will be even more functional.

Acknowledgements. Special thanks to the ONCE, CIDAT and Loquendo for their support.

References

1. Angulo, C., Tellez, R.: Distributed Intelligence for Smart Home Appliances. On: Tendencies of data mining in Spain. In: Spanish Data Mining Network, Barcelona, Spain (2004)
2. Badino, L., Barolo, C., Quazza, S.: Language independent phoneme mapping for foreign TTS. In: IEEE Workshop on Text-to-speech Synthesis (2004)
3. Cesta, A., Bahadori, S., Cortellesa, G., Grisetti, G., Giuliani, M., Locchi, L., Leone, G., Nardo, D., Oddi, A., Pecora, F., Rasconi, R., Saggese, A., Scopelliti, M.: The RoboCare Project, Cognitive Systems for the Care of the Elderly. In: Proceedings of International Conference on Aging, Disability and Independence (ICADI), Washington D.C., USA (2003)
4. Corchado, J.M., Bajo, J., Abraham, A.: GERAmI: Improving the delivery of health care. IEEE Intelligent Systems, Special Issue on Ambient Intelligence (2008)
5. Fernández, F., Rodillo, D., del Pino, F., Bajo, J., Corchado, J.M.: WirePET: Desarrollando Videojuegos para Dispositivos Móviles con Comunicación Bluetooth. In: 6th International Workshop on Practical Applications of Agents and Multiagent Systems, pp. 258–268 (2007)
6. Fraile, J.A., Bajo, J., Corchado, J.M., Abraham, A.: Applying wearable solutions in dependent environments. IEEE Transactions on Information Technology in Biomedicine (2010) (in press)
7. Hansen, T.R., Bardram, J.E., Soegaard, M.: Moving out of the lab: Deploying pervasive technologies in a hospital. IEEE Pervasive Computing 5(3), 24–31 (2006)
8. Hsu, Y., Yang, C., Tsai, T., Cheng, C., Wu, C.: Development of a Decentralized Home Telehealth Monitoring System. Telemedicine and e-Healt 13(1), 69–78 (2007)
9. Moreno, A., Valls, A., Isern, D., Sanchez, D.: Applying Agent Technology to Healthcare: The GruSMA Experience. In: Intelligent Systems IEEE, pp. 63–67 (2006)
10. Tapia, D.I., Corchado, J.M.: An Ambient Intelligence Based Multi-Agent System for Alzheimer Health Care. International Journal of Ambient Computing and Intelligence (IJACI) 1(1), 15–26 (2009)
11. Vergara, M., Díaz-Hellín, P., Fontecha, J., Hervás, R., Sánchez Barba, C., Fuentes, C., Bravo, J.: Mobile Prescription: An NFC-Based Proposal for AAL. In: 2nd International Workshop on Near Field Communication, Monaco (2010)
12. Want, R.: An Introduction to RFID Technology. IEEE Pervasive Computing 5, 25–33 (2006)

Head Tracking System for Wheelchair Movement Control

R. Berjón, M. Mateos, A. Barriuso, I. Muriel, and G. Villarrubia

Abstract. This paper presents an interface to control an electric wheelchair by using the movements of the head, the voice and a mobile device. The system aims to cover the need of independence of severe disabled people that cannot use a common wheelchair joystick. The structure of the system as well as the preliminary results obtained are presented in this paper.

Keywords: Head control, disabled people, mobile devices, motion detection.

1 Introduction

Different electric wheelchairs nowadays are based on a modular system, where the Human Machine Interfaces (HMI) are separate from the electronic circuits that allow the wheelchair to move. These systems allow changes in the HMI so that they can be adapted to cover the largest number of user disabilities.

Severe disabled people are not able to use the usual joystick found on most wheelchairs, as in most cases they are not able to move anything but their head.

There are several HMI systems being developed to control wheelchair, using infrared sensors [1] to detect head movements, for example, based on the SIAMO [2] project, that try to make disabled people's life easier.

Aescolapius is an HMI system that aims at those severe disabled people, as well as any other person it could help, and tries to help them and make them as independent as possible. This system can be adapted to work with most modular based wheelchair systems, like SIAMO [2].

Like [1], our system is able to detect a person's face, but using an RGB camera instead of infrared light, and like [2] Aescolapius is able to interpret voice commands and move the wheelchair accordingly. But what really makes it different is that it also can be controlled by an external person, through a mobile device. Figure 1 shows the architecture of the Aescolapius HMI system and the modules described in this paper.

R. Berjón · M. Mateos · A. Barriuso · I. Muriel · G. Villarrubia
Universidad Pontificia de Salamanca, Compañía 5, 37002, Salamanca, Spain
e-mail: {rberjonga,mmateossa,imurielnu.eui}@upsa.es,
albertolopez_4k@hotmail.com, gabri@grupodeltron.com

J.B. Pérez et al. (Eds.): Highlights in PAAMS, AISC 89, pp. 307–315.
springerlink.com

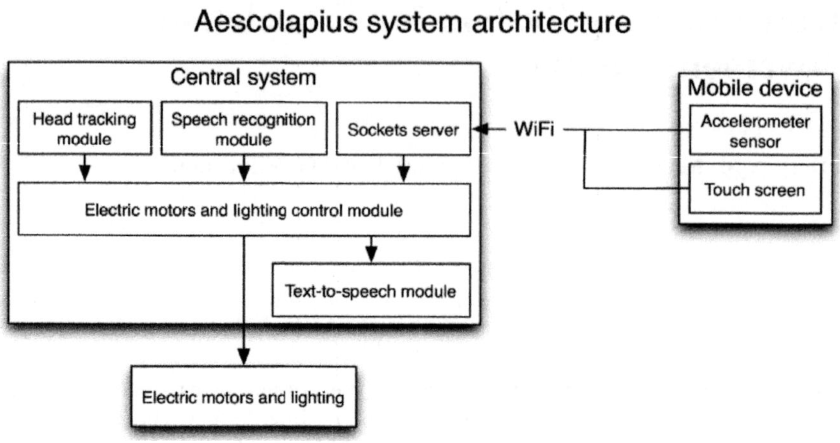

Fig. 1 Shows system's architecture.

The rest of the paper is structured as follows: Section 2 presents the alternative ways of detecting the user face and describes the behavior of the head tracking system. Section 3 shows and describes the voice recognition module, which is part of the system's core. Section 4 and 5 describe the HMI implemented to be used on mobile devices. Section 6, shows the prototype Aescolapius, made for testing purposes. Finally, section 7 presents the preliminary results obtained and the conclusion.

2 Head Movement Tracking

We manage to move the chair by using a camera connected to a portable computer to track the face of the person sitting on the wheelchair. An algorithm called Optical Flow [4] is used to track the face's movements in real time.

There are different ways to track someone's face with a camera, like Haar-like features or Optical flow.

Haar-like features [5] consider adjacent rectangular regions at a specific location in a detection window, it sums up the pixel intensities in these regions and calculates the difference between them. This difference is then used to categorize subsections of an image. For example, let us say we have an image database with human faces. It is a common observation that among all faces the region of the eyes is darker than the region of the cheeks. Therefore a common haar feature for face detection is a set of two adjacent rectangles that lie above the eye and the cheek region. The position of these rectangles is defined relative to a detection window that acts like a bounding box to the target object (the face in this case).

In our system we use a Haar-like features combined with an Optical Flow algorithm (both supplied by the Open Computer Vision libraries created by Intel [3]). Optical flow allows us to get the pattern of apparent motion of objects between an observer (the camera) and the scene (the subject).

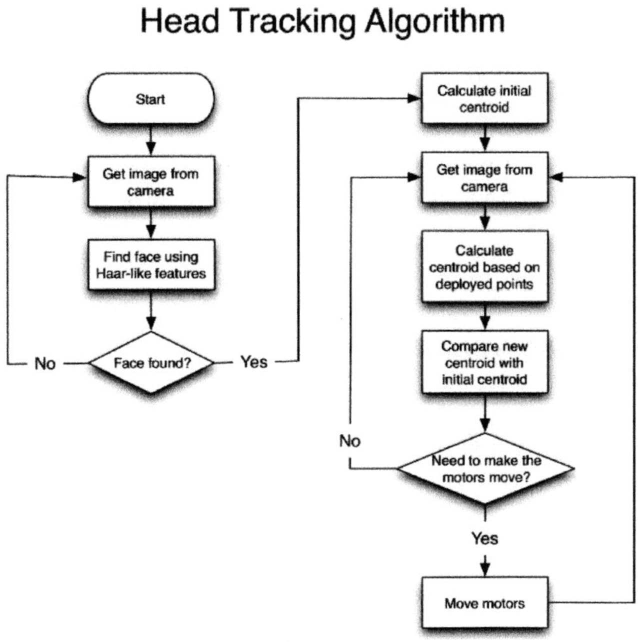

Fig. 2 State diagram for the Head Tracking Algorithm.

As shown in figure 2, when the head tracker module is called, the first thing that is done is find a face in the images received from the camera using Haar-like features, features that are previously stored in a Haar database file. If it is found more than one face, the bigger one is always chosen, and it is assumed that that one is the nearest to the camera.

After that, the previously found face is tracked using the Optical Flow algorithm.

The Haar-like features algorithm returns the area, a rectangle, of the image containing the face. Then the centroid of that area is calculated, also called the geometric center, or barycenter of the area.

After calculating this initial centroid, the Optical Flow algorithm starts tracking some points over the face, which it does by itself. The centroid of the shape created by these points is the same point as the initial centroid.

As the face moves and more images are obtained from the camera, the optical flow algorithm finds those same points inside the new images and calculates the centroid of the new shapes formed by the new points. Then the relative position of those new centroids and the initial centroid is compared. If the new centroids are on the left, right, above or under the initial centroid, the electric motors move will be sent the signal to move.

The sensibility of the system can be adjusted by setting the distance that needs to exist between the first centroid and the current centroid to make the motors move.

By doing this, the system can be adapted to the person sitting on the chair, so it is always sure that the person is able to move the chair properly.

3 Speech Recognition

Using an Android platform and a Windows 7 computer there are two different ways of getting the voice translated to commands. One way is using Google Speech Recognition API. This way, the voice of the person has to be recorded in an audio file, then that audio file has to be sent to the Google Speech Recognition Service and it will return a string containing, in plain text, the spoken words contained in the audio file.

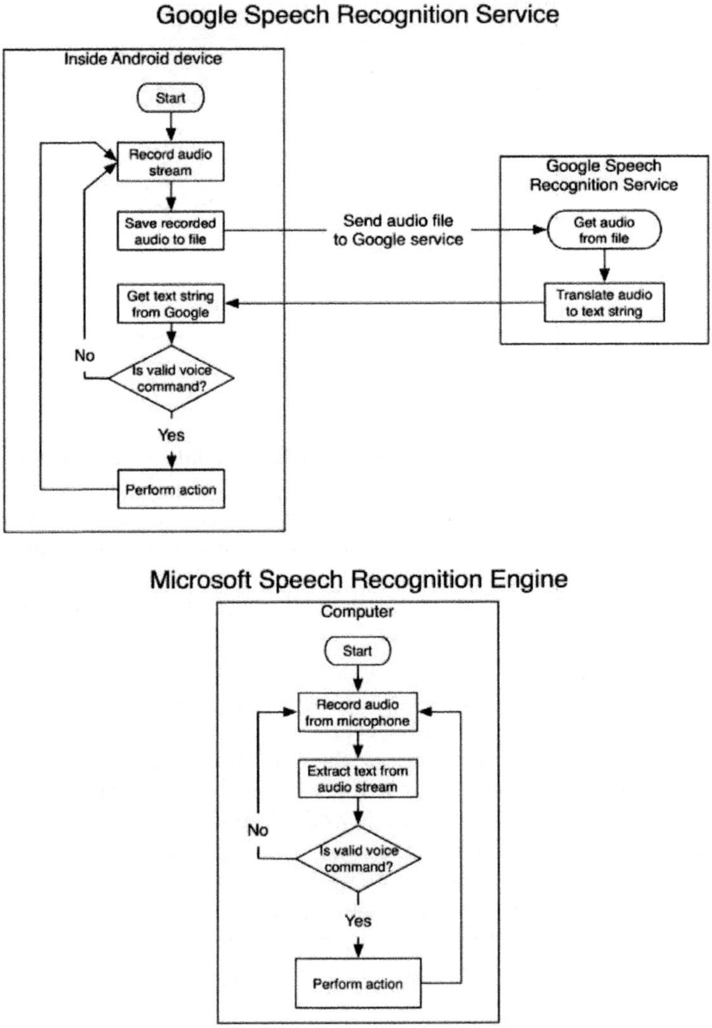

Fig. 3 Differences between Google's and Microsoft's Speech Recognition Services.

Another way is to use Microsoft SAPI [6] (Speech API) 5, bundled inside Windows 7. This way allows us to continuously get audio from a microphone and translate that audio into plain text, without any need to send the audio to any server.

Both ways have been tested and it is been found that Microsoft's one is better for our purpose. It doesn't only allows to continuously get sounds from the audio stream from the microphone, avoiding any command losses, it is also faster because the audio stream is converted locally, and not in a server. And when we are talking about a wheelchair, and that someone could get harmed, fastness becomes a critical factor.

Once chosen the right way, as it is described before, we use the Microsoft Speech API to get the voice commands in plain text. A dictionary of 'allowed commands' is defined and everything that is not in that dictionary is just ignored. Once a valid voice command is obtained, the action attached to that command is performed. The action can be one of this: turning the lights on, making the wheelchair go forward, backwards, left or right, turning on the face tracking system, reading the news, reading the weather forecast or stopping the wheelchair.

Figure 3 shows in a schematic way the main difference in the architecture of the Microsoft SAPI and the Google Speech Recognition Service.

Once all of the voice recognition system was implemented, we found out that any person just talking by the wheelchair could fire a voice action and make the wheelchair start moving. To prevent this, before any command a special word that makes no sense in a normal conversation has been added, so there is no way to make it work by just talking to it.

For example, instead of just saying "forward" to make the wheelchair go forward, you need to add the word "execute" before, so the full command is "execute forward". As this makes the command a little longer to pronounce, the stopping command is the only one that remains the same: "abort". Here is the list of commands that the system can understand, and the action that each fires (this list has the actual commands translated into English, but it is possible to make the system understand English commands):

- "Execute forward": Makes the wheelchair go forward.
- "Execute backwards": Makes it move backwards.
- "Execute left": Makes the chair go left.
- "Execute right": Makes the wheel chair move right.
- "Abort": Makes the wheelchair stop moving.
- "Execute video": Activates the face tracking system, in order to move the chair with the face.
- "Execute news": Makes the chair download and read aloud news from a newspaper's RSS.
- "Execute weather": Makes the system read the weather forecast from an RSS feed.
- "Execute lights": Turns on the wheelchair's lights, in case there isn't enough light to light up the face.

4 Accelerometer of a Mobile Device

Most of the smart phones nowadays in the market, even the cheapest ones, have an acceleration sensor bundled inside. That sensor allows us to measure the inclination of the device with respect to the ground, in other words the gravity vector direction. By doing this we are able to simulate a kind of joystick using the phone, by tilting the phone.

When we were designing the system, we thought about the person sitting on it. But many times another person goes along with him, or her. And we asked ourselves "What about the person walking by the chair?". It is difficult to control a normal electric wheelchair if you walk along with it by using its joystick, because as you walk you move it and make the chair go faster or slower that your walking speed. And if we are talking about the speech recognition system, or the face tracking system it is even more difficult (you cannot put your face in front of the camera as you walk).

So, we thought about it, and realized that nearly every person nowadays has a mobile phone in their pocket, and the number of smart phones is rising exponentially. Why not use that smart phone as an interface to the system, removing this way the physical link between the wheelchair and the person next to it.

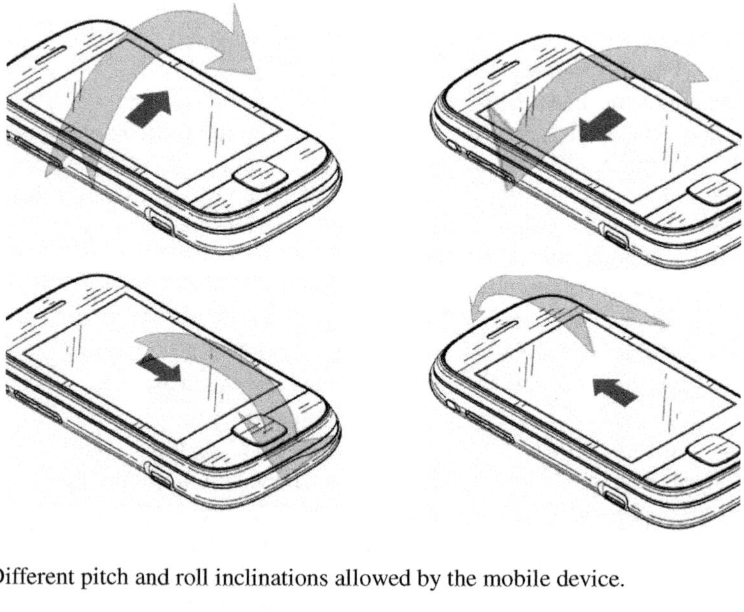

Fig. 4 Different pitch and roll inclinations allowed by the mobile device.

Figure 4 shows how should the smart phone be inclined in order to get the wheelchair moving. Tilting the device as it is shown in the upper-left corner, the system will make the wheel chair move forward.

It was found out that in every person that has tested the system this has been the favorite interaction way. They find that tilting the phone is very usable and comfortable, as they walk beside the wheelchair.

5 Touch Screen

Another input method available on many smart phones is a touch screen. The system has the ability to receive input events from a mobile device provided with a touch screen.

This way, the system is able to aim at an even wider range of people, including many of those who dislike the accelerometer control. But apart from the person beside the wheelchair another use for this method was found. There is severe disabled people that have a little movement in their hands, but not enough strength or movement freedom to use the wheelchair joystick. So it was found that with the mobile device on the chair, right under the hand, the touch screen worked smooth and succeeded moving the chair properly.

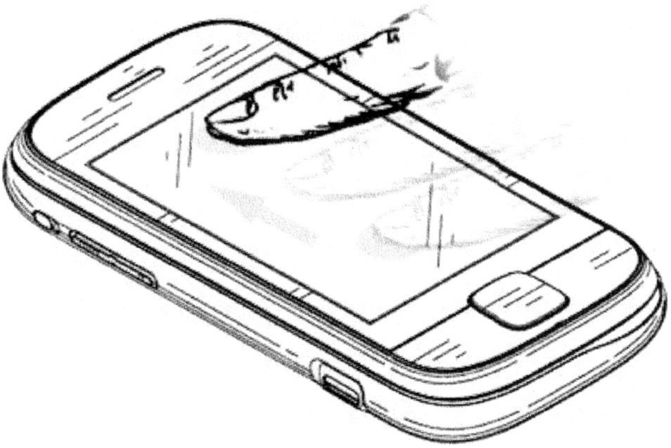

Fig. 5 Shows how the touch screen has to be used.

In figure 5 is shown the way to use the device's touch screen. With a finger, slide it from the bottom to the top of the screen to make the system move the wheelchair forward. The opposite movement, while moving forward, will make the chair stop. And that movement again, while the chair is idle, will make it move backwards. Sliding the finger from left to right will move the wheelchair to the right, and the opposite finger slide will do it to the left.

6 Case Study: Aescolapius

Aescolapius is the name of the wheelchair prototype built to test the system we have presented. It is a handmade wheelchair with all of the components of the system included in it. It features a portable laptop, for processing all of the information, a camera to get images from the face and a microphone in order to listen to voice commands. There was also an Android powered mobile device to test the accelerometer and the touch screen.

Fig. 6 Aescolapius prototype built for testing the system.

7 Conclusions

Aescolapius is a HMI that can be adapted to virtually any wheelchair system. The provided interfaces are: head tracking, speech recognition and two more on the mobile device: accelerometer and touch screen.

By now, Aescolapius is just a prototype but we are working on some specific areas to improve the responsiveness of the system, vary the speed of the motors while running and on the smoothness of the wheelchair's movements.

While testing, the system behaved properly reacting to user commands, whatever voice, face or mobile device commands were sent. Reaction times were short enough not to hurt anybody on the chair or around it, but they need to be improved.

The system also showed robustness against different kinds on faces, with and without glasses, long and short hair or different hair colors and styles. Light conditions have to be good for the head tracking module to work properly, although the Aescolapius prototype has its own lighting system.

Future work involves: obstacle detection using infrared and ultrasonic sensors, route recording, both inside buildings and the street for later autonomous playback. We think route recording is very useful as it can save time in common routes, like going from the living room to the kitchen, for instance. Another useful situation could be in public buildings, like hospitals, where the person on the wheelchair will just need to select where to go, and the wheelchair will do it all by itself.

Acknowledgments. This research has been partially supported by the Junta de Castilla y León project *Estudio y propuesta de un framework para el desarrollo de aplicaciones en plataformas móviles heterogéneas* (PON002B10-2) and by the Spanish Ministry of Science and Innovation project TRA2009_0096.

References

[1] Infrared Non-Contact Head Sensor, for Control of Wheelchair Movements, http://citeseerx.ist.psu.edu/viewdoc/download?doi=10.1.1.86.8316&rep=rep1&type=pdf

[2] SIAMO Wheelchair project, http://www.geintra-uah.org/archivos/1999-HMIs-Wheelcair-UAH.pdf

[3] Bradski, G., Kaehler, A.: Learning OpenCV: Computer Vision with the OpenCV Library (2008)

[4] Optical Flow guide, http://robots.stanford.edu/cs223b05/notes/CS%20223-B%20T1%20stavens_opencv_optical_flow.pdf

[5] Viola, Jones: Rapid object detection using boosted cascade of simple features. In: Computer Vision and Pattern Recognition (2001)

[6] Microsoft site for SAPI, http://www.microsoft.com/speech/default.aspx

Smart m-Learning Reusing Educational Contents

Ana Mª Fermoso García and Alberto Pedrero Esteban

Abstract. Smart mobility tries to help user offering him information adapted to her real context. It would be useful that having this also on mind when the user tries to learn about anything. In order to learn, m-learning can be considered as an evolution of e-learning that allows the user to use the advantages of mobile technologies to support the learning process. There are many educational contents in digital format and it would be very good if they could be reused in mobile contexts. This paper seeks to achieve this objective by providing access to educational content packaged in an e-learning standard. In this way user can reuse in any context the educational contents that already exist for that context using the mobile devices to access them.

Keywords: m-learning, SCORM, reusing, e-learning standards, educational contents.

1 Introduction

The e-learning has enabled that training reach more people through the release of the temporal and spatial barriers. The goal is the user can learn effectively at anytime, anywhere. In this context and taking into account the expansion and acceptance of mobile devices today, there is a trend towards the use of these devices as a basis for new learning models, appearing what is known as m-learning.

The m-learning can be considered as an evolution of e-learning that allows the user to use the advantages of mobile technologies to support the learning process. Mobile phone is today a very usual device, and it is not mandatory permanent Internet connection to work. For all this reasons it is an ideal tool for us to use it at anytime, and also to learn and make better use of our time.

An objective in the e-learning is to achieve a common methodology to ensure the accessibility, interoperability, durability and reuse of materials in digital format. The e-learning standards make it easy.

Smart mobility tries to help user offering him information adapted to her real context. It would be useful that having this also on mind when the user tries to

Ana Mª Fermoso García · Alberto Pedrero Esteban
Computer Sciences Faculty, Pontifical University of Salamanca
e-mail: {afermosoga,apedreroes}@upsa.es

J.B. Pérez et al. (Eds.): Highlights in PAAMS, AISC 89, pp. 317–323.
springerlink.com

learn about anything. In order to learn, m-learning allows user to use the advantages of mobile technologies to support the learning process. There are many educational contents in digital format and it would be very good if they could be reused in mobile contexts. In this sense we can consider "smart" at the m-learning process because it would allow reuse all the digital content that already exist instead of designing new educational contents only for the mobile context. In this paper we try to answer this question applying e-learning standards to the mobile learning context.

In order to improve the reuse of educational resources in digital format appears the concept of learning object. A learning object includes not only educational content, but also metadata descriptions that describe the own object and make easier its use and location.

Another resource to achieve the e-learning objectives is the Sharable Content Object Reference (SCORM) model. SCORM is a common model for learning objects. Its main advantage is to make easy the reuse of educational contents between different systems (learning object repositories and e-learning platforms, for example).

SCORM is now a de facto standard supported by major manufacturers and consumers of e-learning and is supported by the leading learning platforms.

In this paper it will be put on relation the SCORM model with mobile learning to promote the use and reuse of the educational contents also at mobile devices.

Nowadays contents in m-learning are tailored and not reuse the educational resources that already exist. This paper tries to give a solution to this problem.

In the first section it will be contextualized the problem studying on one hand the main trends in m-learning research. On the other hand the fundamentals of the SCORM model. In the third section it will be described the SCORMmobile project which uses the SCORM model in m-learning context. Finally the paper finishes with the conclusions and future work.

2 Research Trends in m-Learning

The m-learning as a teaching platform is increasing their users and utilities. There are different lines of advances in research in this area. It will be discussed the most interesting. In one of them could frame the SCORMmobile project.

One of the lines of research focuses on the analysis of issues involving education and teaching methods through mobile environments. A project which aims to develop innovative teaching methods using the mobile can be found in the [4] e-book.

Another line is the adaptation of contents in order these contents are accessible from mobile devices due to the limitations and special features that have these devices. The idea is to use "converters" of multimedia content, images, web,... to a format that can be viewed in the mobile device. In the market already exist some applications that allow different types of format conversion.

Another trend is to integrate mobility systems in the learning management systems (LMS). This line is mainly oriented to include mobility features in online learning platforms. One of the most extended learning platform is Moodle. There are several studies of this type with Moodle but most of them are in process or are not released: MLE [5], mTouch of iPhone [6] as a native application for Moodle, or MOMO [7], developed specifically for Moodle but it requires download a client for mobile and does not work in version 2.0 of Moodle. However, these adaptations are always partial and specific of a particular learning platform. In addition they are not operating or are not accessible.

Finally, another important line of research, that would fit the project of this paper, is oriented to allow viewing from a mobile device, learning content that follow e-learning standards. This line has a lot of development potential in the future.

By the moment the most important initiative in this area is Pocket SCORM Runtime Environment [8] that allows tailored the SCORM contents to mobile . However the functionality of this project in reference to the SCORM standard has become somewhat outdated and limited, and does not supported sequencing and timing characteristic of the new version of the standard.

Other authors propose using an intermediate layer (framework) to adapt the SCORM object to use it in mobile devices like in [3].

This research is also going on this last line, which is the less explored. It aims to achieve display the contents of a SCORM in mobile phones.

3 E-Learning Standard for Educational Contents Exchange

SCORM model is a de facto standard in e-learning. SCORM is a specification of the ADL (Advanced Distributed Learning) [2][9]. The main objective is to share educational content between different systems to facilitate the reuse of this content.

SCORM is composed by three models:

- Aggregation model (CAM) which defines how the educational content may be packaged in a .zip file. In this way the content will be able to be transferred between different systems.
- Run Time Environment which defines the communication between the e-learning platform (LMS - Learning Management Systems) and the host where is the educational content.
- Sequencing and navigation model (SCORM SN): This model is very important in the new 2004 version [1]. It consist in a set of rules that specifies the order in which learner should acquire the content.

In the project which is proposed in this paper it is necessary to study in detail the Aggregation model.

From the point of view of this model the content is packaged in a .zip file. This file contains a special file. This file is a XML file which contains the information about the structure (organizations) of the object and their resources. Apart from this, the object and their items could also contain metadata about themselves. These metadata are described using another e-learning standard, LOM.

LOM define nine categories to describe a Learning Object. These categories describe for example, general information about the object, technical requirements, copyrights, pedagogical characteristics to use it,...

The LMS who wants to use the SCORM object will have to interpret the content and structure of the XML file.

4 Study Case: M-Learning Application for Reusing Educational Contents

Nowadays most of the educational contents for m-learning are believed tailored. In this sense, on one hand it will not be reusable all the educational contents that already exist. On the other hand all the educational contents specially designed for m-learning will not be able to be used in others environments, may not be reused.

There are many educational contents in different learning object repositories (LOR) or in Learning Management Systems (LMS). SCORM enables content sharing across platforms. If content is packaged using the SCORM model, this content could be interpreted an accessed from all the LMSs and LORs that support this model.

Therefore, if from a mobile phone was possible to interpret a SCORM object, all the contents packaged with SCORM could be accessible from mobile and could be use as resources for m-learning.

SCORMmobile is the application which gives the answer to this question because enable educational contents in digital format that could came from different platforms, could be accessed from the mobile.

The objective of this project is to view and access from a mobile device to educational content in form of learning object packaged under the SCORM model.

Given the advantages of the SCORM standard and the increasing amount of educational materials that follows this pattern, SCORMmobile would make it accessible from the phone a lot of materials already existing in digital format, which follows standard rules for ease of use and reuse. This would be very useful from the point of view of m-learning, where it is common that the materials created are very specific to a particular technology and unable to be reused in other environments than those for which they were created.

SCORMmobile shows the content of a SCORM object in a mobile phone. To achieve it is it is necessary not only obtaining the resources of the educational content, but also finding out the structure and organization of these contents.

The architecture of the application include a software component which one time the user has in his mobile the SCORM object that he wants to access, will show in the screen of the mobile the structure of the resources of this object. From this moment the user only will have to select the resources of the SCORM object which wishes to visualize in his mobile.

To reach this objective internally the software component has to carry out several steps.

The SCORM object appears as a packed file in zip format. Therefore the first activity will be unpacked the zip file.

When the file is unpacked it is obtained, between others, a special XML file which contains all the information about the organization and resources of the object. Therefore it is necessary to explore this XML file in order to find the structure of the object.

The content of the object can be organized in different ways from the pedagogical point of view. Therefore the same object could have many organizations to view its educational resources.

Finally, to show the structure of the object to the user it is used a tree view. This is the same format in what is showed this type of information in an LMS.

One time the user select in the tree a resource of the learning object, its content is showed in the screen of the mobile.

Fig. 1 Visualization of the structure of a SCORM object in a mobile phone

Figure 1 shows the tree associated to a learning object packed in SCORM. The tree contains two organizations for the object. Inside each organization appear the resources of the object. The user only has to select the resource who wants to open and its content will be shown. The process is similar to the opening of a SCORM object in a LMS. Figure 2 shows the content of the resource of the SCORM object selected in Figure 1.

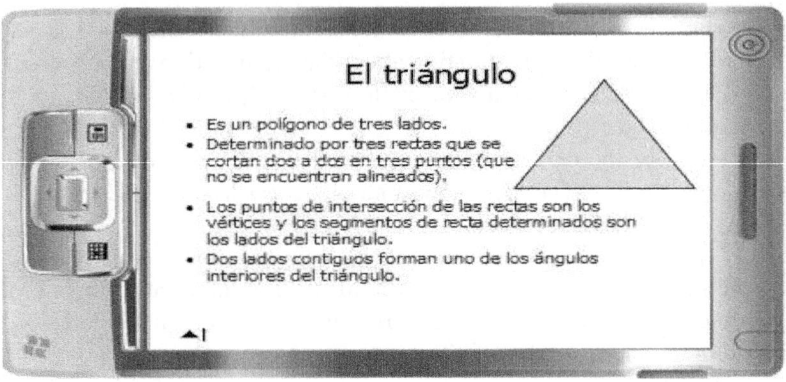

Fig. 2 Visualization of the content of a resource of a SCORM object

5 Conclusions and Future Work

This paper described a case study which enables reused educational contents in mobile environments. In this sense it could be considered as smart.

The implemented application allows visualizing any educational contents packaged in SCORM format. In this way any learning object packed in SCORM would be accessible from a mobile device. Learning platforms allows importing SCORM resources to visualize them. Learning repositories allows importing and exporting SCORM objects. Therefore, the resources in SCORM format stored in learning platforms or learning repositories will be accessible from mobile with the SCORMmobile application. In this way we can access from the mobile for multiple educational contents that already exist.

Nowadays the application is going to be used in an organization whose main activity is e-learning. With SCORMmobile this organization will also teach using mobile devices.

Nowadays the application works in mobiles with Windows Mobile as operating system. The first line to research in the future will be to increase the number of mobile devices that can access the application. To this end the application will be developed also in others operating systems like Android or IPhone.

Acknowledgements. This study has been supported by the Spanish Ministry of Science and Innovation project TRA2009_0096.

References

1. ADL (Advanced Distributed Learning). SCORM 2004 4th edn. Overview (2009), `http://www.adlnet.gov/Technologies/scorm/SCORMSDocuments/2004%204th%20Edition/Overview.aspx` (last access December 21, 2010)
2. ADL (Advanced Distributed Learning). SCORM, `http://www.adlnet.gov/Technologies/scorm/default.aspx` (last access December 21, 2010)

3. Drire, Pirelli, Laurossy, Derick, Benghezala: What We Can fit in Mobile Learning Systems?. In: IMCL 2006, Amman, Jordan (2006)
4. Herrington, J., Herrington, A., Mantei, J., Olney, I., Ferry, B.: New Technologies, new pedagogies: Mobile Learning in Higher Education (2009), http://ro.uow.edu.au/newtech/ (last access December 21, 2010)
5. Mobile Learning Engine (MLE), http://mle.sourceforge.net (last access December 21, 2010)
6. MoodleTouch (mTouch) of iPhone, http://www.pragmasql.com/home/moodletouch.aspx (last access December 21, 2010)
7. Mobile Moodle (MOMO), http://www.mobilemoodle.org/momo18 (last access December 21, 2010)
8. Lin, N.H., Shih, T.K., Hsu, H.-H., Chang, H.-P., Chang, H.-B., Ko, W.C., Lin, L.J.: Pocket SCORM. In: 24th International Conference on Distributed Computing Systems Workshops - W1: NMSA (ICDCSW 2004), vol. 1, pp. 274–279 (2004)
9. Rustici Software. SCORM explained, http://www.scorm.com/scorm-explained/business-of-scorm/ (last access December 21, 2010)

Mobile Assistant for the Elder (MASEL): A Practical Application of Smart Mobility

M.A. Sánchez, E. Beato, D. Salvador, and A. Martín

Abstract. This article introduces the use of a practical application of smart mobility called MASEL (Mobile Assistant for the Elder). This system is more than a typical medical reminder, and it is specially oriented to help to the elder. It is the combination of mobile applications with e-health systems and other web applications and services to connect patients, doctors, pharmacist and families. And this is the main objective of MASEL project. It is based on several technologies to improve the quality of life and facilitate certain tasks to dependent people and to bring a detailed monitoring of medication needed by the patient. MASEL controls the medicines the user has taken and warns him at the right time to be taking medications and monitoring the operation. This is an innovative system especially easy to use. In addition, it is able to manage and keep track of medication that a patient needs regardless of location. Patients will not even need to configure the application with the medicines they need in their treatments. Medical personnel perform this task remotely.

Keywords: Smart mobility, Pervasive computing, Health Care, Geriatric residences.

1 Introduction

We are in times of transition towards a new social order characterized in the continued growth and progressive aging of the population, thus the dependency ratio increases and with it the demand for services that reduce this dependence. In this new scenario, mobile technologies [2][4] are of vital importance and are very important to reach the main objectives of pervasive computing [1][6]. Properly used, these technologies will moderate the impact of this situation [10][11]. Some previous experiences [3][7] show that smart mobility is useful in these cases.

In addition, agent or multi-agent systems [5] and mobile or RFID technology [8] can be combined with health care and pervasive computing [9] to implement practical cases of smart mobility. One of these cases is MASEL project, based on many of these technologies to improve the quality of life and facilitate certain

M.A. Sánchez · E. Beato · D. Salvador · A. Martín
Pontifical University of Salamanca, c/ Compañía 5, 37002 Salamanca, Spain
e-mail: {masanchezvi,ebeatogu,dsalvadorma,amartinro}@upsa.es

J.B. Pérez et al. (Eds.): Highlights in PAAMS, AISC 89, pp. 325–331.
springerlink.com

tasks to dependent people everyday. It is based on the use of a mobile phone with specific features like location and Internet connection. These technologies will allow MASEL to bring a detailed monitoring of medication needed by the patient. MASEL controls the drugs that the user has taken and warns him at the right time to be taking medications and monitoring the operation.

This is an innovative system especially easy to use. In addition, it is able to manage and keep track of medication that a patient needs regardless of location. Patients will not even need to configure the application with the medicines they need in their treatments. Medical personnel perform this task remotely.

Nowadays all this control is centralized in specific locations such as homes, day care centers or private homes and really there are no solutions independent of location.

The main purpose of MASEL is to improve the quality of life of a person who needs regular medication in a personalized way. There is also the possibility of interaction between medical personnel in those services that require it, even family members can be informed at all times. MASEL is looking to be as open as possible to mobile platforms in order to reach as many people as possible regardless of the mobile platform they use.

2 General Description of MASEL

MASEL is based on the use of alarms to remind the user what drug and dose to be taken. By showing the alarm the user can indicate that you have taken the drug, or has delayed in time if possible. All these operations are recorded and monitored by medical staff remotely.

The innovative nature of this idea is particularly highlighted in this point, because today there are not specific mobile applications that allow comprehensive control of medicines to be taken by a patient with the characteristics of MASEL. The most different feature of MASEL is the remote configuration, control and monitoring by medical personnel. At present, the mechanism commonly used to determine which pills to take in every moment are the ``pill boxes" with compartments to store the drugs function to take the time and day of the week.

In addition, there are some mobile medication reminder and guide like pill phone (www.pillphone.com), available for different mobile platforms. These kinds of solutions are simple and based on automatic systems that a patient can use like calendar alarms on the mobile phone or PDA. Generally these applications use information from a generic pharmaceutical database and, for this reason, the treatment is not personalized. This is a very important detail for the elder because each person normally has very different treatments or dosages. MASEL will solve this problem because it is customizable and it has the feature of remote monitoring and medical control.

Other kind of mobile applications are based on location services. For instance, this is useful for families to always know where their old parents are. But this is not enough.

All these applications are supported on the e-health idea of healthcare based on informatics and digital or electronic processes, including diagnosis and monitoring of patients. The use of these technologies makes possible to treat the patient in the distance (tele-assistance), so that medical services the patient needs can be provided remotely. The information managed by e-health systems about patients, could be shared with our application (see section 3, about the architecture) to follow the treatments in detail.

In the area of e-health for geriatric residences we have found and use management applications like Gerosalus (www.gerosalus.com), which enable physicians, among other things, manage the medical record and keep track of medication. This software is used in "Residencial La Vega", a geriatric residence in Salamanca (www.esgrasalamanca.com), our main collaborator in this research. For this reason, we consider a possible connection interface from Gerosalus to provide data to our system MASEL, something that could be a key concept for a practical implementation.

The main features of MASEL are:

- **Ease to use.** The application simply shows an alert to the user and that is the beginning of a simple interaction; the user only must specify if he or she has already taken the medication or postponed.
- **Independent of geographical location.** It uses a permanent Internet connection in the mobile phone.
- **Registry of operations and remote monitoring.** From any Internet connected computer could check the registry for treatment of a patient.
- **Flexible.** The architecture of MASEL allows easy adaptation to various mobile platforms through the use of web services. This facilitates the extension to other mobile platforms, even those that appear in the future.

3 Architecture, Profiles and Main Applications of MASEL

As shown in figure 1, MASEL is an integrated system formed by a central server with a medical and patient database, specific web applications to be used for the doctors, families or MASEL administrators and connections to specific smartphone applications to be used for the patients. In addition it can be connected to e-health systems like Gerosalus. This option is very important in residences like "Residencial La Vega" because all data of each patient is stored in this system. Using this type of connection, MASEL can be quickly configured and will be easily integrated with e-health systems. Gerosalus is one of them, but MASEL could be integrated with any other similar software.

The final user uses a mobile application that facilitates access to services for older people regardless of their location. Today we have versions for Windows Mobile 6.x and in future for iPhone platforms. Besides, the application in future can be developed for any other mobile platform.

Therefore, MASEL architecture is configurable and adaptable to any mobile platform that exists today. The use of web services makes it adaptable to any e-health system.

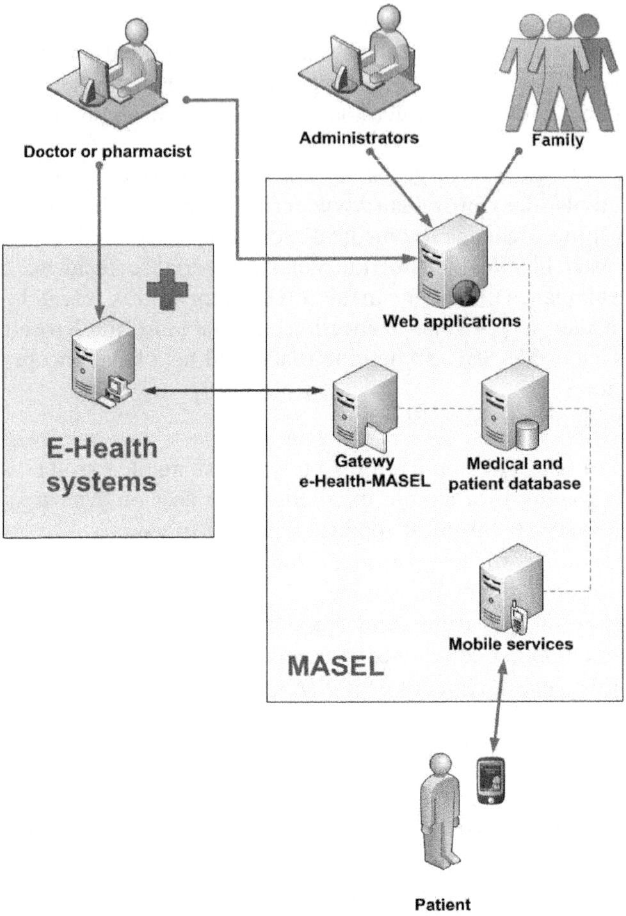

Fig. 1 Main profiles and components of MASEL architecture

As described in figure 1, there are three different profiles, but not exclusive. The types of application profiles are: medical, patient, family and administrator.

The medical profile is related to doctors and pharmaceutics. They can use a web application included in the MASEL system to manage the data of the patient. In addition they can use this application to add and edit patient treatments information like medicines, duration and personal dosage. For this reason, the treatment is customizable for each person. Finally, this application can also report the doctor about the course of the treatment because it can manage the information received from the mobile application used by the patients.

The patient profile is associated to persons that use a smartphone application to receive medication alerts and to confirm when they have taken each medicine. This application does not need to initialize the data of medical alerts because MASEL system can provide automatically this data obtained in the medical profile by the medical web application. So, for the final user, the application is really easy to use. This is very important for the elder. In addition, the information about when the patient takes medicines is reported to doctors and families using MASEL architecture.

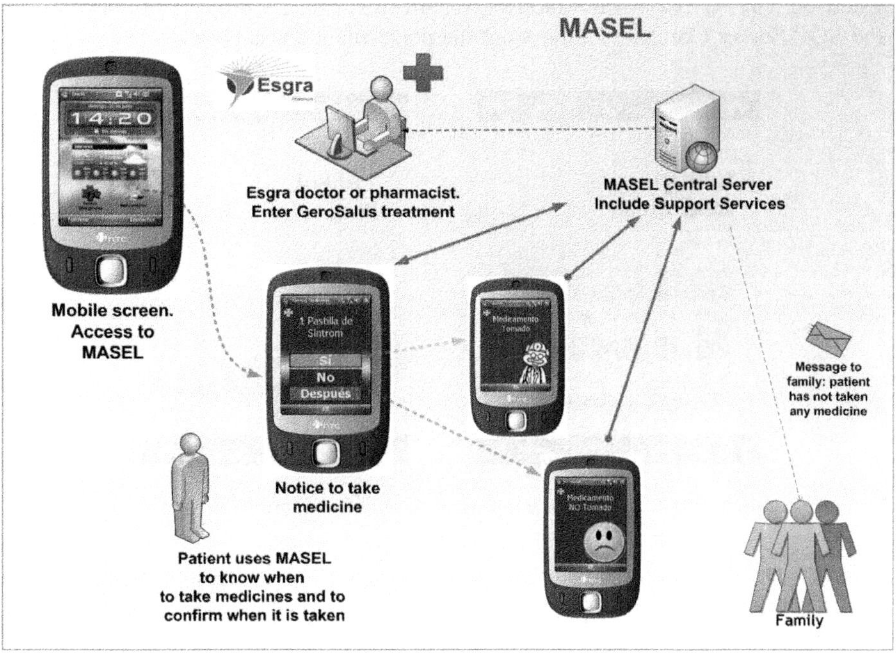

Fig. 2 Functionalities Scheme Information System Profiles

The family profile has been created to report information to family or friends who care the patients. This is possible because MASEL provides a web application for these persons that give them customizable reports. In addition, they can receive specific alerts if the patient does not take a medicine in a specific period of time. This way, family can be reported about the course of the treatment.

Finally, administration profile is necessary for MASEL administrators to create users in each profile and to give the adequate permissions for each application. These users also need special application to interact with e-health systems.

In figure 2 we describe an example to illustrate how MASEL works with the profiles and applications of the system.

First, a doctor or a pharmacist of Esgra, the residence of the elder, adds a treatment for a specific old person. For this he or she will use Gerosalus software instead the web application. In our central server this information will be added to the medical database.

Using a specific web service, the patient will receive the personal treatment in the mobile phone. This way, he or she will have the medication alerts added by the doctor or pharmacist of the residence.

The scheduled alerts will use the mobile calendar or a specific reminder depending on the operative system of the smartphone. The mobile application that we have developed for Windows Mobile 6.x uses a specific mobile database and the alerts are not translated to the calendar. For this reason the application needs to be running constantly. This could be a problem in some terminals. We have it tested on a HTC HD with Windows Mobile 6.1, .NET Compact Framework 2.5 and SQL Server Compact Edition and the performance was good.

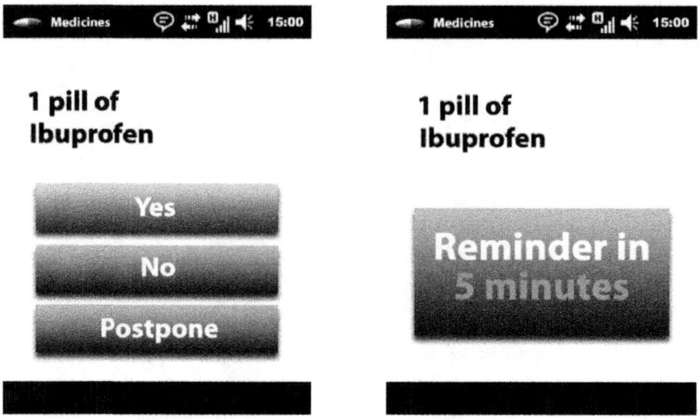

Fig. 3 Example of use postponing the taking of medication (option "Postpone")

When an alert prompts the patient will see a simple message on the mobile screen similar to the left part of the figure 3. So, the objective of this alert is to remember the old person to take a pill of ibuprofen. But we need the interaction of the user to know if finally the pill has been taken. So the application gives three options: yes, no or postpone. If the person cannot take the pill in this moment can press postpone button. Five minutes after, the same alert will appear. This information will be send to the central server to control the treatment by the doctors and to report to the families.

4 Conclusions and Future Work

The idea to use mobile phones to help the elder is not new. There are real applications like pillphone that shows how to perform this task for all kind of people.

But MASEL is more than a typical medical reminder, and it is specially oriented to help to the elder. It is the combination of mobile applications with e-health systems and other web applications and services to connect patients, doctors, pharmacist and families. And this is the main objective of our project.

To make MASEL a real support system for the elder we are testing it at "Residencial La Vega" from Esgra at Salamanca. The present results are not enough to conclude if this is a final system to be used in a real context. We still have more software to develop and to test. Right now, we have tested the mobile application only in a terminal with Windows Mobile. This kind of mobile phone is not very used by the elder. And for this reason we want to test more mobile platforms to know which of them is well accepted in this range of population. In addition there are differences depending on the range of age.

Acknowledgements. This study has been supported by the Spanish Ministry of Science and Innovation project TRA2009_0096.

References

1. Bardram, J.E., Baldus, H., Favela, J.: Pervasive computing in hospitals. In: Bardram, J.E., Mihailidis, A., Boca Raton, D. (eds.) Pervasive Computing in Healthcare, pp. 49–77. CRC Press, Boca Raton (2007)
2. Beato, M.E., et al.: Mobile devices: Practical cases on this new applications platform. In: Intl. Workshop on Practical Applications of Agents and Multiagent Systems, Salamanca, pp. 231–240 (2007)
3. Farmer, A., Gibson, O., Hayton, P., Bryden, K., Dudley, C., Neil, A., Tarassenko, L.: A real-time, mobile phone-based telemedicine system to support young adults with type 1 diabetes. Informatics in Primary Care. 13(3), pp. 171-177 (2005)
4. Fraile, J.A., Delgado, M., Sánchez, M.A., Beato Gutiérrez, M.E.: The UPSA Mobile Information System – MoviUPSA. In: Intl. Workshop on Practical Applications of Agents and Multiagent Systems, pp. 221–230 (2007)
5. Fraile, J.A., Tapia, D.I., Sánchez, M.A.: Hybrid multi-agent architecture (HoCa) applied to the control and supervision of patients in their homes. In: Corchado, E., Abraham, A., Pedrycz, W. (eds.) HAIS 2008. LNCS (LNAI), vol. 5271, pp. 54–61. Springer, Heidelberg (2008)
6. Hansen, T.R., Bardram, J.E., Soegaard, M.: Moving Out of the Lab: Deploying Pervasive Technologies in a Hospital. IEEE Pervasive Computing 5(3), 24–31 (2006)
7. Mattila, E., Korhonen, I., Saranummi, N.: Mobile and personal health and wellness management systems. In: Bardram, J.E., Mihailidis, A., Boca Raton, D. (eds.) Pervasive Computing in Healthcare, pp. 105–134. CRC Press, Boca Raton (2007)
8. Siegemun, F., et al.: Interaction in pervasive computing settings using Bluetooth-enabled active tags and passive RFID technology together with mobile phones. In: Proceedings of The First IEEE International Conference on Pervasive Computing And Communications (2003)
9. Orwat, C., Graefe, A., Faulwasser, T.: Towards pervasive computing in health care – A literature review. In: BMC Medical Informatics and Decision Making, pp. 8–26 (2008)
10. Stanford, V.: Using pervasive computing to deliver elder care. IEEE Pervasive Computting 1(1), 10–13 (2002)
11. Zhibo, P., et al.: A pervasive and preventive healthcare solution for medication noncompliance and daily monitoring. In: Applied Sciences in Biomedical and Communication Technologies, pp. 1–6 (2009)

Interaction Mechanism for Language Learning for Elderly People through Mobile Devices

Antonia Macarro, Emma Villafaina, Ana María Pinto, and Javier Bajo

Abstract. Education for elderly people has acquired a growing relevance during the last decade. Elderly people represent a high percentage of the current population and require special education. This paper present a personal agent specifically designed to provide advanced interfaces for elderly people that can be executed on mobile devices. The agent is aimed at providing language learning facilities that con complement the traditional techniques. The mechanism has been initially tested for French language learning courses for elderly people. The preliminary results obtained are presented within this papers.

Keywords: Context-Aware, Adaptive Algorithms, Machine Learning, Elderly People, Language Learning.

1 Introduction

During the recent years there is a growing need for adapting the educational environments to the new requirements of the information technologies. One of the segments of the population which will benefit with the advent of systems based on Ambient Intelligence will be the elderly and people with disabilities 8, contributing to improve their quality of life 10. Ambient Intelligence evolves from the ubiquitous computing 4, and constitutes the most promising technological approach to meet the challenge of developing strategies in dependency environments 17.

This work presents an innovative method, based on the Ambient Intelligence (AmI) paradigm 5 15, for formal teaching of languages oriented to elderly people. Recently, teaching oriented to elderly people has acquired a great importance in the universities, with the so-called "Universidades de la Experiencia" studies. The aim of these studies is to facilitate a learning method for elderly peole (up to 55 years old), and their integration in the information society. However, the special characteristics of elderly people require new solutions, since the learning methods used in early ages are not appropriated for this sector of the society most of the times. It is necessary to investigate in new techniques and methods to satisfy the learning needs of the elderly people.

Antonia Macarro · Emma Villafaina · Ana María Pinto · Javier Bajo
Universidad Pontificia de Salamanca, c/ Compañía 5, 37002 Salamanca, Spain
e-mail:{amacarroal,evillafainamu,ampintoll,jbajope}@upsa.es

J.B. Pérez et al. (Eds.): Highlights in PAAMS, AISC 89, pp. 333–340.
springerlink.com

This paper focuses in the combination of the new information technologies along with the traditional teaching. In this way it will be possible to combine the advantages of the face to face teaching with the advantages of distance learning. It will be necessary to upgrade the systems of evaluation/accreditation to assess the knowledge or skills acquired during the learning process. To achieve this objective, we propose the use of mobile devices, intelligent systems and wireless communications.

Different authors 3 9 18 consider that in the near future, the educational institutions will be provided with intelligent systems specifically designed to facilitate the interaction with the human users. These intelligent systems will be able to personalize the services offered to the users, depending on their concrete profiles. It is necessary to improve the supply of services, as well as the way to offer them. Trends situate the user surrounded of technology that facilitates new and personalized services. Techniques and architectures based on mobile devices have been recently explored as a system of interaction with the elderly and dependent 1. These systems can provide support in the daily lives of dependent people 20, providing a cognitive and physical support for the assisted person 2. They can also provide mechanisms for establishing new strategies for learning interaction, which greatly facilitates the teaching, particularly languages learning.

The rest of the paper is structured as follows: Next section introduces the problem that motivates most of this research. Section 3 presents the proposed user agent for mobile devices. Section 4 describes a case study to test the proposal and, finally, Section 5 presents the results and conclusions obtained.

2 Related Work

There is an ever growing need to supply constant care and support to the disabled and elderly and the drive to find more effective ways to provide such care has become a major challenge for the scientific community 4. During the last three decades the number of Europeans over 60 years old has risen by about 50%. Today they represent more than 25% of the population and it is estimated that in 20 years this percentage will rise to one third of the population, meaning 100 millions of citizens 4. In the USA, people over 65 years old are the fastest growing segment of the population 1 and it is expected that in 2020 they will represent about 1 of 6 citizens totaling 69 million by 2030. Furthermore, over 20% of people over 85 years old have a limited capacity for independent living, requiring continuous monitoring and daily care 2. Some estimations of the World Health Organization show that in 2025 there will be more than 1000 million people aged over 60 in the world, so if this trend continues, by 2050 will be double, with about the 80% concentrated in developed countries 18.

Education is the cornerstone of any society and it is the base of most of the values and characteristics of that society. The new knowledge society offers significant opportunities for AmI applications, especially in the fields of education and learning 17. The new communication technologies propose a new paradigm focused on integrating learning techniques based on active learning (learning by doing things, exchange of information with other users and the sharing of resources), with techniques based on passive learning (learning by seeing and hearing, Montessori, etc.) 7. While the traditional paradigm, based on a model focused on face to face education, sets as fundamental teaching method the role of the teachers and their knowledge, the paradigm based on a learning model highlights the role of the students. In this second paradigm the students play an active role, and build, according to a personalized action plan, their own knowledge. Moreover, they can establish their own work rhythm and style. The active methodology proposes learning with all senses (sight, hearing, touch, smell and taste), learn through all possible methods (school, networking, etc.), and have access to knowledge without space or time restrictions (anywhere and at any time).

There are different studies that have used the Ambient Intelligence to facilitate learning. In 4, Bomsdorf shows the need to adapt intelligent environments to changes depending on the educational context and the characteristics of users. Morken *et al.* 18 analyze the characteristics of intelligent environments for learning. They focus on the role of mobility in educational environments and the role that acquire the mobile devices. Naismith *et al.* 19 conducted a detailed study describing the role of mobile devices in education, analyzing the characteristics of the devices and their capacity for learning in educational environments. All these approaches are focused on the role of learning in Ambient Intelligence environments, but none of them is oriented on learning for dependents or elderly people. The following section presents a multiagent architecture that facilitates learning methodology using an active through mobile devices.

3 Proposed Mechanism

Information technologies can notably improve the learning process in educational environments, providing the users with new technological resources 7. At present there are various tools to facilitate the active learning, such as forums, wikis, email, chat, virtual campuses, and so on. However, none of them is focused to language learning for elderly people. This paper presents an interactive system specifically designed for language learning for elderly people and people with visual disabilities. Multiagent architecture specially designed to facilitate active learning in dependence environments. Agents can be characterized through their capacities in areas such as autonomy, reactivity, pro-activity, social abilities, reasoning, learning and mobility 12 21, which make them particularly suitable for use in the design and development of intelligent environments for dependency. That way, each student can have a personal agent, able to adapt to his personal needs taking into account the characteristics of his profile.

The personal user agent includes new interaction techniques adapted to be applied in mobile devices. Interaction techniques have been used to design Ambient Intelligence interaction mechanisms suitable for use in teaching languages to people involved in courses at the University of Experience. Therefore, interfaces have been made simple and straightforward and have facilitated interaction through touch devices. The application developed for mobile devices contain a series of tests that individuals can complete to carry out language learning. Thus, the student Agent is a type of interface agent that allows a student to interact with the system. The student agents adapt to the needs of the human students, taking into consideration their profiles and customizing their access to the system. Figure1 shows a simple example of the student agent being executed in a mobile device. As can be seen in Figure 1, the test are very simple.

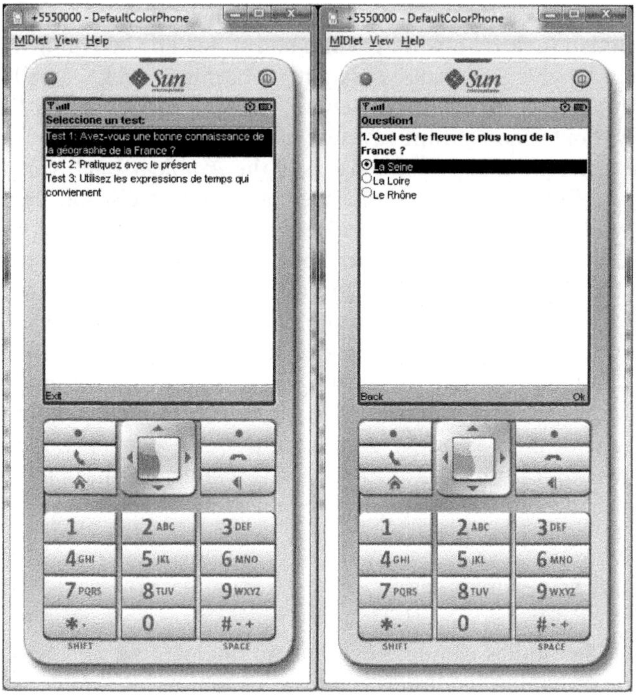

Fig. 1 Student agent executed in a mobile device.

The proposed architecture has great capacity for learning and adaptation to the user needs. Some works 14, which proposes an e-mail system adapted to elderly people without experience in the use of computers, show that not all users have the same difficulties. Even when an application was developed specifically for them, may require different versions for different users depending on their characteristics. The customization of the tasks and the interface can solve this problem because it allows different versions depending on the needs and particularities of each user 18. One of the advantages of the mechanism proposed in this paper is the ability of the agents to communicate through wireless channels, which facilitate independence of location or time constraints.

4 Case Study Special Courses for Elderly People

Concern about the education of elderly people has led the universities to set up, under different names, studies for elderly people. These initiatives began to be put into practice in Spain in the early 90's, and nowadays all the Spanish universities have an educational program for elderly. In general, elderly people are inexperienced in the use of the Internet, finding problems to understand and use some of the conventions used in user interfaces: double-clicking with the mouse, movements between text fields to be filled in, bar displacement 16. Moreover, the problems are common vision 15, cognitive problems 10, manual dexterity 20. In order to benefit this group from the advantages of the digital environment, it is necessary to provide new methods and techniques. The usability is a key factor, given that a difficult-to-use system will be probably rejected by the users, increasing the fear and, perhaps, a sense of frustration in the use of these technologies. The work of Marqui and Todman 18 presents the difficulties and negative experiences of the elderly people in the use of computers (mainly the anxiety it causes).

Among the agencies that investigate new proposals to improve the daily life of elderly and dependent people, we can find the European Educational Programs Autonomous Agency, whose mission is to manage the Spanish participation in the Lifelong Learning Program (PAP) of the European Union. Within the adult education program is included the Grundtvig program. Encourage learning throughout life has become a political priority for all European countries. Language training for dependents is common and widely accepted. In this way, under the frame of the Grundtvig program, we defined a case study at the University of the Experience at the Pontifical University of Salamanca. Tests consisted of 2 tests for two differnt groups, one performed in Group 1 and another made to Group 2. The first test involved 4 individuals, while in the second test involved 5 individuals. The sex of individuals was not taken into consideration to perform this test, because this parameter has not been considered as significant for this study. Each of the individuals completed one of the tests proposed in the mobile application, and noted by 10 questions. After the test, each individual completed a form on which he was asked about the evaluation of the test. The tests assesses 5 items:

- Usefulness of the test. This item is valued feedback from users about the usefulness of the test as a learning tool.
- Easy to use mobile phone. This item assesses the usability of the proposed application.
- Utility as a tool for teaching languages. This item is valued feedback from users about the usefulness of the tool for language learning.
- Using mobile phone before for similar activities. This item is valued the ease and frequency of use of mobile devices.
- Overall assessment of the test. This item assesses the overall opinion about the test users.

5 Results and Conclusions

This paper has presented a methodology for teaching languages to dependents, using techniques of Ambient Intelligence. The new multi-architecture presented in this paper provides an active learning through simple and intuitive interfaces, installed on mobile devices. This requires the integration of intelligent agents with innovative strategies of teaching languages and mobile devices. In this way we have obtained:

- An active method of teaching and learning of languages for dependents and elderly people.
- An interaction system based on the AmI paradigm for language education.
- The obtained language learning system was adapted to be executed on mobile devices, facilitating the adaptation to the needs of the dependent and elderly people.
- The multiagent architecture presented in this work has been tested by means of the case study presented in Section 4.

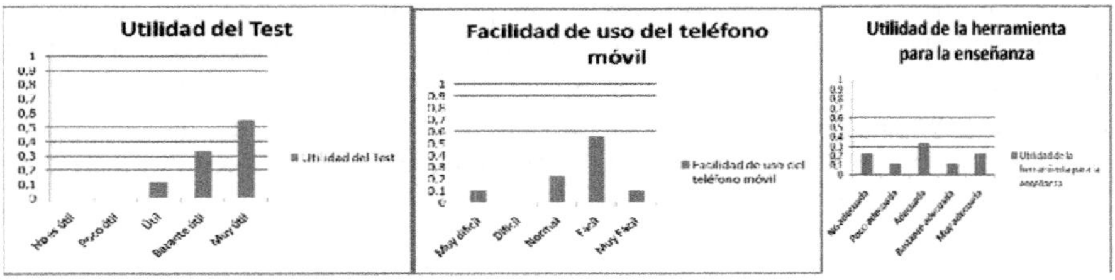

Fig. 2 Results obtained.

Figure 2 shows the results of the tests performed in the case study presented in Section 4, specifically designed for elderly students. Figure 2 a shows the utility of the tests. As can be seen, the user positively evaluate the new mechanism. Figure 2 b evaluates the usability of the application in the mobile device. The users appreciate the new application, however, some of them had problems to interact with the system. Figure 3c presents the appropriateness of the application to be applied in educational environments for elderly people.

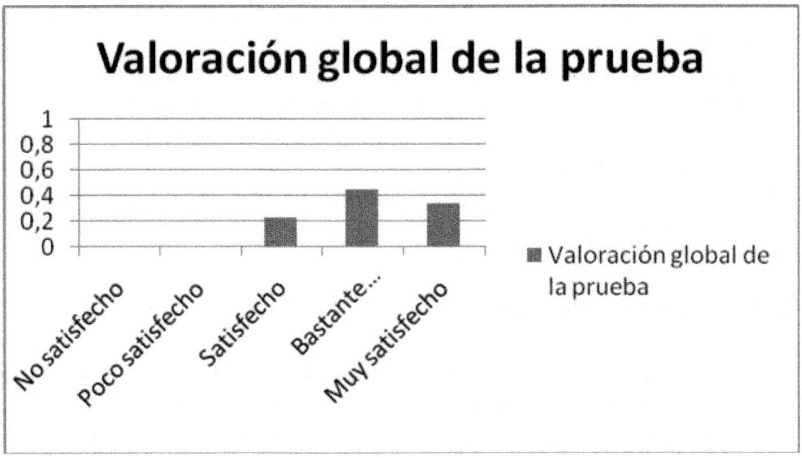

Fig. 3 Satisfaction degree for the users.

Figure 3 shows the overall assessment of the appropriateness of the tests for the individuals. The 44.44% of individuals are quite satisfied after testing. 33.33% of individuals are very pleased and 22.22% re-stant of individuals is very pleased. The results are promising and to demonstrate the good agreement obtained by the proposed application. However, there is still much work to do, especially work aimed at improving the system of interaction and the achievement of a greater number of tests in different scenarios.

Acknowledgements. This study has been supported by the Spanish Ministry of Science and Innovation project TRA2009_0096.

References

1. Anastasopoulos, M., Niebuhr, D., Bartelt, C., Koch, J., Rausch, A.: Towards a Reference Middleware Architecture for Ambient Intelligence Systems. In: ACM Conference on Object-Oriented Programming, Systems, Languages, and Applications (2005)
2. Angulo, C., Tellez, R.: Distributed Intelligence for smart home appliances. Tendencias de la minería de datos en España. Red Española de Minería de Datos, Barcelona, España (2004)
3. Bajo, J., Molina, J.M., Corchado, J.M.: Ubiquitous computing for mobile environments. In: Issues in Multi-Agent Systems: The AgentCities. ES Experience. Whitestein Series in Software Agent Technologies and Autonomic Computing, pp. 33–58. Birkhäuser, Basel (2007)
4. Bajo, J., Corchado, J.M., de Paz, Y., de Paz, J.F., Rodríguez, S., Martín, A., Abraham, A.: SHOMAS: Intelligent Guidance and Suggestions in Shopping Centres. Applied Soft Computing 9(2), 851–862 (2009)
5. Bajo, J., de Paz, J.F., de Paz, Y., Corchado, J.M.: Integrating Case-based Planning and RPTW Neural Networks to Construct an Intelligent Environment for Health Care. Expert Systems with Applications 36(2) (part 2), 5844–5858 (2009)
6. Bomsdorf, B.: Adaptation of Learning Spaces: Supporting Ubiquitous Learning in Higher Distance Education. In: Davies, N., Kirste, T., Schumann, H. (eds.) Dagstuhl Seminar Proceedings. Mobile Computing and Ambient Intelligence: The Challenge of Multimedia, Dagstuhl, Germany (2005)
7. Brown, T.H.: Beyond constructivism: Exploring future learning paradigms. Education Today (2) (2005)
8. Carretero, N., Bermejo, A.B.: Inteligencia Ambiental. CEDITEC: Centro de Difusión de Tecnologías, Universidad Politécnica de Madrid, España (2005)
9. Corchado, J.M., Laza, R.: Constructing Deliberative Agents with Case-based Reasoning Technology. International Journal of Intelligent Systems 18(12), 1227–1241 (2003)
10. Corchado, J.M., Bajo, J., Abraham, A.: GERAmI: Improving the delivery of health care. IEEE Intelligent Systems 23(2), 19–25 (2008)
11. Corchado, J.M., Bajo, J., de Paz, Y., Tapia, D.: Intelligent Environment for Monitoring Alzheimer Patients, Agent Technology for Health Care. Decision Support Systems 34(2), 382–396 (2008), ISSN 0167-9236
12. Corchado, J.M., Glez-Bedia, J., de Paz, Y., Bajo, J., de Paz, J.F.: Replanning mechanism for deliberative agents in dynamic changing environments. Computational Intelligence 24(2), 77–107 (2008)

13. Czaja, S.J., Lee, C.C.: Designing computer systems for older adults. In: Jacko, J.A., Sears, J. (eds.) The Human-Computer Interaction Handbook: Fundamentals, Evolving Technologies and Emerging Applications, pp. 413–427. Lawrence Erlbaum Associates, Mahwah (2003)
14. Dickinson, A., Newell, A.F., Smith, M.J., Hill, R.L.: Introducing the Internet to the over-60s: developing an email system for older novice computer users. Interacting with Computers 17, 621–642 (2005)
15. Echt, K.V.: Designing web-based health information for older adults: visual considerations and design directives. In: Morrell, R.W. (ed.) Older adults, Health information and the World Wide Web, pp. 61–87. Lawrence Erlbaum Associates, Mahwah (2002)
16. Ellis, R.D., Kurniawan, S.H.: Increasing the usability of online information for older users: a case study in participatory design. International Journal of Human-Computer Interaction 2(12), 263–276 (2000)
17. Friedewald, M., Da Costa, O.: Science and Technology Roadmapping: Ambient Intelligence in Everyday Life (AmI@Life). Working Paper. Institute for Prospective Technology Studies IPTS, Seville (2003)
18. Kurniawan, S.H., King, A., Evans, D.G., Blenkhorn, P.L.: Personalising web page presentation for older people. Interacting with Computers 18, 457–477 (2006)
19. Naismith, L., Lonsdale, P., Vavoula, G., Sharples, M.: Futurelab Literature Review in Mobile Technologies and Learning, Technical Report for Futurelab (2004), `http://www.futurelab.org.uk/research/reviews/reviews_11_and12/11_01.htm`
20. Ranganathan, V.K., Siemionow, V., Sahgal, V., Yue, G.H.: Effects of aging on hand function. Journal of the American Geriatrics Society 49, 1478–1484 (2001)
21. Wooldridge, M., Jennings, N.R.: Agent Theories, Architectures, and Languages: a Survey. In: Wooldridge, M.J., Jennings, N.R. (eds.) ECAI 1994 and ATAL 1994. LNCS, vol. 890, pp. 1–22. Springer, Heidelberg (1995)

Author Index